D0279774

McCORMACK
2 KNUTSFORD DRIVE
CLIFTONVILLE ROAD
BELFAST BT14 6LZ

# CLINICAL ANATOMY

## A REVISION AND APPLIED ANATOMY
## FOR CLINICAL STUDENTS

## HAROLD ELLIS

M.A., M.Ch., D.M., F.R.C.S.
*Professor of Surgery*
*Westminster Medical School, London*
*Formerly Examiner in Anatomy*
*Primary F.R.C.S. (Eng.)*

### SIXTH EDITION

SECOND PRINTING

BLACKWELL SCIENTIFIC PUBLICATIONS

OXFORD LONDON EDINBURGH MELBOURNE

# To my wife and parents

©1962, 1966, 1969, 1971 and 1977 by
Blackwell Scientific Publications
Osney Mead, Oxford
8 John Street, London, WC1
9 Forrest Road, Edinburgh
P.O. Box 9, North Balwyn, Victoria, Australia

All rights reserved. No part of this publication
may be reproduced, stored in a retrieval system,
or transmitted, in any form or by any means,
electronic, mechanical, photocopying, recording
or otherwise without the prior permission of
the copyright owner.

First published 1960
Second edition 1962, reprinted 1963
Third edition 1966
Fourth edition 1969
Fifth edition 1971, reprinted 1972 1973 1974
Sixth edition 1977, reprinted 1978

Distributed in the United States of America by
J. B. Lippincott Company, Philadelphia
and in Canada by
J. B. Lippincott Company of
Canada Ltd, Toronto

**British Library Cataloguing in Publication Data**

Ellis, Harold, b. 1926
  Clinical anatomy.—6th ed.
  1. Anatomy, Pathological
I. Title
616.07                    RB25

ISBN 0-632-00446-0

Filmset in Ireland by Doyle Photosetting Ltd., Tullamore
and printed and bound in Great Britain by
Billing & Sons Ltd., Guilford and Worcester

# Contents

iii

# *Contents*

# Contents

PART THREE

# THE UPPER LIMB

PART FOUR

# THE LOWER LIMB

PART FIVE

# THE HEAD AND NECK

# Contents

x        *Contents*

# Preface to the sixth edition

Experience of teaching clinical students at three medical schools and of examining them in ten cities and in seven countries has convinced me that there is still an unfortunate hiatus between the anatomy which the student learns in his pre-clinical years and that which he later encounters in the wards and operating theatres.

This book attempts to counter this situation. It does so by highlighting those features of anatomy which are of clinical importance, in medicine and midwifery as well as in surgery. It presents the facts which a student might reasonably be expected to carry with him during his years on the wards, through his final examinations and into his postgraduate years; it is designed for the clinical student.

Anatomy is a vast subject and therefore, in order to achieve this goal, I have deliberately carried out a rigorous selection of material so as to cover only those of its thousands of facts which I consider form the necessary anatomical scaffolding for the clinician. Wherever possible practical applications are indicated throughout the text—they cannot, within the limitations of a book of this size, be exhaustive, but I hope that they will act as signposts to the student and indicate how many clinical phenomena can be understood and remembered on simple anatomical grounds.

In this sixth edition a complete revision of the text has been carried out. Many pages have been redrafted and entirely new sections have been added on the segmental anatomy of the liver and on the autonomic nervous system. Twenty-six new figures have been added and many older illustrations replaced or enlarged.

The continued success of this volume owes much to the helpful comments which the author has received from readers all over the world. Every suggestion is given the most careful consideration in an attempt to keep the material abreast of the needs of today's medical students.

*Westminster Medical School, 1976*                    HAROLD ELLIS

# Acknowledgments

I wish to thank my colleagues—the registrars, house surgeons and dressers at Westminster Hospital—who have kindly perused and commented on the text and have given valuable help in proof-reading.

The majority of the illustrations are by Miss Margaret McLarty and Miss Audrey Arnott; I must thank them sincerely for all their care.

I am grateful to the following authors for permission to reproduce illustrations:

Lord Brock for Figs. 20 and 21 (from *Lung Abscess*);

Professor R. G. Harrison for Figs. 12, 32 and 67 (from *A Textbook of Human Embryology*);

and Professor Sheila Sherlock for Fig. 69 (from *Diseases of the Liver and Biliary System*).

To my sister, Mrs L. Witte, go my grateful thanks for invaluable secretarial assistance. Finally, I wish to express my debt to Mr Per Saugman and the staff of Blackwell Scientific Publications for their continued and unfailing help.

H. E.

# PART I

# THE THORAX

## Surface anatomy and surface markings
## Figs. 1, 2 and 3

The experienced clinician spends much of his working life relating the surface anatomy of his patients to their deep structures.

The following bony prominences can usually be palpated in the living subject (corresponding vertebral levels are given in brackets):

- superior angle of the scapula (T2);
- upper border of the manubrium sterni, the suprasternal notch (T2/3);
- spine of the scapula (T3).
- sternal angle (of Louis)—the transverse ridge at the manubrio–sternal junction (T4/5);
- inferior angle of scapula (T8);
- xiphisternal joint (T9);
- lowest part of costal margin—10th rib (L3);

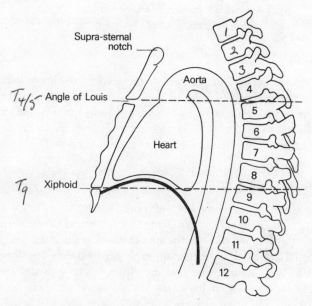

**Fig. 1.** Lateral view of the thorax—its surface markings and vertebral levels. (Note that the angle of Louis (T4/5) demarcates the superior mediastinum, the upper margin of the heart and the beginning and end of the aortic arch.)

Note from Fig. 1 that the manubrium corresponds to the 3rd and 4th thoracic vertebrae and overlies the aortic arch, and that the sternum corresponds to the 5th to 8th vertebrae and neatly overlies the heart.

Since the 1st and 12th ribs are difficult to feel, the ribs should be enumerated from the 2nd costal cartilage, which articulates with the sternum at the angle of Louis.

The spinous processes of all the thoracic vertebrae can be palpated in the mid-line posteriorly, but it should be remembered that the first spinous process which can be felt is that of C7 (the vertebra prominens).

The position of the *nipple* varies considerably in the female, but in the male it usually lies in the 4th intercostal space about 4 in (10 cm) from the mid-line. The *apex beat*, which marks the lowest and outermost point at which the cardiac impulse can be palpated, is normally in the 5th intercostal space $3\frac{1}{2}$ in from the mid-line. (Just below and medial to the nipple.)

The *trachea* is palpable in the suprasternal notch midway between the heads of the two clavicles.

**Surface markings of the more important
thoracic contents (Figs. 2–4)**

### The trachea

The trachea commences in the neck at the level of the lower border of the cricoid cartilage (C6) and runs vertically downwards to end at the level of the sternal angle of Louis (T4/5), just to the right of the mid-line, by dividing to form the right and left main bronchi. In the erect position and in full inspiration the level of bifurcation is at T6.

### The pleura

The *cervical pleura* can be marked out on the surface by a curved line drawn from the sterno-clavicular joint to the junction of the medial and middle thirds of the clavicle; the apex of the pleura is about 1 in (2.5 cm) above the clavicle. This fact is easily explained by the oblique slope of the first rib. It is important because the pleura can be wounded (with consequent pneumothorax) by a stab wound—and this includes the surgeon's knife and the anaesthetist's needle—above the clavicle.

The lines of pleural reflexion pass from behind the sterno-clavicular joint on each side to meet in the mid-line at the 2nd costal cartilage (the

angle of Louis). The pleural edge then passes vertically downwards to the 6th costal cartilage and then crosses:
- the 8th rib in the mid-clavicular line;
- the 10th rib in the mid-axillary line; and
- the 12th rib at the lateral border of the erector spinae.

The pleura actually descends just below the 12th rib margin at its medial extremity—or even below the edge of the 11th rib if the 12th is unusually short; obviously in this situation the pleura may be opened accidentally in making a loin incision to expose the kidney, perform an adrenalectomy or to drain a subphrenic abscess.

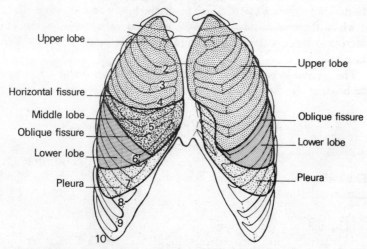

Fig. 2.   The surface markings of the lungs and pleura—anterior view.

## The lungs

The surface projection of the lung is somewhat less extensive than that of the parietal pleura as outlined above, and in addition it varies quite considerably with the phase of respiration. The *apex* of the lung closely follows the line of the cervical pleura and the surface marking of the *anterior border of the right lung* corresponds to that of the right mediastinal pleura. On the left side, however, the *anterior border* has a distinct notch (the *cardiac notch*) which passes behind the 5th and 6th costal cartilages. The *lower border* of the lung has an excursion of as much as 2–3 in (5–8 cm) in the extremes of respiration, but in the neutral position (midway between inspiration and expiration) it lies along a line

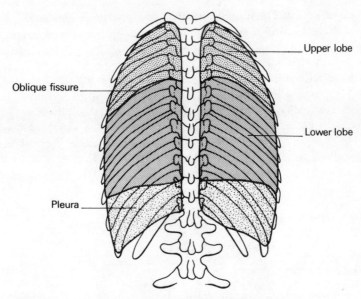

**Fig. 3.** The surface markings of the lungs and pleura—posterior view.

which crosses the 6th rib in the mid-clavicular line, the 8th rib in the mid-axillary line, and reaches the 10th rib adjacent to the vertebral column posteriorly.

The *oblique fissure*, which divides the lung into upper and lower lobes, is indicated on the surface by a line drawn obliquely downwards and outwards from 1 in (2.5 cm) lateral to the spine of the 5th thoracic vertebra to the 6th costal cartilage about $1\frac{1}{2}$ in (4 cm) from the mid-line. This can be represented approximately by abducting the shoulder to its full extent; the line of the oblique fissure then corresponds to the position of the medial border of the scapula.

The surface marking of the *transverse fissure* (separating the middle and upper lobes of the right lung) is a line drawn horizontally along the 4th costal cartilage and meeting the oblique fissure where the latter crosses the 5th rib.

### The heart

The outline of the heart can be represented on the surface by the irregular quadrangle bounded by the following four points (Fig. 4):

1. the 2nd *left* costal cartilage $\frac{1}{2}$ in (12 mm) from the edge of the sternum;

2.   the 3rd *right* costal cartilage $\frac{1}{2}$ in (12 mm) from the sternal edge;
3.   the 6th right costal cartilage $\frac{1}{2}$ in (12 mm) from the sternum; and
4.   the 5th left intercostal space $3\frac{1}{2}$ in (9 cm) from the mid-line (corresponding to the apex beat).

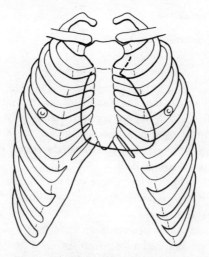

Fig. 4.   The surface markings of the heart.

The *left border* of the heart (indicated by the curved line joining points 1 and 4) is formed almost entirely by the left ventricle, the *lower border* (the horizontal line joining points 3 and 4) corresponds to the right ventricle and the apical part of the left ventricle; the *right border* (marked by the line joining points 2 and 3) is formed by the right atrium.

The surface markings of the vessels of the thoracic wall are of importance if these structures are to be avoided in performing aspiration of the chest. The *internal mammary vessels* run vertically downwards behind the costal cartilages half an inch from the lateral border of the sternum. The intercostal vessels lie immediately below their corresponding ribs (the vein above the artery) so that it is safe to pass a needle immediately *above* a rib, dangerous to pass it immediately *below* (Fig. 7).

# The thoracic cage

The thoracic cage is formed by the vertebral column behind, the ribs and intercostal spaces on either side and the sternum and costal cartilages in front. Above it communicates through the 'thoracic inlet' with the root of the neck; below it is separated from the abdominal cavity by the diaphragm (Fig. 1).

## The thoracic vertebrae

See 'vertebral column', pages 346–7.

## The ribs

The greater part of the thoracic cage is formed by the twelve pairs of ribs. Of these, the first seven are connected anteriorly by way of their costal cartilages to the sternum, the cartilages of the 8th, 9th and 10th articulate each with the cartilage of the rib above and the last two ribs are free anteriorly ('floating ribs').

Each typical rib (Fig. 5) has a *head* bearing two articular facets, for articulation with the numerically corresponding vertebra and the vertebra above, a stout *neck*, a *tubercle* with a rough non-articular portion and a smooth facet, for articulation with the transverse process of the corresponding vertebra, and a long shaft flattened from side to side and divided into two parts by the 'angle' of the rib.

The following are the significant features of the 'atypical' ribs.

*1st Rib* (Fig. 6). This is flattened from above downwards. It is not only the flattest but also the shortest and most curvaceous of all the ribs.

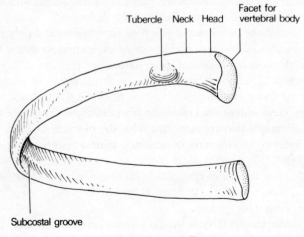

Fig. 5.  A typical rib.

It has a prominent *tubercle* on the inner border of its upper surface for the insertion of scalenus anterior. In front of this tubercle, the subclavian vein crosses the rib; behind the tubercle is the *subclavian groove* where the subclavian artery and lowest trunk of the brachial plexus lie in relation to the bone. It is here that the anaesthetist can infiltrate the plexus with local anaesthetic.

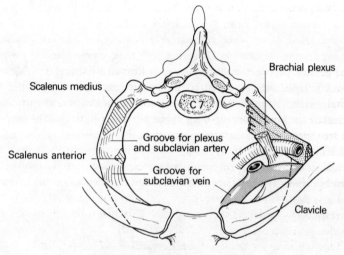

Fig. 6.  Structures crossing the first rib.

The *2nd rib* is much less curved than the 1st and about twice as long. The *10th rib* has only one articular facet on the head.

The *11th* and *12th ribs* are short, have no tubercles and only a single facet on the head. The 11th rib has a slight angle and a shallow subcostal groove; the 12th has neither of these features.

CLINICAL FEATURES

1.  The chest wall of the child is highly elastic and therefore fractures of the rib in children are rare. In adults, the ribs may be fractured by direct violence or indirectly by crushing injuries; in the latter the rib tends to give way at its weakest part in the region of its angle. Not unnaturally, the upper two ribs, which are protected by the clavicle, and the lower two ribs, which are unattached and therefore swing free, are the least commonly injured.

In a severe crush injury to the chest several ribs may fracture in front and behind so that a whole segment of the thoracic cage becomes torn free ('stove-in chest'). With each inspiration this loose flap sucks in, with each expiration it blows out, thus undergoing paradoxical respiratory movement. The associated swinging movements of the mediastinum produce severe shock and this injury calls for urgent treatment by insertion of a chest drain with under-water seal, followed by endotracheal intubation, or tracheostomy, combined with positive pressure respiration.

**2.** *In coarctation of the aorta*, the intercostal arteries derived from the aorta receive blood from the superior intercostals (from the costocervical trunk of the subclavian artery) and from the arteries anastomosing around the scapula. Together with the communication between the internal mammary and inferior epigastric arteries, they provide the principle collaterals between the aorta above and below the block. In consequence, the intercostal arteries undergo dilatation and tortuosity and erode the lower borders of the corresponding ribs to give the characteristic irregular *notching of the ribs* which is very useful in the radiographic confirmation of this lesion.

**3.** *A cervical rib* (Fig. 7) occurs in 0.5 per cent of subjects and is bilateral in half of these. It is attached to the transverse process of the 7th cervical

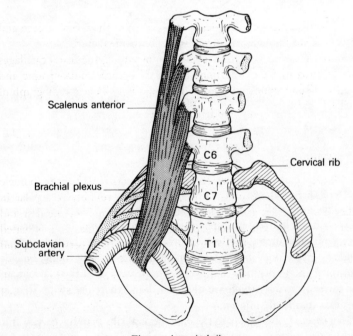

Fig. 7.   A cervical rib.

vertebra and articulates with the 1st (thoracic) rib or, if short, has a free distal extremity. Pressure of such a rib on the lowest trunk of the brachial plexus arching over it may produce paraesthesiae along the ulnar border of the forearm and wasting of the small muscles of the hand (T1). Less commonly vascular changes, even gangrene, may be caused by pressure of the rib on the overlying subclavian artery. This results in a post-stenotic dilatation of the vessel distal to the rib in which thrombus forms from which emboli are thrown off.

## The costal cartilages

These bars of hyaline cartilage serve to connect the upper seven ribs directly to the side of the sternum and the 8th, 9th and 10th ribs to the cartilage immediately above. The cartilages of the 11th and 12th ribs merely join the tapered extremities of these ribs and end in the abdominal musculature.

CLINICAL FEATURES
1. The cartilages add considerable resilience to the thoracic cage and protect the sternum and ribs from more frequent fracture.
2. In old age (and sometimes also in young adults) the costal cartilages undergo progressive ossification; they then become radio-opaque and may give rise to some confusion when examining a chest radiograph of an elderly patient.

## The sternum

This dagger-shaped bone, which forms the anterior part of the thoracic cage, consists of three parts. The *manubrium* is roughly triangular in outline and provides articulation for the clavicles and for the first and upper part of the 2nd costal cartilages on either side. It is situated opposite the 3rd and 4th vertebrae. Opposite the disc between T4 and T5 it articulates at an oblique angle with the *body of the sternum* (placed opposite T5 to T9). This is composed of four parts or 'sternebrae' which fuse between puberty and 25 years of age. Its lateral border is notched to receive part of the 2nd and the 3rd to the 7th costal cartilages. The *xiphoid process* is the smallest part of the sternum and usually remains cartilaginous well into adult life.

CLINICAL FEATURES

1. The attachment of the elastic costal cartilages largely protects the sternum from injury, but indirect violence accompanying fracture dislocation of the thoracic spine may be associated with a sternal fracture. Direct violence to the sternum may lead to displacement of the relatively mobile body of the sternum backwards from the relatively fixed manubrium.

2. In a sternal puncture a wide-bore needle is pushed through the thin layer of cortical bone covering the sternum into the highly vascular spongy bone beneath, and a specimen of bone marrow aspirated with a syringe.

3. In operations on the thymus gland, and occasionally for a retrosternal goitre, it is necessary to split the manubrium in the mid-line in order to gain access to the superior mediastinum. A complete vertical split of the whole sternum is one of the approaches to the heart and great vessels used in modern cardiac surgery.

## The intercostal spaces

There are slight variations between the different intercostal spaces, but typically each space contains three muscles, comparable to those of the abdominal wall, and an associated neuro-vascular bundle (Fig. 8). The muscles are:

1. *The external intercostal*, the fibres of which pass downwards and forwards from the rib above to the rib below and reach from the vertebrae behind to the costo-chondral junction in front, where muscle is replaced by the *anterior intercostal membrane*.

2. *The internal intercostal*, which runs downwards and backwards from the sternum to the angles of the ribs where it becomes the *posterior intercostal membrane*.

3. *The innermost intercostal* is only incompletely separated from the internal intercostal muscle by the neuro-vascular bundle.

Just as in the abdomen, the nerves and vessels of the thoracic wall lie between the middle and innermost layers of muscles. This neurovascular bundle consists, from above downwards, of vein, artery and nerve, the vein lying in a groove on the under-surface of the corresponding rib (remember—V,A,N).

The vessels comprise the posterior and anterior intercostals.

The *posterior intercostal arteries* of the lower nine spaces are branches of the thoracic aorta while the first two are derived from the superior intercostal branch of the costo-cervical trunk, the only branch of the second part of the subclavian artery. Each runs forward in the subcostal groove just beneath its concomitant vein (a tributary of the azygos or hemi-azygos system) to anastomose with the anterior intercostal artery. Each has a number of branches to adjacent muscles, to the skin and to the spinal cord.

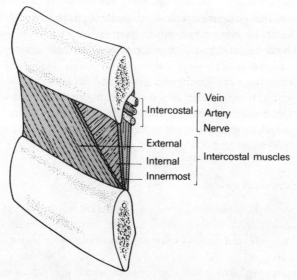

**Fig. 8.** The relationship of an intercostal space.
(Note that a needle passed into the chest immediately above a rib will avoid the neuro-vascular bundle.)

The *anterior intercostal arteries* are branches of the internal mammary artery (1st–6th space) or of its musculo-phrenic branch (7th–9th spaces). Perforating branches pierce the upper 5 or 6 intercostal spaces; those of the 2nd–4th spaces are large and supply the breast.

The *intercostal nerves* are the anterior primary rami of the thoracic nerves, each of which gives off a collateral muscular branch and lateral and anterior cutaneous branches for the innervation of the thoracic and abdominal walls (Fig. 9).

CLINICAL FEATURES
1. Local irritation of the intercostal nerves by such conditions as Pott's disease of the thoracic vertebrae (tuberculosis) may give rise to pain which

is referred to the front of the chest or abdomen in the region of the peripheral termination of the nerves.

**2.** Local anaesthesia of an intercostal space is easily produced by in-filtration around the intercostal nerve trunk and its collateral branch—a procedure known as *intercostal nerve block*.

**3.** In a conventional postero-lateral *thoractomy* (e.g. for a pulmonary lobectomy) an incision is made along the line of the 5th or 6th rib; the periosteum over a segment of the rib is elevated, thus protecting the neurovascular bundle, and the rib excised. Access to the lung or medi-astinum is then gained through the intercostal space which can be opened out considerably owing to the elasticity of the thoracic cage.

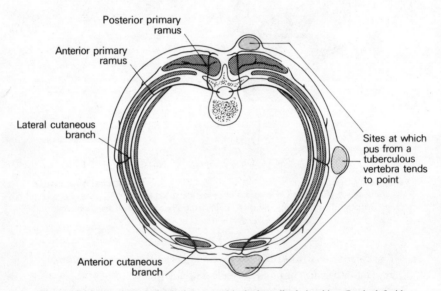

**Fig. 9.** Diagram of a typical spinal nerve and its body-wall relationships. On the left side the sites of eruption of a cold abscess tracking forwards from a diseased vertebra is shown—these occur at the points of emergence of the cutaneous branches.

**4.** Pus from the region of the vertebral column tends to track around the thorax along the course of the neuro-vascular bundle and to 'point' to the three sites of exit of the cutaneous branches of the intercostal nerves, which are lateral to erector spinae, in the mid-axillary line and just lateral to the sternum (Fig. 9).

## THE DIAPHRAGM

The diaphragm is the dome-shaped septum dividing the thoracic from the abdominal cavity. It comprises two portions: a peripheral muscular part which arises from the margins of the thoracic outlet and a centrally-placed aponeurosis (Fig. 10).

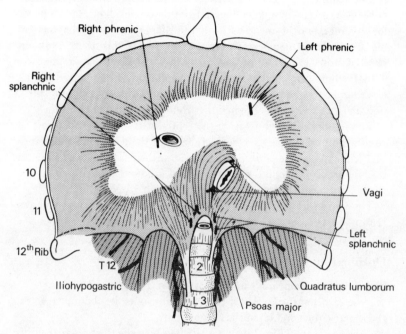

**Fig. 10.** The diaphragm—inferior aspect. The three major orifices, from above downwards, transmit the i.v.c., oesophagus and aorta.

The muscular fibres are arranged in three parts.

1. A *vertebral part* arising from the crura and from the arcuate ligaments. The *right crus* arises from the front of the bodies of the upper three lumbar vertebrae and intervertebral discs; the *left crus* is only attached to the first two vertebrae. The arcuate *ligaments* are a series of fibrous arches, the *medial* being a thickening of the fascia covering psoas major and the *lateral* of fascia overlying quadratus lumborum. The fibrous medial borders of the two crura form a *median arcuate ligament* over the front of the aorta.

2. A *costal part* attached to the inner aspect of the lower six ribs and costal cartilages; and

3.   *A sternal portion* consisting of two small slips from the deep surface of the xiphisternum.

The central tendon, into which the muscular fibres are inserted, is trefoil in shape and is partially fused with the under-surface of the pericardium.

The diaphragm receives its entire motor supply from the phrenic nerve (C3, 4, 5) whose long course from the neck follows the embryological migration of the muscle of the diaphragm from the cervical region (see below). Injury or operative division of this nerve results in paralysis and elevation of the corresponding half of the diaphragm.

Radiographically, paralysis of the diaphragm is recognized by its elevation and paradoxical movement; instead of descending on inspiration it is forced upwards by pressure from the abdominal viscera.

The sensory nerve fibres from the central part of the diaphragm also run in the phrenic nerve, hence irritation of the diaphragmatic pleura (in pleurisy) or of the peritoneum on the under surface of the diaphragm by subphrenic collections of pus or blood produce referred pain in the corresponding cutaneous area, the shoulder-tip.

The peripheral part of the diaphragm, including the crura, receives sensory fibres from the lower intercostal nerves.

### Openings in the diaphragm

The Three main openings in the diaphragm (Figs. 10 and 11) are:
1.   the *aortic* (at the level of T12) which transmits the abdominal aorta, the thoracic duct and often the azygos vein;
2.   the *oesophageal* (T10) which is situated between the muscular fibres of the right crus of the diaphragm and transmits, in addition to the oesophagus, branches of the left gastric artery and vein and the two vagi;
3.   *the opening for the inferior vena cava* (T8) is placed in the central tendon and also transmits the right phrenic nerve.

In addition to these structures, the greater and lesser splanchnic nerves piece the crura and the sympathetic chain passes behind the diaphragm deep to the medial arcuate ligament.

### The development of the diaphragm and the anatomy of diaphragmatic herniae

The diaphragm is formed (Fig. 12) by fusion in the embryo of:
1.   the septum transversum (forming the central tendon);

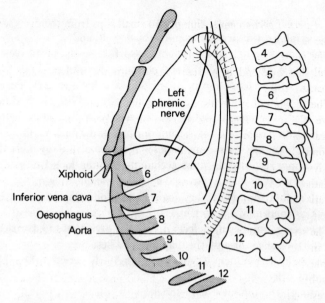

**Fig. 11.** Schematic lateral view of the diaphragm to show the levels at which it is pierced by major structures.

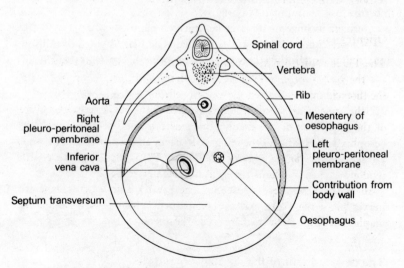

**Fig. 12.** The development of the diaphragm.

This drawing shows the four elements contributing to the diaphragm—(1) the septum transversum, (2) the dorsal mesentery of the oesophagus, (3) the body wall, and (4) the pleuro-peritoneal membrane.

2. the dorsal oesophageal mesentery;
3. a peripheral rim derived from the body wall; and
4. the pleuro-peritoneal membranes, which close the primitive communication between the pleural and peritoneal cavities.

The septum transversum is the mesoderm which, in early development, lies in front of the head end of the embryo. With the folding off of the head, this mesodermal mass is carried ventrally and caudally, to lie in its definitive position at the anterior part of the diaphragm. During this migration, the cervical myotomes and nerves contribute muscle and nerve supply respectively, thus accounting for the long course of the phrenic nerve (C3, 4 and 5) from the neck to the diaphragm.

With such a complex embryological story, one may be surprised to know that congenital abnormalities of the diaphragm are unusual.

However, a number of defects may occur giving rise to a variety of congenital herniae through the diaphragm. These may be:
1. through the foramen of Morgagni; anteriorly between the xiphoid and costal origins;
2. through the foramen of Bochdalek—the pleuro-peritoneal canal—lying posteriorly;
3. through a deficiency of the whole central tendon (occasionally such a hernia may be traumatic in origin);
4. through a congenitally large oesophageal hiatus.

Far more common are the *acquired* hiatus herniae (divided by Allison into sliding and rolling herniae). These occur in patients usually of

**(a) "Sliding hernia"**    **(b) "Rolling hernia"**

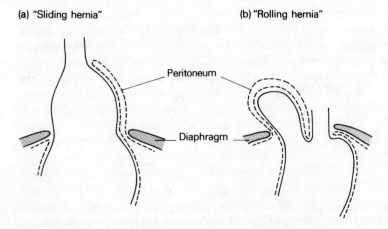

Peritoneum

Diaphragm

Fig. 13.   (a) A sliding hiatus hernia.   (b) A rolling hiatus hernia.

middle age where weakening and widening of the oesophageal hiatus has occurred (Fig. 13).

In the *sliding herniae* the upper stomach and lower oesophagus slide upwards into the chest through the lax hiatus when the patient lies down or bends over; the competence of the cardia is disturbed and peptic juice can therefore regurgitate into the gullet. This may be followed by oesophagitis with consequent heartburn, bleeding, and, eventually, stricture formation.

In the *rolling hernia* (which is far less common) the cardia remains in its normal position and the cardio-oesophageal junction is intact, but the fundus of the stomach rolls up through the hiatus in front of the oesophagus, hence the alternative term of paraoesophageal hernia. In such a case there may be epigastric discomfort, flatulence and even dysphagia, but *no* regurgitation because the cardiac mechanism is undisturbed.

## The movements of respiration

During inspiration the movements of the chest wall and diaphragm result in an increase in all diameters of the thorax. This in turn brings about an increase in the negative intra-pleural pressure and an expansion of the lung tissue. Conversely, in expiration the relaxation of the respiratory muscles and the elastic recoil of the lung reduce the thoracic capacity and force air out of the lungs.

In quiet *inspiration* the first rib remains relatively fixed, but contraction of the external intercostals elevates and, at the same time, everts the succeeding ribs. In the case of the 2nd–7th ribs this principally increases the antero-posterior diameter of the thorax (by the forward thrust of the sternum). The corresponding movement of the lower ribs raises the costal margin and leads mainly to an increase in the transverse diameter of the thorax. The depth of the thorax is increased by the contraction of the diaphragm which draws down its central tendon. Normal quiet *expiration* is aided by the tone of the abdominal musculature which, acting through the contained viscera, forces the diaphragm upwards.

In deep and in forced inspiration additional muscles are called into play (e.g. scalenus anterior, sternomastoid, serratus anterior and pectoralis major) to increase further the capacity of the thorax. Similarly, in deep expiration, forced contraction of the abdominal muscles aids the normal expulsive factors described above.

## THE PLEURAE

The two pleural cavities are totally separate from each other (Fig. 2). Each *pleura* consists of two layers: a *visceral layer* intimately related to the surface of the lung, and a *parietal layer* lining the inner aspect of the chest wall, the upper surface of the diaphragm and the sides of the pericardium and mediastinum. The two layers are continuous in front and behind the root of the lung, but below this the pleura hangs down in a loose fold, the *pulmonary ligament*, which forms a 'dead-space' for distension of the pulmonary veins. The surface markings of the pleura and lungs have already been described in the section on surface anatomy.

Notice that the lungs do not occupy all the available space in the pleural cavity; it is only in forced inspiration that the basal parts of the lung reach the costo-diaphragmatic line of pleural reflexion.

CLINICAL FEATURES

1.   Normally the two pleural layers are in close apposition and the space between them is only a potential one. It may, however, fill with air (pneumothorax), blood (haemothorax) or pus (empyema).

2.   Fluid can be drained from the pleural cavity by inserting a wide-bore needle through an intercostal space (usually the 7th posteriorly). The needle is passed along the superior border of the lower rib, thus avoiding the intercostal nerves and vessels (see Fig. 8). Below the 7th intercostal space there is danger of penetrating the diaphragm.

3.   Since the parietal pleura is segmentally innervated by the intercostal nerves, inflammation of the pleura results in pain referred to the cutaneous distribution of these nerves (i.e. to the thoracic wall or, in the case of the lower nerves, to the anterior abdominal wall).

# The lower respiratory tract

## THE TRACHEA (Figs. 14 and 15)

The trachea is about four and a half inches in length and nearly one inch in diameter. It commences at the lower border of the cricoid cartilage (C6) and terminates by bifurcating at the level of the sternal angle of Louis (T4/5) to form the right and left main bronchi. (In the living subject, the level of bifurcation varies slightly with the phase of respiration; in deep inspiration it descends to T6 and in expiration it rises to T4.)

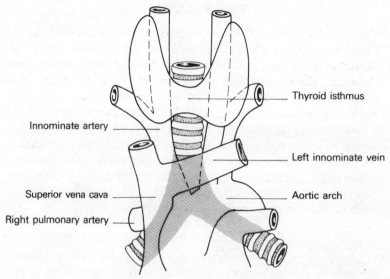

**Fig. 14.**  The trachea and its anterior relationships.

Lying partly in the neck and partly in the thorax, its relations are:

*Cervical*
●anteriorly—the isthmus of thyroid gland, inferior thyroid veins, sterno-hyoid and sternothyroid muscles;
●laterally—the lobes of thyroid gland and the common carotid artery;
●posteriorly—the oesophagus with the recurrent laryngeal nerve lying in the groove between oesophagus and trachea (Fig. 16).

*Thoracic.*
In the superior mediastinum its relations are:
●anteriorly—commencement of the innominate artery (brachiocephalic trunk) and left carotid artery, from the arch of the aorta, and the left innominate vein;
●posteriorly—oesophagus.
●to the left—arch of the aorta, left common carotid and left subclavian arteries, left recurrent laryngeal nerve and pleura;
●to the right—innominate artery, right vagus and pleura.
(Fig. 17.)

*Structure.*
The patency of the trachea is maintained by a series of 15–20 U-shaped cartilages. Posteriorly, where the cartilage is deficient, the trachea is

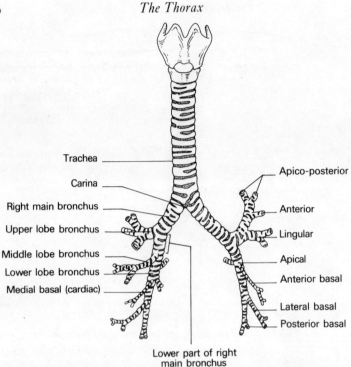

Trachea

Apico-posterior

Carina

Anterior

Right main bronchus

Upper lobe bronchus

Lingular

Middle lobe bronchus

Apical

Lower lobe bronchus

Anterior basal

Medial basal (cardiac)

Lateral basal

Posterior basal

Lower part of right
main bronchus

Fig. 15.   The trachea and main bronchi viewed from the front.

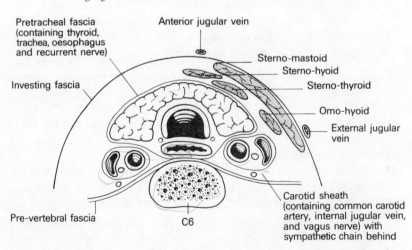

Pretracheal fascia
(containing thyroid,
trachea, oesophagus
and recurrent nerve)

Anterior jugular vein

Sterno-mastoid

Sterno-hyoid

Investing fascia

Sterno-thyroid

Omo-hyoid

External jugular
vein

Pre-vertebral fascia

C6

Carotid sheath
(containing common carotid
artery, internal jugular vein,
and vagus nerve) with
sympathetic chain behind

Fig. 16.   The cervical part of the trachea and its environs in transverse section (through
the 6th cervical vertebra).

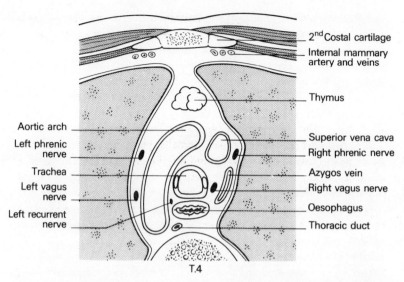

Fig. 17. The thoracic part of the trachea and its environs in transverse section (through the 4th thoracic vertebra).

flattened and its wall completed by fibrous tissue and a sheet of smooth muscle (the trachealis). Within, it is lined by a ciliated columnar epithelium with many goblet cells.

CLINICAL FEATURES

1. Since it contains air, the trachea is more radio-translucent than the neighbouring structures and is seen in P-A and lateral radiographs as a dark area passing downwards, backwards and slightly to the right. In the elderly, calcification of the tracheal rings may be a source of radiological confusion.

2. The trachea may be compressed or displaced by pathological enlargement of the neighbouring structures, particularly the thyroid gland and the arch of the aorta.

3. The intimate relationship between the arch of the aorta and the trachea and left bronchus is responsible for the physical sign known as 'tracheal-tug', characteristic of aneurysms of the aortic arch.

4. *Tracheostomy* may be required for laryngeal obstruction (diphtheria, tumours, inhaled foreign bodies), for the evacuation of excessive secretions (severe post-operative chest infection in a patient who is too weak to cough adequately), and for long-continued artificial respiration (polio-

myelitis, severe chest injuries). It is important to note that respiration is further assisted by considerable reduction of the dead-space air.

The neck is extended and the head held exactly in the mid-line by an assistant. A vertical incision is made downwards from the cricoid cartilage, passing between the anterior jugular veins. Alternatively a more cosmetic transverse skin crease incision, placed half-way between the cricoid and suprasternal notch, is employed. A hook is thrust under the lower border of the cricoid to steady the trachea and pull it forward. The pretracheal fascia is split longitudinally, the isthmus of the thyroid either pushed upwards or divided between clamps and the cartilages of the trachea clearly exposed. A circular opening is then made into the trachea to admit the tracheostomy tube.

In children the neck is relatively short and the left innominate vein may come up above the suprasternal notch so that dissection is rather more difficult and dangerous. This difficulty is made greater because the child's trachea is softer and more mobile than the adult's and therefore not so readily identified and isolated. Its softness means that care must be taken, in incising the child's trachea, not to let the scalpel plunge through and damage the underlying oesophagus.

In contrast, the trachea may be ossified in the elderly and small bone shears requred to open into it.

The golden rule of tracheostomy—based entirely on anatomical considerations—is *'stick exactly to the mid-line'*.

If this is not done, major vessels are in jeopardy and it is possible, although the student may not credit it, to miss the trachea entirely.


## THE BRONCHI (Fig. 15)

*The right main bronchus* is wider, shorter and more vertical than the left. It is about 1 in (2.5 cm) long and passes directly to the root of the lung at T5. Before joining the lung it gives off its *upper lobe branch*, and then passes below the pulmonary artery to enter the hilum of the lung. It has two important relations: the azygos vein, which arches over it from behind to reach the superior vena cava, and the pulmonary artery which lies first below and then anterior to it.

*The left main bronchus* is nearly 2 in (5 cm) long and passes downwards and outwards below the arch of the aorta, in front of the oesophagus and descending aorta. It reaches the hilum of the lung opposite T6 and lies at first behind and then below the left pulmonary artery.

*Surface markings of lung . 6 8 10.*
*pleura . 8 10 12.*

CLINICAL FEATURES

1. The greater width and more vertical course of the right bronchus accounts for the greater tendency for foreign bodies and aspirated material to pass into the right bronchus (and thence especially into the middle and lower lobes of the ring lung) rather than into the left.

2. The inner aspect of the whole of the trachea, the main and lobar bronchi and the commencement of the first segmental divisions can be seen at bronchoscopy.

3. Widening and distortion of the angle between the bronchi (the carina) as seen at bronchoscopy is a serious prognostic sign, since it usually indicates carcinomatous involvement of the tracheo-bronchial lymph nodes around the bifurcation of the trachea.

## THE LUNGS (Figs. 18 and 19)

Each lung is conical in shape, having a blunt apex which reaches above the sternal end of the 1st rib, a concave base overlying the diaphragm, an extensive costo-vertebral surface moulded to the form of the chest wall and a mediastinal surface which is concave to accommodate the pericardium.

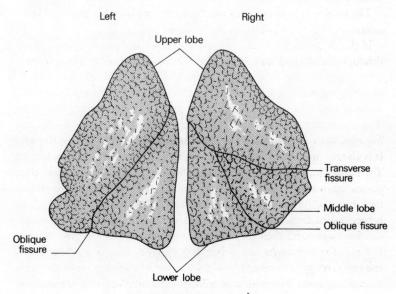

Fig. 18. The lungs, lateral aspects.

The right lung is slightly larger than the left and is divided into three lobes—upper, middle and lower, by the oblique and horizontal fissures. The left lung has only an oblique fissure and hence only two lobes.

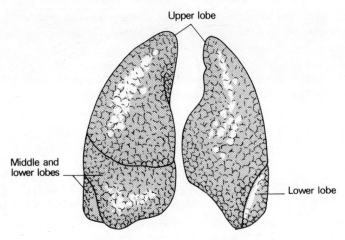

Upper lobe

Middle and lower lobes

Lower lobe

Fig. 19.   The lungs, anterior aspects.

### The broncho-pulmonary segments of the lungs (Figs. 20 and 21)

Recent advances in thoracic surgery have emphasized the importance of a knowledge of the finer arrangement of the bronchial tree. Each lobe of the lung is subdivided into a number of broncho-pulmonary segments each of which is supplied by a segmental bronchus, artery and vein. These segments are wedge-shaped with their apices at the hilum and bases at the lung surface; if excised accurately along their boundaries (which are marked by intersegmental veins) there is little bleeding or alveolar air leakage from the raw lung surface.

The names and arrangement of the bronchi are given in Table 1; each broncho-pulmonary segment takes its title from that of its supplying segmental bronchus (listed in the right-hand column of the table).

The left upper lobe has a *lingular segment*, supplied by the *lingular bronchus* from the main upper lobe bronchus. This lobe is equivalent to the *right middle lobe* whose bronchus arises as a branch from the main bronchus. Apart from this, differences between the two sides are very slight; on the left, the upper lobe bronchus gives off a combined apico-

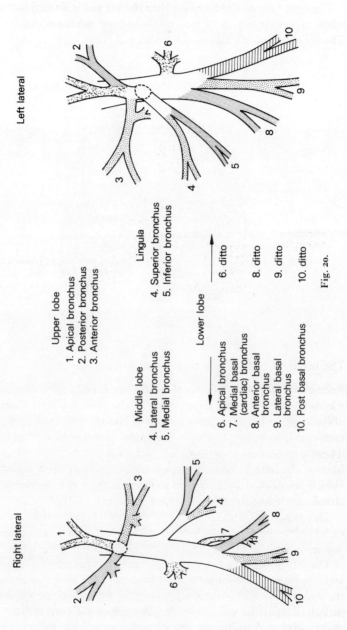

Left lateral

Right lateral

Upper lobe
1. Apical bronchus
2. Posterior bronchus
3. Anterior bronchus

Lingula
4. Superior bronchus
5. Inferior bronchus

Middle lobe
4. Lateral bronchus
5. Medial bronchus

Lower lobe

6. Apical bronchus
7. Medial basal (cardiac) bronchus
8. Anterior basal bronchus
9. Lateral basal bronchus
10. Post basal bronchus

6. ditto

8. ditto

9. ditto

10. ditto

Fig. 20.

**Table 1.** The named divisions of the main bronchi

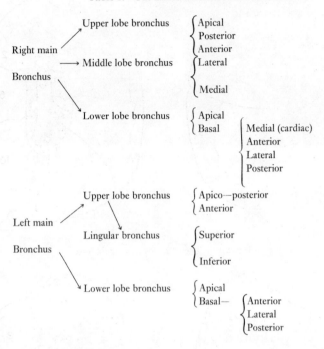

| | | |
|---|---|---|
| | Upper lobe bronchus | Apical / Posterior / Anterior |
| Right main Bronchus | Middle lobe bronchus | Lateral / Medial |
| | Lower lobe bronchus | Apical / Basal — Medial (cardiac) / Anterior / Lateral / Posterior |
| | Upper lobe bronchus | Apico—posterior / Anterior |
| Left main Bronchus | Lingular bronchus | Superior / Inferior |
| | Lower lobe bronchus | Apical / Basal — Anterior / Lateral / Posterior |

posterior segmental bronchus and an anterior branch, whereas all three branches are separate on the right side.

On the right also there is a small medial (or cardiac) lower lobe bronchus which is absent on the left, the lower lobes being otherwise mirror images of each other.

*Blood supply.*

Mixed venous blood is returned to the lungs by the pulmonary arteries; the air passages are themselves supplied by the bronchial arteries which are small branches of the descending aorta.

*Lymphatic drainage.*

The lymphatics of the lung drain centripetally from the pleura towards the hilum. From the *broncho-pulmonary lymph nodes* in the hilum, efferent lymph channels pass to the *tracheo-bronchial nodes* at the bifurcation of the trachea, thence to the *paratracheal nodes* and the mediastinal

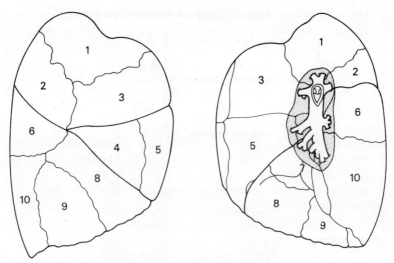

Fig. 21a.    The segments of the right lung.

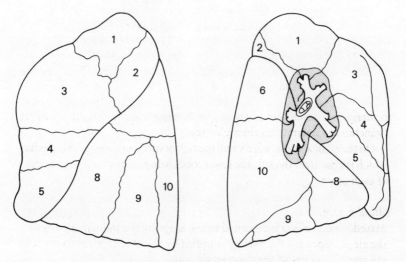

Fig. 21b.    The segments of the left lung.

lymph trunks to drain usually directly into the innominate veins or, rarely, indirectly via the thoracic or right lymphatic duct.

*Nerve supply.*
The pulmonary plexuses derive fibres from both the vagi and the sym-

pathetic trunk. They supply efferents to the bronchial musculature and receive afferents from the mucous membrane of the bronchioles and from the alveoli.

# The mediastinum

For descriptive purposes the mediastinum is divided by a line drawn horizontally from the sternal angle to the lower border of T4 (angle of Louis) into superior and inferior mediastina. The latter is further subdivided into an anterior mediastinum in front of the pericardium, a middle mediastinum containing the pericardium itself with the heart and great vessels, and a posterior mediastinum between the pericardium and the lower eight thoracic vertebrae (Fig. 22).

## THE PERICARDIUM

The heart and the roots of the great vessels are contained within the conical *fibrous pericardium* the apex of which is fused with the adventitia of the great vessels and the base with the central tendon of the diaphragm. Anteriorly it is related to the body of the sternum, the 3rd–6th costal cartilages and the anterior borders of the lungs; posteriorly, to the oesophagus, descending aorta, and vertebrae T5–T8, and on either side to the roots of the lungs, the mediastinal pleura and the phrenic nerves.

The inner aspect of the fibrous pericardium is lined by the *parietal layer of serous pericardium*. This, in turn, is reflected around the roots of the great vessels to become continuous with the *visceral layer* or *epicardium*. The lines of pericardial reflexion are marked on the posterior surface of the heart (Fig. 23) by the *oblique sinus*, bounded by the inferior vena cava and the four pulmonary veins, which form a recess between the left atrium and the pericardium, and the *transverse sinus* between the superior vena cava and left atrium behind and the pulmonary trunk and aorta in front.

## THE HEART (Fig. 24)

Because of its great importance, no excuse need be offered for dealing with the heart in considerable detail.

The heart is irregularly conical in shape, and it is placed obliquely in the middle mediastinum. Viewed from the front, portions of all the

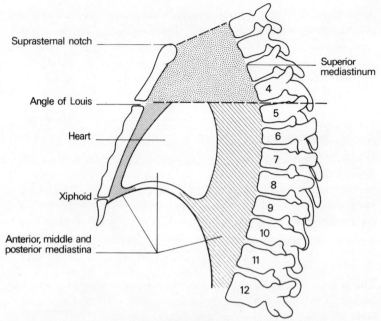

**Fig. 22.** The subdivisions of the mediastinum.

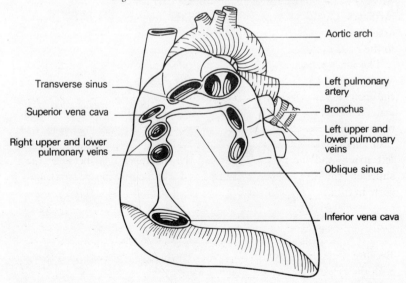

**Fig. 23.** The transverse and oblique sinuses of the pericardium. In this illustration the heart has been removed from the pericardial sac, which is seen in anterior view.

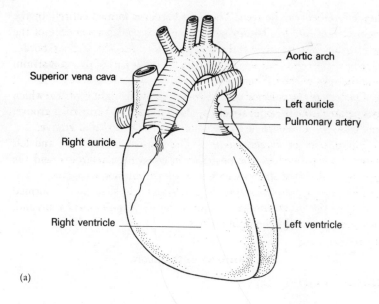

Aortic arch

Superior vena cava

Left auricle

Pulmonary artery

Right auricle

Right ventricle

Left ventricle

(a)

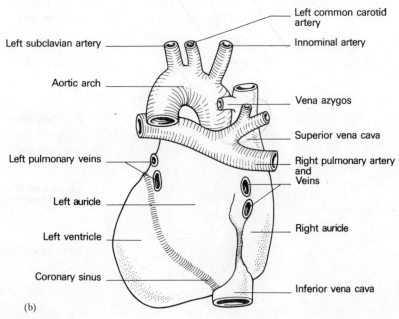

Left common carotid artery

Left subclavian artery

Innominal artery

Aortic arch

Vena azygos

Superior vena cava

Left pulmonary veins

Right pulmonary artery and Veins

Left auricle

Left ventricle

Right auricle

Coronary sinus

Inferior vena cava

(b)

Fig. 24.  The heart, (a) anterior and (b) posterior aspects.

heart chambers can be seen. The right border is formed entirely by the right atrium, the left border partly by the auricular appendage of the left atrium but mainly by the left ventricle, and the inferior border chiefly by the right ventricle but also by the lower part of the right atrium and the apex of the left ventricle.

The bulk of the anterior surface is formed by the right ventricle which is separated from the right atrium by the vertical atrio-ventricular groove, and from the left ventricle by the anterior inter-ventricular groove.

The inferior or *diaphragmatic surface* consists of the right and left ventricles separated by the posterior inter-ventricular groove and the portion of the right atrium which receives the inferior vena cava.

The base or *posterior surface* is quadrilateral in shape and is formed mainly by the left atrium with the openings of the pulmonary veins and, to a lesser extent, by the right atrium.

## Chambers of the heart

### Right atrium (Fig. 25)

The right atrium receives the superior vena cava in its upper and posterior part, the inferior vena cava and coronary sinus in its lower part,

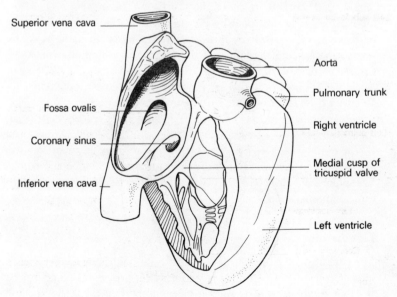

**Fig. 25.** The interior of the right atrium and ventricle.

and the anterior cardiac vein (draining much of the front of the heart) anteriorly. Running more or less vertically downwards between the venae cavae is a distinct muscular ridge, the *crista terminalis* (indicated on the outer surface of the atrium by a shallow groove—the *sulcus terminalis*). This ridge separates the smooth-walled posterior part of the atrium, derived from the sinus venosus, from the rough-walled anterior portion which is prolonged into the *auricular appendage* and which is derived from the true fetal atrium.

The openings of the inferior vena cava and the coronary sinus are guarded by rudimentary valves; that of the inferior vena cava being continuous with the *annulus ovalis* around the shallow depression on the atrial septum, the *fossa ovalis*, which marks the site of the fetal *foramen ovale*.

### Right ventricle (Fig. 25)

The right ventricle is joined to the right atrium by way of the vertically disposed tricuspid valve, and with the pulmonary trunk through the pulmonary valve. A muscular ridge, the *infundibulo-ventricular crest*, between the atrio-ventricular and pulmonary orifices, separates the 'inflow' and 'outflow' tracts of the ventricle. The inner aspect of the inflow tract path is marked by the presence of a number of irregular muscular elevations (*trabeculae carneae*) from some of which the *papillary muscles* project into the lumen of the ventricle and find attachment to the free borders of the cusps of the tricuspid valve by way of the *chordae tendineae*. The *moderator band* is a muscular bundle crossing the ventricular cavity from the interventricular septum to the anterior wall and is of some importance since it conveys the right branch of the atrio-ventricular bundle to the ventricular muscle.

The outflow tract of the ventricle or *infundibulum* is smooth-walled and is directed upwards and to the right towards the pulmonary trunk. The pulmonary orifice is guarded by the *pulmonary valves* comprising three semilunar cusps.

### Left atrium

The left atrium is rather smaller than the right but has somewhat thicker walls. On the upper part of its posterior wall it presents the openings of the four pulmonary veins and on its septal surface there is a shallow depression corresponding to the fossa ovalis of the right atrium. As on the right side, the main part of the cavity is smooth-walled but the surface of the auricle is marked by a number of ridges due to the underlying pectinate muscles.

### Left ventricle (Fig. 26)

The left ventricle communicates with the left atrium by way of the *mitral valve*, which possesses a large anterior and a smaller posterior cusp attached to papillary muscles by chordae tendineae. With the exception of the fibrous *vestibule* immediately below the aortic orifice, the wall of the left ventricle is marked by thick trabeculae carneae.

The aortic orifice is guarded by the three semilunar cusps of the *aortic valve*, immediately above which are the dilated *aortic sinuses*. The mouths of the right and left coronary arteries are seen in the anterior and left posterior sinus respectively.

### Systems of the heart

### The conducting system of the heart

This consists of specialized cardiac muscle found in the sinu-atrial node and in the atrio-ventricular node and bundle. The heart-beat is initiated in the *sinu-atrial* node (the 'pacemaker of the heart'), situated in the upper part of the crista terminalis just to the right of the opening of the superior vena cava into the right atrium. From there the cardiac impulse spreads throughout the atrial musculature to reach the *atrio-ventricular node* lying in the atrial septum immediately above the opening of the coronary sinus. The impulse is then conducted to the ventricles by way of the specialized tissue of the *atrio-ventricular bundle (of His)*. This bundle divides at the junction of the membranous and muscular parts of the interventricular septum into its right and left branches which run immediately beneath the endocardium to activate all parts of the ventricular musculature.

### The blood supply of the heart (Fig. 27)

The heart's blood supply is derived from the right and left coronary arteries whose main branches lie in the inter-ventricular and atrio-ventricular grooves.

The *right coronary artery* arises from the anterior aortic sinus and

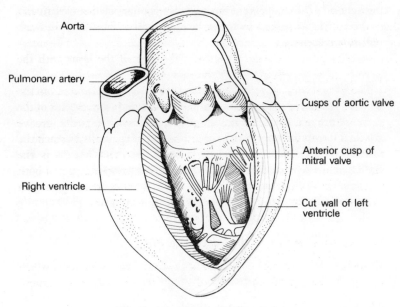

Aorta

Pulmonary artery

Right ventricle

Cusps of aortic valve

Anterior cusp of
mitral valve

Cut wall of left
ventricle

Fig. 26.   The interior of the left ventricle.

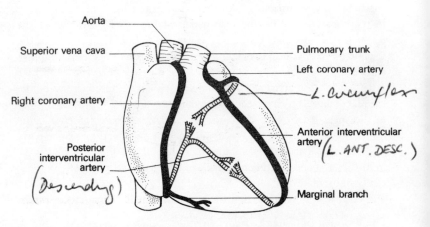

Aorta

Superior vena cava

Right coronary artery

Posterior
interventricular
artery

*(Descending)*

Pulmonary trunk

Left coronary artery

*L. Circumflex*

Anterior interventricular
artery *(L. ANT. DESC.)*

Marginal branch

Fig. 27.   The coronary arteries.

passes forwards between the pulmonary trunk and the right atrium to
descend in the right part of the atrio-ventricular groove. At the inferior
border of the heart it continues along the atrio-ventricular groove to

anastomose with the left coronary at the posterior inter-ventricular groove. It gives off a *marginal branch* along the lower border of the heart and an *interventricular branch* which runs forward in the inferior inter-ventricular groove to anastomose near the apex of the heart with the corresponding branch of the left coronary artery.

The *left coronary artery*, which is larger than the right, arises from the left posterior aortic sinus. Passing first behind and then to the left of the pulmonary trunk, it reaches the left part of the atrio-ventricular groove in which it runs laterally round the left border of the heart to reach the posterior interventricular groove. Its most important branch is the *anterior interventricular artery* which supplies the anterior aspect of both ventricles and passes around the apex of the heart to anastomose with the posterior interventricular branch of the right coronary.

### The venous drainage of the heart (Fig. 28)

About two-thirds of the venous drainage of the heart is by veins which accompany the coronary arteries and which open into the right atrium.

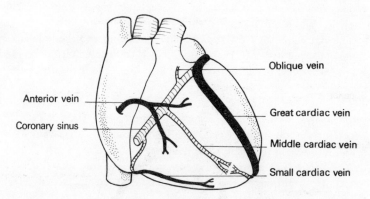

Anterior vein

Coronary sinus

Oblique vein

Great cardiac vein

Middle cardiac vein

Small cardiac vein

Fig. 28.   The coronary veins.

The rest of the blood drains by means of small veins *(venae cordis minimae)* directly into the cardiac cavity.

The *coronary sinus* lies in the posterior atrio-ventricular groove and opens into the right atrium just to the left of the mouth of the inferior vena cava.

It receives:

1.   the *great cardiac vein* in the anterior interventricular groove;

2. the *middle cardiac vein* in the inferior interventricular groove; and
3. the *small cardiac vein*—accompanying the marginal artery along the lower border of the heart.

The *anterior cardiac vein* lies in the anterior atrio-ventricular groove, drains much of the anterior surface of the heart and opens directly into the right atrium.

## Nerve supply

The nerve supply of the heart is derived from the vagus (cardio-inhibitor) and the cervical and upper thoracic sympathetic ganglia (cardio-accelerator) by way of the superficial and deep cardiac plexuses.

## The development of the heart

The primitive heart is a single tube which soon shows grooves demarcating the *sinus venosus, atrium, ventricle* and *bulbus cordis* from behind forward. As this tube enlarges it kinks so that its caudal end, receiving venous blood, comes to lie behind its cephalic end with its emerging arteries (Fig. 29).

The sinus venosus later absorbs into the atrium and the bulbus becomes incorporated into the ventricle so that, in the fully developed heart, the atria and great veins come to lie posterior to the ventricles and the roots of the great arteries.

The boundary tissue between the primitive single atrial cavity and single ventricle grows out as a *dorsal* and a *ventral endocardial cushion* which meet in the mid-line thus dividing the common atrio-ventricular orifice into a right (tricuspid) and left (mitral) orifice.

The division of the primitive atrium into two is a complicated process but an important one in the understanding of congenital septal defects (Fig. 30). A partition, the *septum primum*, grows downwards from the posterior and superior walls of the primitive common atrium to fuse with the endocardial cushions. Before fusion is complete, however, a hole appears in the upper part of this septum which is termed *the foramen secundum in the septum primum.*

A second membrane, the *septum secundum*, then develops to the right of the primum but this is never complete; it has a free lower edge which does, however, extend low enough for this new septum to overlap the foramen secundum in the septum primum and hence to close it.

The two overlapping defects in the septa form the valve-like *foramen*

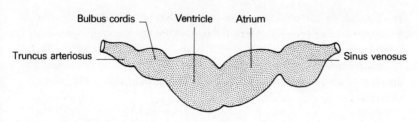

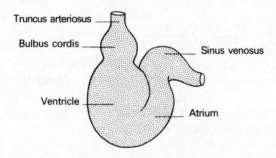

Fig. 29.   The coiling of the primitive heart tube into its definitive form.

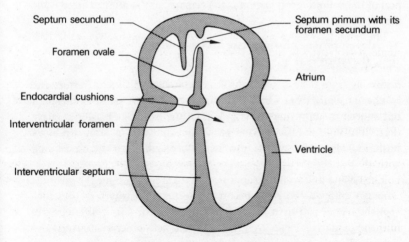

Fig. 30.   The development of the chambers of the heart.
(Note the septum primum and septum secundum which form the inter-atrial septum,
leaving the foramen ovale as a valve-like opening passing between them.)

*ovale* which shunts blood from the right to left heart in the foetus (see 'fetal circulation' below). After birth, this foramen usually becomes completely fused leaving only the fossa ovalis on the septal wall of the right atrium as its memorial. In about 10 per cent of adult subjects, however, a probe can still be insinuated through an anatomically patent, although functionally sealed, foramen.

The division of the ventricle into two is a much simpler affair. The bulk of the interventricular septum is an upgrowth from the apex of the heart, but the small pars membranacea septi in the uppermost part of this partition is a downgrowth from the inter-atrial septum. The single *truncus arteriosus*, emerging from the bulbus cordis and the common ventricle, is divided into the aorta and pulmonary trunk by a spiral septum, hence the spiral relationship of these vessels to each other.

The primitive *sinus venosus* absorbs into the right atrium so that the venae cavae draining into the sinus come to open separately into this atrium. The smooth-walled part of the adult atrium represents the contribution of the sinus venosus, the pectinate part represents the portion derived from the primitive atrium.

Rather similarly the adult left atrium has a double origin. The original single pulmonary venous trunk entering the left atrium becomes absorbed into it, and donates the smooth-walled part of this chamber with the pulmonary veins entering as four separate openings; the trabeculated part of the definitive left atrium is the remains of the original atrial wall.

### The development of the aortic arches and their derivatives (Fig. 31)

Emerging from the bulbus cordis is a common arterial trunk termed the *truncus arteriosus* from which arise six pairs of aortic arches, equivalent to the arteries supplying the gill clefts of the fish. These arteries curve dorsally around the pharynx on either side and join to form two longitudinally placed dorsal aortae which fuse distally into the descending aorta.

The 1st and 2nd arches disappear; the 3rd arches become the carotids. The 4th arch on the right becomes the innominate artery (brachiocephalic trunk) and right subclavian artery; on the left, it differentiates into the definitive aortic arch, gives off the left subclavian artery and links up distally with the descending aorta.

The 5th arch artery is rudimentary and disappears.

When the truncus arteriosus splits longitudinally to form the ascending

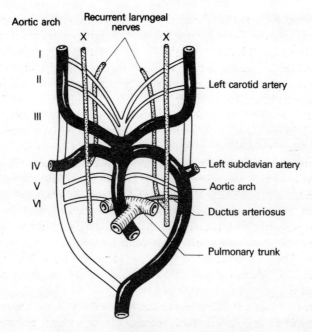

**Fig. 31.** The aortic arches and their derivatives.

This diagram explains the relationship of the right recurrent laryngeal nerve to the right subclavian artery and the left nerve to the aortic arch and the ligamentum arteriosum (or to a patent ductus arteriosus).

aorta and pulmonary trunk, the 6th arch unlike the others remains linked with the latter and forms the right and left pulmonary arteries. On the left side this arch retains its connection with the dorsal aorta to form the *ductus arteriosus* (the ligamentum arteriosum of adult anatomy).

This asymmetrical development of the aortic arches accounts for the different course taken by the recurrent laryngeal nerve on each side. In the early fetus the vagus nerve lies lateral to the primitive pharynx, separated from it by the aortic arches. What are to become the recurrent laryngeal nerves pass medially, caudal to the aortic arches, to supply the developing larynx. With elongation of the neck and caudal migration of the heart, the recurrent nerves are caught up and dragged down by the descending aortic arches. On the right side the 5th and distal part of the 6th arch absorb, leaving the nerve to hook round the 4th arch (i.e. the right subclavian artery). On the left side, the nerve remains looped around the persisting distal part of the 6th arch (the ligamentum arteriosum) which is overlapped and dwarfed by the arch of the aorta.

## The fetal circulation (Fig. 32)

The circulation of the blood in the embryo is a remarkable example of economy in Nature and results in the shunting of well-oxygenated blood from the placenta to the brain and heart, leaving relatively desaturated blood for less essential structures.

Blood is returned from the placenta by the umbilical vein to the inferior vena cava and thence the right atrium, most of it by-passing the liver in the *ductus venosus*. Relatively little mixing of oxygenated and de-

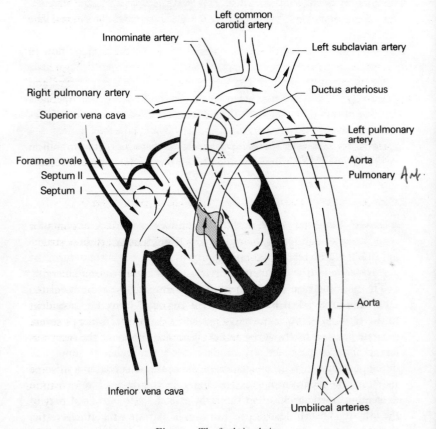

Fig. 32. The fetal circulation.

oxygenated blood occurs in the right atrium since the valve overlying the orifice of the inferior vena cava serves to direct the flow of oxygenated blood from that vessel through the *foramen ovale* into the left atrium, while the de-oxygenated stream from the superior vena cava is directed through the tricuspid valve into the right ventricle. From the left atrium the oxygenated blood (together with a small amount of de-oxygenated blood from the lungs) passes into the left ventricle and hence into the ascending aorta for the supply of the brain and heart via the vertebral, carotid and coronary arteries.

As the lungs of the fetus are inactive, most of the de-oxygenated blood from the right ventricle is short circuited by way of the *ductus arteriosus* from the pulmonary trunk into the descending aorta. This blood supplies the abdominal viscera and the lower limbs and is shunted to the placenta, for oxygenation, along the *umbilical arteries* arising from the internal iliac arteries.

At birth, expansion of the lungs leads to an increased blood flow in the pulmonary arteries; the resulting pressure changes in the two atria bring the *septum primum* and *septum secundum* into apposition and effectively close off the foramen ovale. At the same time active contraction of the muscular wall of the ductus arteriosus results in a functional closure of this arterial shunt and, in the course of the next 2–3 months, its complete obliteration. Similarly, ligature of the umbilical cord is followed by thrombosis and obliteration of the umbilical vessels.

### Congenital abnormalities of the heart and great vessels

The complex development of the heart and major arteries accounts for the multitude of congenital abnormalities which may affect these structures, either alone or in combination.

*Dextro-rotation of the heart* means that this organ and its emerging vessels lie as a mirror-image to the normal anatomy. It may be associated with reversal of all the intra-abdominal organs; I have seen a student correctly diagnose acute appendicitis as the cause of a patient's severe *left* iliac fossa pain because he found that the apex beat of the heart was on the right side!

*Septal defects* include a patent foramen ovale (which occurs in some 10 per cent of subjects) and atrial or ventricular septal defects. An ostium secundum defect lies high up in the atrial wall and is relatively easy to close surgically. An ostium primum defect lies immediately above the atrio-ventricular boundary and may be associated with a defect of the pars membranacea septi of the ventricular septum; it thus presents a more serious surgical problem.

Occasionally the ventricular septal defect is so huge that the ventricles form a single cavity, giving a *trilocular heart*.

*Congenital pulmonary stenosis* may affect the trunk of the pulmonary artery, its valve or the infundibulum of the right ventricle. If stenosis occurs in conjunction with a septal defect, the compensatory hypertrophy of the right ventricle (developed to force blood through the pulmonary obstruction) develops a sufficiently high pressure to shunt blood through the defect into the left heart; this mixing of the de-oxygenated right heart blood with the oxygenated left-sided blood results in the child being *cyanosed at birth*.

The commonest congenital abnormality combination causing cyanosis is *Fallot's tetralogy* (Fig. 33) comprising pulmonary stenosis, right ventricular hypertrophy (as a result of this), a septal defect and an overriding aorta, whose mouth comes to lie over the septal defect and receives blood from both ventricles.

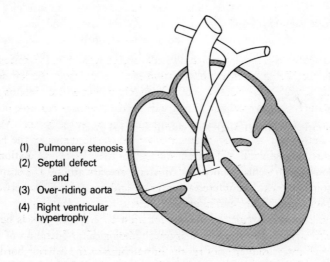

(1) Pulmonary stenosis
(2) Septal defect
    and
(3) Over-riding aorta
(4) Right ventricular
    hypertrophy

Fig. 33.   The tetralogy of Fallot.

*A patent ductus arteriosus* (Fig 34a) is a relatively common congenital defect. If left uncorrected it causes progressive work hypertrophy of the left heart and pulmonary hypertension.

*Aortic coarctation* (Fig. 34b) is thought to be due to an abnormality

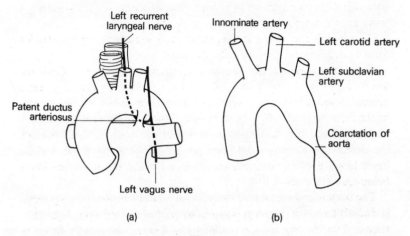

Fig. 34.
a.   Patent ductus arteriosus—showing its close relationship to the left recurrent laryngeal-
nerve.
b.   Coarctation of the aorta.

of the obliterative process which normally occludes the ductus arteriosus.
There may be an extensive obstruction of the aorta from the left subclavian artery to the ductus, which is widely patent and maintains the
circulation to the lower part of the body; often there are multiple other
defects and frequently infants so afflicted die at an early age. More
commonly there is a short segment involved in the region of the ligamentum arteriosum or still patent ductus. In these cases, circulation to
the lower limb is maintained via collateral arteries around the scapula
anastomosing with the intercostal arteries, and via the link up between
the internal mammary and inferior epigastric arteries.

Clinically this circulation may be manifest by enlarged vessels being
palpable around the scapular margins; radiologically, dilatation of the
engorged intercostal arteries results in notching of the inferior borders
of the ribs.

Abnormal development of the primitive aortic arches may result in
the aortic arch being on the right or actually being double. An abnormal
right subclavian artery may arise from the dorsal aorta and pass behind
the oesophagus—a rare cause of difficulty in swallowing (dysphagia
lusoria).

Rarely the division of the truncus into aorta and pulmonary artery is incomplete, leaving an *aorto-pulmonary window*, the most unusual congenital fistula between the two sides of the heart.


## THE SUPERIOR MEDIASTINUM

This is bounded in front by the manubrium sterni and behind by the first 4 thoracic vertebrae. Above, it is in direct continuity with the root of the neck and below it is continuous with the three compartments of the inferior mediastinum. Its principal contents are: the great vessels, trachea, oesophagus, thoracic duct, vagi, left recurrent laryngeal nerve and the phrenic nerves (Fig. 17).

The *arch of the aorta* is directed from before backwards, its three great branches, the *innominate, left carotid* and *left subclavian arteries*, ascend to the thoracic inlet, the first two forming a V around the trachea. The *innominate (brachiocephalic) veins* lie in front of the arteries, the left running almost horizontally across the superior mediastinum and the right vertically downwards; the two unite to form the *superior vena cava*. Separating the veins from the arteries are the phrenic and vagus nerves. Posteriorly lies the trachea with the oesophagus immediately behind it lying against the vertebral column.


## THE OESOPHAGUS

The oesophagus, which is 10 in (25.5 cm) long, extends from the lower broder of the cricoid cartilage to the cardiac orifice of the stomach (Fig. 35). It has the following course and relations:

*In the neck* it commences in the median plane and deviates slightly to the left as it approaches the thoracic inlet. The trachea and the thyroid gland are its immediate anterior relations, the lower cervical vertebrae and prevertebral fascia are behind it and on either side it is related to the common carotid arteries and the recurrent laryngeal nerves. On the left side it is also related to the subclavian artery and the terminal part of the thoracic duct (Fig. 16).

*The thoracic part* traverses first the superior and then the posterior mediastinum. From being somewhat over to the left, it returns to the mid-line at T5 then passes downwards, forwards and to the left to reach the oesophageal opening in the diaphragm (T10). For convenience, the relations of this part are given in sequence from above downwards.

*Anteriorly*, it is crossed by the trachea, the left bronchus (which constricts

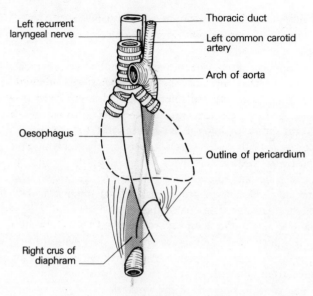

Left recurrent laryngeal nerve

Thoracic duct

Left common carotid artery

Arch of aorta

Oesophagus

Outline of pericardium

Right crus of diaphram

Fig. 35. The oesophagus and its relations.

it), the pericardium (separating it from the left atrium) and the diaphragm.

*Posteriorly*, lie the thoracic vertebrae, the thoracic duct, the azygos vein and its tributaries and, near the diaphragm, the descending aorta.

*On the left side* it is related to the terminal part of the aortic arch, the left recurrent laryngeal nerve, left subclavian artery, the thoracic duct and the left pleura.

*On the right side* there is the pleura and the azygos vein.

Below the root of the lung the vagi form a plexus on the surface of the oesophagus, the left vagus lying anteriorly, the right posteriorly.

*In the abdomen*. Passing forwards through the opening in the right crus of the disphragm, the oesophagus comes to lie in the oesophageal groove on the posterior surface of the left lobe of the liver, covered by peritoneum on its anterior and left aspects. Behind it is the left crus of the diaphragm.

*Structure*. The oesophagus is made up of:

1.   an outer connective tissue sheath of areolar tissue;
2.   a muscular layer of external longitudinal and internal circular fibres which are striated in the upper two-thirds and smooth in the lower one-third;

3. a submucous layer containing mucous glands;
4. a mucosa of stratified epithelium passing abruptly into the columnar epithelium of the stomach.

*Blood supply* is from the inferior thyroid artery, branches of the descending thoracic aorta and the left gastric artery. *The veins* from the cervical part drain into the inferior thyroid veins, from the thoracic portion into the azygos vein and from the abdominal portion partly into the azygos and partly into the left gastric veins.

The *lymphatic* drainage is from a peri-oesophageal lymph plexus into the posterior mediastinal nodes which drain both into the supra-clavicular nodes and into nodes around the left gastric vessels. It is not uncommon to be able to palpate hard, fixed supraclavicular nodes in patients with advanced oesophageal cancer.

*Radiographically* the oesophagus is studied by X-rays taken after a barium swallow, in which it is seen lying in the retro-cardiac space just in front of the vertebral column. Anteriorly the normal oesophagus is indented from above downwards by the three most important structures that cross it, the arch of the aorta, the left bronchus and the left atrium.

CLINICAL FEATURES

1. For oesophagoscopy, measurements are made from the upper incisor teeth; the three important levels 7 in (17.5 cm), 11 in (28 cm) and 17 in (43 cm) corresponding to the commencement of the oesophagus, the point at which it is crossed by the left bronchus and its termination respectively.

2. These three points also indicate the narrowest parts of the oesophagus; the sites at which, as might be expected, swallowed foreign bodies are most likely to become impacted and strictures to occur after swallowing corrosive fluids.

3. The anastomosis between the azygos (systemic) and left gastric (portal) venous tributaries in the oesophageal veins is of great importance. In portal hypertension these veins distend into large collateral channels which may then rupture with severe haemorrhage (probably as a result of peptic ulceration of the overlying mucosa).

4. Use is made of the close relationship between the oesophagus and the left atrium in determining the degree of left atrial enlargement in mitral stenosis; a barium swallow may show marked backward displacement of the oesophagus caused by the dilated atrium.

## Development of the oesophagus

The oesophagus develops from the distal part of the primitive fore-gut. From the floor of the fore-gut also differentiate the larynx and trachea, first as a groove (the laryngo-tracheal groove) which then converts into a tube, a bud on each side of which develops and ramifies into the lung.

This close relationship between the origins of the oesophagus and trachea accounts for the relatively common malformation in which the upper part of the oesophagus ends blindly while the lower part opens into the lower trachea at the level of T4 *(oesophageal atresia with oesophago-tracheal fistula)*. Less commonly the upper part of the oesophagus opens into the trachea, or oesophageal atresia occurs without concomitant fistula into the trachea. Rarely there is a tracheo-oesophageal fistula without atresia (Fig. 36).

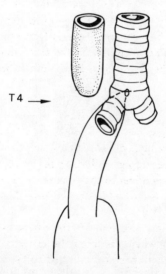

T4 ⟶

Fig. 36. The usual form of oesophageal stenosis. The upper oesophagus ends blindly; the lower oesophagus communicates with the trachea at the level of the 4th thoracic vertebra.

## THE THORACIC DUCT (Fig. 37)

The *cisterna chyli* lies between the abdominal aorta and the right crus of the diaphragm. It drains lymphatics from the abdomen and the lower limbs, then passes upwards through the aortic opening to become the *thoracic duct*. This ascends behind the oesophagus, inclines to the left

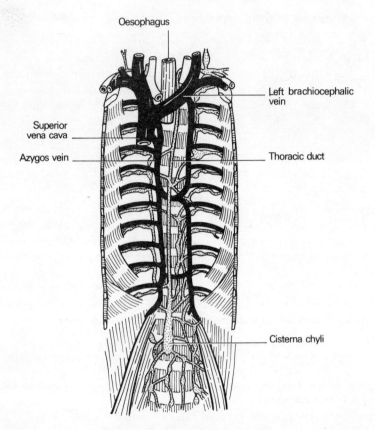

Oesophagus

Left brachiocephalic vein

Superior vena cava

Azygos vein

Thoracic duct

Cisterna chyli

Fig. 37. The course of the thoracic duct.

of the oesophagus at the level of T5, then runs upwards on its left side into the neck. The duct then passes laterally behind the carotid sheath, descends over the subclavian artery and drains into the commencement of the left innominate vein.

The left *jugular*, *subclavian* and *mediastinal lymph trunks*, draining the left side of the head and neck, upper limb and thorax respectively, usually join the thoracic duct, although they may open directly into the adjacent large veins at the root of the neck.

The thoracic duct thus usually drains the whole lymphatic field below the diaphragm and the left half of the lymphatics above it.

On the right side, the right subclavian jugular and mediastinal trunks may open independently into the great veins. Usually the subclavian and jugular trunks first join into a *right lymphatic duct* and this may be joined by the mediastinal trunk so that all three then have a common opening into the origin of the right innominate vein.

CLINICAL FEATURES

1. The lymphatics may become blocked by infection and fibrosis due to the Microfilaria bancrofti. This usually results in lymphoedema of the legs and scrotum but occasional involvement of the main channels of the trunk and thorax is followed by chylous ascites, chyluria and chylous pleural effusion.

2. The thoracic duct may be damaged during block dissection of the neck. If noticed at operation, the injured duct should be ligated; lymph then finds its way into the venous system by anastomosing channels. If the accident is missed, there follows an unpleasant chylous fistula in the neck.

Tears of the thoracic duct have also been reported as a complication of fractures of the thoracic vertebrae to which, in its lower part, the duct is closely related. Such injuries are followed by a chylothorax.

THE THORACIC SYMPATHETIC TRUNK (Fig. 38)

The sympathetic chain is the most laterally placed structure in the posterior mediastinum.

Descending from the cervical chain, it crosses:
● the neck of the 1st rib,
● the heads of the 2nd to 10th ribs, and
● the bodies of the 11th and 12th thoracic vertebrae.

It then passes behind the medial arcuate ligament of the diaphragm to continue as the lumbar sympathetic trunk.

The thoracic chain bears a ganglion for each spinal nerve; the first frequently joins the inferior cervical ganglion to form the *stellate ganglion*. Each ganglion receives a white ramus communicans containing pre-ganglionic fibres from its corresponding spinal nerve and donates back a grey ramus, bearing post-ganglionic fibres.

*Branches*

1. Sympathetic fibres are distributed to the skin with each of the thoracic spinal nerves.

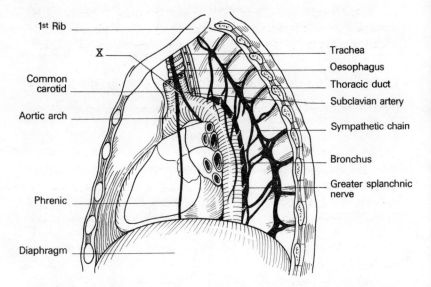

1st Rib

X

Common
carotid

Aortic arch

Phrenic

Diaphragm

Trachea
Oesophagus
Thoracic duct
Subclavian artery

Sympathetic chain

Bronchus

Greater splanchnic
nerve

**Fig. 38.**    The left thoracic sympathetic trunk with a display of the left mediastinum.

2.    Post-ganglionic fibres from T1–5 are distributed to the thoracic viscera—the heart and great vessels, the lungs and the oesophagus.

3.    Mainly pre-ganglionic fibres from T5–12 form the *splanchnic nerves* which pierce the crura of the diaphragm and pass to the coeliac and renal ganglia from which they are relayed as post-ganglionic fibres to the abdominal viscera.

These splanchnic nerves are the:
- *greater splanchnic* (T5–10),
- the *lesser splanchnic* (T10–11) and
- *least splanchnic* (T12).

They lie medial to the sympathetic trunk on the bodies of the thoracic vertebrae and are quite easily visible through the parietal pleura.

CLINICAL FEATURES

1.    Bilateral thoraco-abdominal sympathectomy is still occasionally performed in cases of severe essential hypertension, although the majority of patients can be controlled medically. The sympathetic chain is removed from T5 to L2 and the splanchnic nerves excised. This results in considerable dilatation of the splanchnic vascular bed, released from sympathetic control, with consequent fall in blood pressure.

2. A high spinal anaesthetic will produce temporary hypotension by paralysing the sympathetic (vasoconstrictor) pre-ganglionic outflow from spinal segment T5 downwards, passing to the abdominal viscera.

## ON THE EXAMINATION OF A CHEST RADIOGRAPH

The following features should be examined in every radiograph of the chest.

### Centring and density of film

The sternal ends of the two clavicles should be equidistant from the shadow of the vertebral spines. The assessment of the density of the film can only be learned by experience, but in a 'normal' film the bony cage should be clearly outlined and the larger vessels in the lung fields clearly visible.

### General shape

Any abnormalities in the general form of the thorax (scoliosis, kyphosis and the barrel chest of emphysema, for example) should always be noted before other abnormalities are described.

### Bony cage

The thoracic vertebrae should be examined first, then each of the ribs in turn (counting conveniently from their posterior ends and comparing each one with its fellow of the opposite side) and finally clavicles and scapulae. Unless this procedure is carried out systematically, important diagnostic clues (e.g. the presence of a cervical rib, or notching of the ribs by enlarged anastomotic vessels) are liable to be missed.

### The domes of the diaphragm

These should be examined for·height and symmetry and the nature of the cardio-phrenic and costophrenic angles observed.

### The mediastinum

The outline of the mediastinum should be traced systematically. Special note should be made of the size of the heart, of mediastinal shift and of the vessels and glands at the hilum of the lung.

### Lung fields

Again, systematic examination of the lung fields visible in each intercostal space is necessary if slight differences between the two sides are not to be overlooked.

## Abnormalities

When this scheme has been carefully followed, any abnormalities in the bony cage, the mediastinum or lung fields should now be apparent. They should then be defined anatomically as accurately as possible and checked, where necessary, by reference to a film taken from a different angle.

## Radiographic appearance of the heart

For the appearance of the heart as seen at fluoroscopy, reference should be made to a standard work on radiology or cardiology. In the present account, only the more important features of the heart and great vessels which can be seen in standard postero-anterior and oblique lateral radiographs of the chest will be described.

## The heart and great vessels in A–P radiographs (Fig. 39)

The greater part of the 'mediastinal shadow' in a A–P film of the chest is formed by the heart and great vessels. These should be examined as follows:

### Size and shape of the heart

Normally the transverse diameter should not exceed half the total width of the chest, but since it varies widely with bodily build and the position of the heart, these factors must also be assessed. The *shape* of the cardiac shadow also varies a good deal with the position of the heart, being long and narrow in a vertically disposed heart and broad and rounded in the so-called horizontal heart.

### The cardiac outline

Each 'border' of the cardiac shadow should be examined in turn. The *right border* of the mediastinal shadow is formed from above downwards by the right innominate vein, the superior vena cava and the right atrium. Immediately above the heart, the *left border* of the mediastinal shadow presents a well-marked projection, the *aortic knuckle*, which represents the arch of the aorta seen 'end-on'. Beneath this there are successively the shadows due to the pulmonary trunk (or the infundibulum of the right ventricle), the auricle of the left atrium, and the left ventricle. The

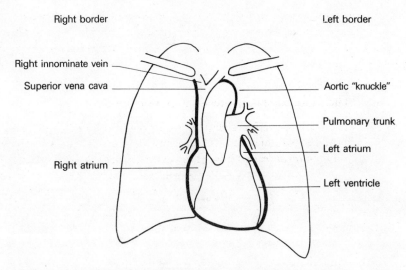

Right border

Left border

Right innominate vein

Superior vena cava

Aortic "knuckle"

Pulmonary trunk

Left atrium

Right atrium

Left ventricle

**Fig. 39.** A tracing of a chest radiograph to show the composition of the right and left borders of the mediastinal shadow.

shadow of the inferior border of the heart blends centrally with that of the diaphragm, but on either side the two shadows are separated by the well-defined cardio-phrenic angles.

## The heart and great vessels in anterior oblique radiographs

### The left oblique view
The greater part of the mediastinal shadow in this view is formed by the right and left ventricles, above which the relation of the arch of the aorta and the pulmonary trunk to the translucent trachea can be seen.

### The right oblique view
Almost all of the cardiac shadow in this view is due to the right ventricle. It is particularly useful for the assessment of the size of the left atrium since its posterior wall forms the upper half of the posterior border of the cardiac shadow. This border can be defined more accurately by giving the patient barium paste to swallow; the outlined oesophagus is indented by an enlarged left atrium.

# PART 2
# THE ABDOMEN AND PELVIS

# Surface anatomy and surface markings

Be able to identify these landmarks on yourself or the patient (Fig. 40).

The *xiphoid*. The *costal margin*, extending from the 7th costal cartilage at the xiphoid to the tip of the 12th rib (although the latter is often difficult to feel); this margin bears a distinct step, which is the tip of the 9th costal cartilage.

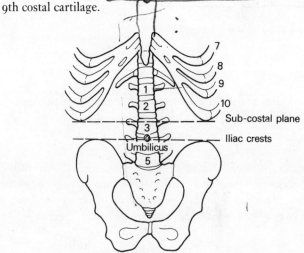

**Fig. 40.** Boundaries, bony landmarks and vertebral levels of the abdomen.

The *iliac crest* ends in front at the *anterior superior spine* from which the *inguinal ligament* (Poupart's ligament) passes downwards and medially to the *pubic tubercle*. Identify this tubercle by direct palpation and also by running the fingers along the adductor longus tendon (tensed by flexing, abducting and externally rotating the thigh) to its origin at the tubercle.

Feel the firm *vas deferens* between the finger and thumb as it lies within the spermatic cord at the scrotal neck. Trace the vas upwards and note that it passes medially to the pubic tubercle and thence through the *external inguinal ring*, which can be felt by invaginating the scrotal skin with the finger-tip.

57

**Vertebral levels** (Fig. 40)

*T9*—the xiphoid.

*L1*—*the transpyloric plane of Addison* lies half-way between the supra-sternal notch and the pubis, or approximately a handsbreadth below the xiphoid. This plane passes through the pylorus, the pancreatic neck, the duodeno-jejunal flexure and the hila of the kidneys.

*L3*—*the subcostal plane*, a line joining the lowest point of the thoracic cage on each side, which is the inferior margin of the 10th rib.

*L4*—the plane of the iliac crests.

The umbilicus is an inconstant landmark. In the healthy adult it lies at the junction of L3 and L4 vertebrae. It is lower in the infant and, naturally, when the abdomen is pendulous.


**Surface markings**

The abdominal viscera are inconstant in their position but the surface markings of the following structures are of clinical value.


*Liver*

The lower border of the liver extends along a line from the tip of the right 10th rib to just below the left nipple; it may just be palpable in the normal subject, especially on deep inspiration. The upper border follows a line passing just below the nipple on each side.


*Spleen*

This underlies the 9th, 10th and 11th ribs posteriorly on the left side.


*Gall bladder*

The fundus of the gall bladder corresponds to the point where the lateral border of the rectus abdominis cuts the costal margin; this is at the tip of the 9th costal cartilage, easily detected as a distinct "step" when the fingers are run along the costal margin.


*Pancreas*

The transpyloric plane defines the level of the neck of the pancreas which overlies the vertebral column. From this landmark, the head can be imagined passing downward and to the right, the body and tail passing upwards and to the left.

*Aorta*

This terminates just to the left of the mid-line at the level of the iliac crests; a pulsatile swelling below this level may thus be an iliac, but cannot be an aortic, aneurysm.

*Kidneys*

The lower pole of the normal right kidney may be felt in the thin subject. Anteriorly, the hilum of the kidney lies on the transpyloric plane 4 finger-breadths from the mid-line. Posteriorly, the upper pole of the kidney lies deep to the 12th rib. The right kidney normally extends about 1 in (2.5 cm) lower than the left. Using these landmarks, the kidney outlines can be projected on to either the anterior or posterior aspects of the abdomen.

In some perfectly normal thin people, especially women, it is possible to palpate the lower border of the liver, the lower pole of the right kidney, the aorta, the caecum and the sigmoid colon.

# The fasciae and muscles of the abdominal wall

### Fasciae of the abdominal wall

There is no deep fascia over the trunk, only the superficial fascia. (If there were, we would presumably be unable to take a deep breath or enjoy a large meal!) This, in the lower abdomen, forms a *superficial fatty layer* (of Camper) and a deeper *fibrous layer* (of Scarpa). The fatty layer is continuous with the superficial fat of the rest of the body, but the fibrous layer blends with the deep fascia of the upper thigh, extends into the penis and scrotum (or labia majora), and into the perineum as *Colles' fascia*. In the perineum it is attached behind to the perineal body and posterior margin of the perineal membrane and, laterally, to the rami of the pubis and ischium. It is because of these attachments that a rupture of the urethral bulb may be followed by extravasation of blood and urine into the scrotum, perineum and penis and then into the lower abdomen deep to the fibrous fascial plane, but not by extravasation downwards into the lower limb, from which the fluid is excluded by the attachment of the fascia to the deep fascia of the upper thigh.

*Nerve supply*

The segmental nerve supply of the abdominal muscles and the overlying skin is derived from T7 to L1. The distribution can be mapped out approximately if it is remembered that the umbilicus is supplied by T10 and the groin and scrotum by L1 (via the ilio-inguinal and ilio-hypogastric nerves). (See Fig. 135.)

## The muscles of the anterior abdominal wall

These are of considerable practical importance because their anatomy forms the basis of abdominal incisions.

*The rectus abdominis* (Fig. 41) arises on a 3 inch (7.5 cm) horizontal line from the 5th, 6th and 7th costal cartilages and is inserted for a length of 1 in (2.5 cm) into the crest of the pubis. At the tip of the xiphoid, at the umbilicus and half-way between are three constant *transverse tendinous*

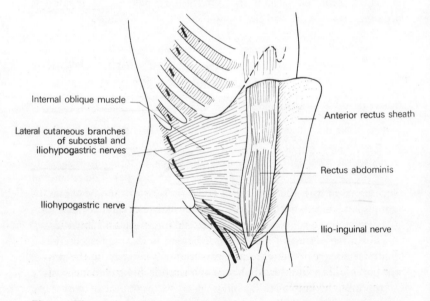

Fig. 41. The anterior rectus sheath has been opened, the external oblique removed, and the chest wall dissected down to the internal intercostal muscles.

*intersections*; below the umbilicus there is sometimes a fourth. These intersections are seen only on the anterior aspect of the muscle and here they adhere to the anterior rectus sheath. Posteriorly they are not in evidence and, in consequence, the rectus muscle is completely free behind. At each intersection, vessels from the superior epigastric artery and vein pierce the rectus.

The sheath in which the rectus lies is formed, to a large extent, by the aponeurotic expansions of the lateral abdominal muscles (Fig. 42).

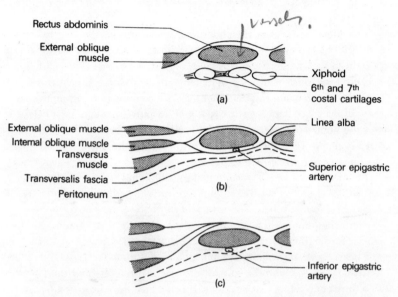

Fig. 42.   The composition of the rectus sheath shown in transverse section.
a. Above the costal margin.   b. Above the arcuate line.   c. Below the arcuate line.

*(a)* Above the costal margin, the anterior sheath comprises the external oblique aponeurosis only; posteriorly lie the costal cartilages.

*(b)* From the costal margin to a point half-way between umbilicus and pubis, the external oblique and the anterior part of the internal oblique aponeurosis form the anterior sheath. Posteriorly lie the posterior part of this split internal oblique aponeurosis and the aponeurosis of transversus abdominis.

*(c)* Below a point half-way between umbilicus and pubis, all the aponeuroses pass in front of the rectus so that the anterior sheath here comprises the tendinous expansions of all three oblique muscles blended together. The posterior wall at this level is made up of the only other structures available—the transversalis fascia and peritoneum.

The posterior junction between *b* and *c* is marked by the *arcuate line of Douglas*, which is the lower border of the posterior aponeurotic part of the rectus sheath. At this point the inferior epigastric artery and vein (from the external iliac vessels) enter the sheath, pass upwards and anastomose with the superior epigastric vessels which are terminal branches of the internal mammary artery and vein. The rectus sheaths fuse in the mid-line to form the *linea alba* stretching from the xiphoid to the pubic symphysis.

The lateral muscles of the abdominal wall comprise the external and internal oblique and the transverse muscles. They are clinically important in making up the rectus sheath and the inguinal canal, and also because they must be divided in making lateral abdominal incisions.

There attachments can be remembered when one bears in mind that they fill the space between the costal margin above, the iliac crest below, and the lumbar muscles covered by lumbar fascia behind. Medially, as already noted, they constitute the rectus sheath and thence blend into the linea alba from xiphoid to pubic crest.

*The obliquus externus abdominis* (external oblique) arises from the outer surfaces of the lower 8 ribs and fans out into the xiphoid, linea alba, the pubic crest, pubic tubercle and the anterior half of the iliac crest.

From the pubic tubercle to the anterior superior iliac spine its lower border forms the aponeurotic *inguinal ligament* of Poupart.

The *obliquus internus abdominis* (internal oblique) arises from the lumbar fascia, the anterior two-thirds of the iliac crest and the lateral two-thirds of the inguinal ligament. It is inserted into the lowest 6 costal cartilages, linea alba and the pubic crest.

The *transversus abdominis* arises from the lowest 6 costal cartilages (interdigitating with the diaphragm), the lumbar fascia, the anterior two-thirds of the iliac crest and the lateral one-third of the inguinal ligament; it is inserted into the linea alba and the pubic crest.

Note that the external oblique passes downwards and forwards, the internal oblique upwards and forwards and the transversus transversely. Note also that the external oblique has its posterior border free but the deeper two muscles both arise posteriorly from the lumbar fascia.

## The anatomy of abdominal incisions

Incisions to expose the intraperitoneal structures represent a compromise on the part of the operator. On the one hand he requires maximum access, on the other hand he wishes to leave a scar which lies, if possible, in an unobtrusive crease, and which will have done minimal damage to the muscles of the abdominal wall and to their nerve supply.

The nerve supply to the lateral abdominal muscles forms a richly communicating network so that cuts across the lines of fibres of these muscles, with division of one or two nerves, produce no clinical ill-effects. The segmental nerve supply to the rectus, however, has little cross-communication and damage to these nerves must, if possible, be avoided.

The copious anastomoses between the blood vessels supplying the abdominal muscles makes damage to these by operative incisions of no practical importance.

### Mid-line incision

The mid-line incision is made through the linea alba. Superiorly, this is a realtively wide fibrous structure, but below the umbilicus it becomes almost hair-line and the surgeon may experience difficulty in finding the exact point of cleavage between the recti at this level. Being made of fibrous tissue only, it provides an almost bloodless line along which the abdomen can be opened rapidly and, if necessarily, from Dan in the North to beer Sheba in the South!

### Paramedian incision

The paramedian incision is placed 1 in (2.5 cm)–1½ in (4 cm) lateral, and parallel to, the mid-line; the anterior rectus sheath is opened, the rectus displaced laterally and the posterior sheath, together with peritoneum, then incised. This incision has the advantage that, on suturing the peritoneum, the rectus slips back into place to cover and protect the peritoneal scar.

Because of the adherence of the anterior sheath to the rectus muscle at its tendinous intersections, the sheath must be dissected off the muscle at each of these sites, at each of which a segmental vessel requires division. Having done this, the rectus is easily slid laterally from the posterior sheath from which it is quite free. The posterior sheath and the peritoneum form a tough membrane down to half-way between pubis and umbilicus, but it is much thinner and more fatty below this where, as we have seen, it loses its aponeurotic component and is made up of only transversalis fascia and peritoneum. The inferior epigastric vessels may be seen passing under the arcuate line of Douglas in the posterior sheath and usually require division in a low paramedian incision.

### The trans-rectus incision

Ocassionally, the rectus muscle is split in the line of the paramedian incision. The rectus receives its nerve supply laterally and the muscle medial to the incision must, in consequence, be deprived of its innervation and undergo atrophy; it is an incision therefore best avoided.

### Para-rectus incision

The para-rectus (Battle) incision opens the rectus sheath and displaces the rectus medially. It was once popular but has the disadvantage that extending this incision longitudinally may damage the nerves running across segmentally to supply the rectus.

### Subcostal incision

The subcostal (Kocher) incision is used on the right side in biliary surgery and, on the left, in exposure of the spleen. The skin incision commences at the mid-line and extends parallel to, and 1 in (2.5 cm) below, the costal margin.

The anterior rectus sheath is opened, the rectus cut and the posterior sheath with underlying adherent peritoneum incised. The small 8th intercostal nerve branch to the rectus is sacrificed but the larger and more important 9th nerve, in the lateral part of the wound, preserved. The divided rectus muscle is held by the intersections above and below and retracts very little. It subsequently heals by fibrous tissue. This incision is valuable in the patient with the wide subcostal angle. Where this angle is narrow, the paramedian incision is usually preferred.

### The muscle split or gridiron approach to the appendix

The oblique skin incision centred at *McBurney's point* (two-thirds of the way laterally along the line from the umbilicus to the anterior superior iliac spine) is now less popular than an almost transverse incision in the line of the skin crease forwards from, and 1 in (2.5 cm) above, the anterior spine.

The aponeurosis of the external oblique is incised in the line of its fibres (obliquely downwards and medially); the internal oblique and transversus muscles are then split in the line of their fibres, and retracted without their having to be divided. On closing the incision, these muscles snap together again leaving a virtually undamaged abdominal wall.

### Transverse and oblique incisions

Incisions cutting through the lateral abdominal muscles do not damage their richly anastomosing nerve supply and heal without weakness. They are useful, for example, in exposing the sigmoid colon or the caecum or, by displacing the peritoneum medially, extraperitoneal structures such as the ureter, sympathetic chain and the external iliac vessels.

*Thoraco-abdominal incisions*

An upper paramedian or upper oblique abdominal incision can be extended through the 8th or 9th intercostal space, the diaphragm incised and an extensive exposure achieved of both upper abdomen and thorax. This is used, for example, in removing growths of the upper stomach or lower oesophagus and in resection of the right lobe of the liver.

*Paracentesis abdominis*

Intraperitoneal fluid collections can be evacuated via a cannula inserted through the abdominal wall. The bladder having been first emptied with a catheter, the cannula is introduced on a trocar either through the mid-line (where the linea alba is relatively bloodless) or lateral to McBurney's point (where there is no danger of wounding the inferior epigastric vessels). The coils of gut are not in danger in this procedure because they are mobile and are pushed away by the tip of the trocar.

## The inguinal canal (Fig. 43)

This canal represents the oblique passage taken through the lower abdominal wall by the testis and cord (the round ligament in the female).

Because of its importance in diagnosis and treatment of hernias, the anatomy of this region is probably asked more often than any other in examinations.

The canal is $1\frac{1}{2}$ in (4 cm) long. It passes downwards and medially from the internal to the external inguinal rings and lines parallel to, and immediately above, the inguinal ligament.

*Relations*

• Anteriorly—the skin, superficial fascia and the external oblique aponeurosis cover the full length of the canal; the internal oblique covers its lateral one-third.

• Posteriorly—the conjoint tendon forms the posterior wall of the canal medially, the transversalis fascia laterally. (The *conjoint tendon* represents the fused common insertion of the internal oblique and transversus into the pubic crest.)

• Above—arch the lowest fibres of the internal oblique and transversus abdominis.

• Below—lies the inguinal ligament.

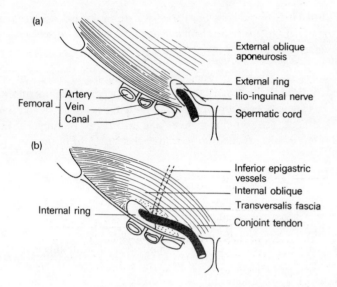

**Fig. 43.** The right inguinal canal.
a. with the external oblique aponeurosis intact.   b. with the aponeurosis layed open.

*The internal ring* represents the point at which the spermatic cord pushes through the transversalis fascia, dragging from it a covering which forms the *internal spermatic fascia*. This ring is demarcated medially by the inferior epigastric vessels passing upwards from the external iliac artery and vein.

*The external ring* is a V-shaped defect in the external oblique aponeurosis and lies immediately above and medial to the pubic tubercle. As the cord traverses this opening, it carries the *external spermatic fascia* from the ring's margins.

The inguinal canal transmits the spermatic cord and the ilio-inguinal nerve in the male and the round ligament and ilio-inguinal nerve in the female.

*The spermatic cord* comprises (Fig. 44):

3 layers of fascia—the *external spermatic*, from the external oblique aponeurosis;

the *cremasteric*, from the internal oblique aponeurosis (containing muscle fibres termed the *cremaster muscle*);

the *internal spermatic*, from the transversalis fascia.

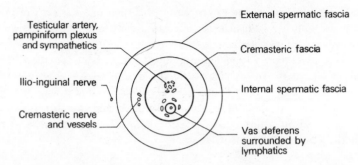

Fig. 44. Scheme of the spermatic cord and its contents, in transverse section.

3 arteries—the testicular (from the aorta);
the cremasteric (from the inferior epigastric artery);
the artery of the vas (from the inferior vesical artery).
3 nerves—the nerve to the cremaster (from the genito-femoral nerve);
sympathetic fibres;
the ilio-inguinal nerve.
3 other structures—the vas deferens;
the pampiniform plexus of veins (draining the right testis into the
inferior vena cava and the left testis into the renal vein);
lymphatics which drain the testis to the aortic lymph nodes.

## CLINICAL FEATURES

An *indirect inguinal hernia* passes through the internal ring, along the canal and then, if large enough, emerges through the external ring and descends into the scrotum. If reducible, such a hernia can be completely controlled by pressure with the finger-tip over the internal ring, which lies $\frac{1}{2}$ in (12 mm) above the point where the femoral artery passes under the inguinal ligament, i.e. $\frac{1}{2}$ in (12 mm) above the femoral pulse. This pulse can be felt at the mid-inguinal point, half-way between the anterior superior iliac spine and the symphysis pubis (Fig. 148).

If the hernia protrudes through the external ring, it can be felt to lie above and medial to the pubic tubercle, and is thus differentiated from a femoral hernia emerging from the femoral canal which lies below and lateral to this landmark (Fig. 171).

A *direct inguinal hernia* pushes its way directly forwards through the posterior wall of the inguinal canal. Since it lies medial to the internal ring it is not controlled by digital pressure applied immediately above the femoral pulse. Occasionally a direct hernia becomes large enough to push its way through the external ring and then into the neck of the scrotum. This is so unusual that one can usually assume that a scrotal hernia is an indirect hernia.

The only certain way of determining the issue is at operation; the inferior epigastric vessels demarcate the medial edge of the internal ring, therefore an indirect hernia sac will pass lateral and a direct hernia medial to these vessels. Quite often both a direct and an indirect hernia co-exist; they bulge through on each side of the inferior epigastric vessels like the legs of a pair of pantaloons.

# Peritoneal cavity

The endothelial lining of the primitive coelomic cavity of the embryo becomes the thoracic pleura and the abdominal peritoneum. Each is invaginated by ingrowing viscera which thus come to be covered by a serous membrane and to be packed snugly into a serous-lined cavity.

In the male, the peritoneal cavity is completely closed, but in the female it is perforated by the openings of the uterine tubes which constitute a possible pathway of infection from the exterior.

To revise the complicated attachments of the peritoneum, it is best to start at one point and trace this membrane in an imaginary round-trip of the abdominal cavity, aided by Figs. 45 and 46. A convenient point of departure is the parietal peritoneum of the anterior abdominal wall below the umbilicus. At this level the membrane is smooth apart from the shallow ridges formed by the median umbilical cord (the obliterated fetal urachus passing from the bladder to the umbilicus) and the lateral umbilical cords (the obliterated umbilical arteries passing to the umbilicus from the internal iliac arteries).

A cicatrix can usually be felt and seen at the posterior aspect of the umbilicus, and from this the *falciform ligament* sweeps upwards and slightly to the right of the mid-line to the liver. In the free border of this ligament lies the *ligamentum teres* (the obliterated fetal umbilical vein) which passes into the groove between the quadrate lobe and left lobe of the liver.

Elsewhere, the peritoneum sweeps over the inferior aspect of the diaphragm, to be reflected onto the liver (leaving a bare area demarcated by the *upper* and *lower coronary ligaments* of the liver) and onto the right margin of the abdominal oesophagus. After enclosing the liver, the peritoneum descends from the porta hepatis as a double sheet, the *lesser*

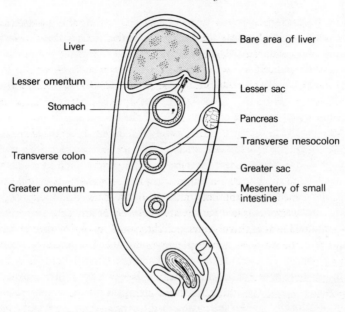

**Fig. 45.** The peritoneal cavity in longitudinal section (female).

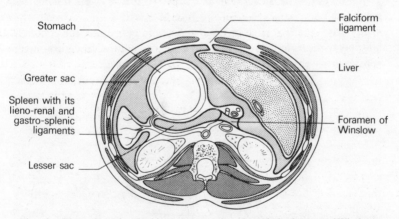

**Fig. 46.** The peritoneal cavity in transverse section (through the foramen of Winslow).

*omentum*, to the lesser curve of the stomach. Here it again splits to enclose this organ, reforms at its greater curve, then loops downwards, then up again to attach to the length of the transverse colon, forming the apron-like *greater omentum*.

The transverse colon, in turn, is enclosed within this peritoneum which then passes upwards and backwards as the *transverse mesocolon* to the posterior abdominal wall, where it is attached along the anterior aspect of the pancreas.

At the base of the transverse mesocolon, this double peritoneal sheet divides once again; the upper leaf passes upwards over the posterior abdominal wall to reflect onto the liver (at the bare area), the lower leaf passes over the lower part of the posterior abdominal wall to cover the pelvic viscera and to link up once again with the peritoneum of the anterior wall. This posterior layer is, however, interrupted by its being reflected along an oblique line running from the duodeno-jejunal flexure to the ileo-caecal junction to form the *mesentery of the small intestine.*

The mesentery of the small intestine, the lesser and greater omenta and mesocolon all carry the vascular supply and lymph drainage of their contained viscera.

*The lesser sac* (Fig. 46) is the extensive pouch lying behind the lesser omentum and the stomach and projecting downwards (although usually this space is obliterated) between the layers of the greater omentum. Its left wall is formed by the spleen attached by the *gastro-splenic* and *lieno-renal* ligaments. The right extremity of the sac opens into the main peritoneal cavity via the *foramen of Winslow* (Fig. 47), whose boundaries are as follows:

•anteriorly—the free edge of lesser omentum, containing the common bile duct to the right, hepatic artery to the left and portal vein posteriorly;
•posteriorly—the inferior vena cava;

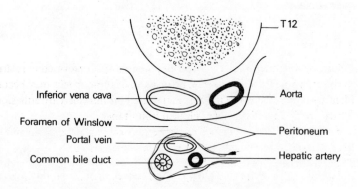

Fig. 47.   The foramen of Winslow in transverse section.

•inferiorly—the 1st part of the duodenum, over which runs the hepatic artery before this ascends into the anterior wall of the foramen;
•superiorly—the caudate process of the liver.

## CLINICAL FEATURES

1. Occasionally a loop of intestine passes through the foramen of Winslow into the lesser sac and becomes strangulated by the edges of the foramen. Notice that none of these boundaries can be incised to release the strangulation; the bowel must be decompressed by a needle to allow its reduction.

2. It is important to the surgeon that the hepatic artery can be compressed between his index finger within the foramen of Winslow and his thumb on its anterior wall. If the cystic artery is torn during cholecystectomy, haemorrhage can be controlled by this manoeuver which then enables the damaged vessel to be identified and secured.

## Intraperitoneal fossae

A number of fossae occur within the peritoneal cavity into which loops of bowel may become caught and strangulated. Those of importance are:

1. *the lesser sac* via the foramen of Winslow, which we have just described;

2. *paraduodenal fossa*—between the duodeno-jejunal flexure and the inferior mesenteric vessels;

3. *retrocaecal fossa*—in which the appendix frequently lies;

4. *intersigmoid fossa*—formed by the inverted V attachment of the meso-sigmoid.

## The subphrenic spaces (Fig 48)

Below the diaphragm are a number of potential spaces formed in relation to the attachments of the liver. One or more of these spaces may become filled with pus (a subphrenic abscess) walled off inferiorly by adhesions. There are five subdivisions of clinical importance:

*The right and left subphrenic spaces* lie between the diaphragm and the liver, separated from each other by the falciform ligament.

*The right and left subhepatic spaces* lie below the liver. The right is the pouch of Morison and is bounded by the posterior abdominal wall behind and by the liver above. It communicates anteriorly with the right subphrenic space around the anterior margin of the right lobe of the

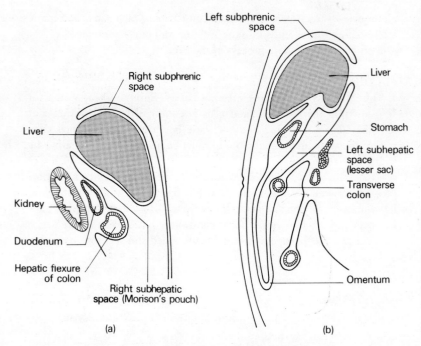

**Fig. 48.** The anatomy of the subphrenic spaces in sagittal section.    a. Right.    b. Left.

liver and below both open into the general peritoneal cavity from which infection may track, for example, from a perforated appendix or a perforated peptic ulcer. The left subhepatic space is the lesser sac which communicates with the right through the foramen of Winslow. It may fill with fluid as a result of a perforation in the posterior wall of the stomach or from an inflamed or injured pancreas to form a pseudocyst of the pancreas.

*The right extraperitoneal space* lies between the bare area of the liver and the diaphragm. It may become involved in retroperitoneal infections or directly from a liver abscess.

Posterior and anterior subphrenic abscesses are drained by an incision below, or through the bed of, the 12th rib. A finger is then passed upwards and forwards between liver and diaphragm to open into the abscess cavity. An anteriorly-placed collection of pus below the diaphragm can alternatively be drained via an incision placed below and parallel to the costal margin.

# The gastrointestinal tract

## THE STOMACH

The stomach is roughly J-shaped, although its size and shape vary considerably. It tends to be high and transverse in the obese short subject and to be elongated in the asthenic individual; even in the same person, its shape depends on whether it is full or empty, on the position of the body and on the phase of respiration. The stomach has two surfaces—the anterior and posterior, two curvatures—the greater and lesser, and two orifices—the cardia and pylorus (Fig. 49).

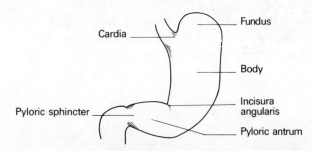

**Fig. 49.** The stomach and its subdivisions.

The stomach projects to the left, above the level of the *cardia*, to form the dome-like gastric *fundus*. Between the cardia and the pylorus lies the *body* of the stomach leading to a narrow portion, immediately preceding the pylorus, which is termed the *pyloric antrum*. The junction of pylorus with duodenum is marked by a constriction externally and also by a constant vein (of Mayo) which crosses it at this level.

The thickened *pyloric sphincter* is easily felt and surrounds the lumen of the *pyloric canal*. The pyloric sphincter is an anatomical structure as well as a physiological mechanism. The cardia, on the other hand, although competent, is not demarcated by a distinct anatomical sphincter. The exact nature of the cardiac sphincter action is still not fully understood, but the following mechanisms have been suggested, each supported by some experimental and clinical evidence.

1.  Mucosal folds at the oesophago-gastric junction act as a valve.
2.  The acute angle of entry of oesophagus into stomach produces a valve-like effect.
3.  The circular muscle of the lower oesophagus is a physiological, as distinct from an anatomical, sphincter.

4.   The arrangement of the muscle fibres of the stomach around the cardia acts either as a sphincter or else maintains the acute angle of entry of oesophagus into stomach.

5.   The right crus of the diaphragm acts as a 'pinch-cock' to the lower oesophagus as it pierces this muscle.

6.   The positive intra-abdominal pressure compresses the walls of the short segment of intra-abdominal oesophagus.

### Relations of the stomach (Fig. 50)

•Anteriorly—the abdominal wall, the left costal margin, the diaphragm and the left lobe of the liver.

•Posteriorly—the lesser sac, which separates the stomach from the pancreas, transverse mesocolon, left kidney, left suprarenal, the spleen and the splenic artery.

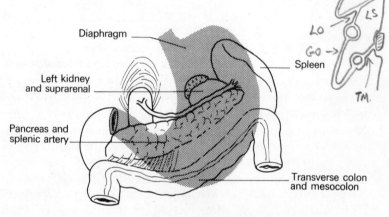

Fig. 50.   The posterior relations of the stomach; the stomach (stippled) is superimposed upon its bed.

•Superiorly—the left dome of the diaphragm.

The lesser omentum is attached along the lesser curvature of the stomach, the greater omentum along the greater curvature. These omenta contain the vascular and lymphatic supply of the stomach.

*The arterial supply* (Fig. 51) to the stomach is extremely rich and comprises:

•the left gastric artery—from the coeliac axis;

•the right gastric artery—from the hepatic artery;

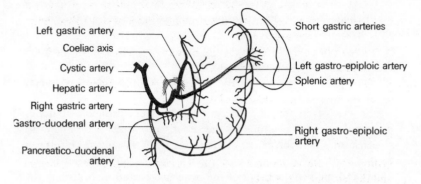

**Fig. 51.** The arterial supply of the stomach.

•the right gastro-epiploic artery—from the gastro-duodenal branch of the hepatic artery;
•the left gastro-epiploic artery—from the splenic artery;
•the short gastric arteries—from the splenic artery.

The corresponding veins drain into the portal system.

*The lymphatic drainage* of the stomach accompanies its blood vessels. The stomach can be divided into three drainage zones (Fig. 52).

•Area I—the superior two-thirds of the stomach drain along the left and right gastric vessels to the aortic nodes.

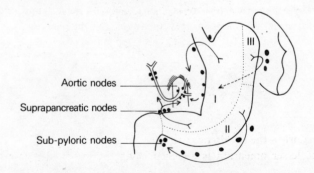

**Fig. 52.** The lymph drainage of the stomach.
•Area I drains along the right and left gastric vessels to the aortic nodes.
•Area II drains to the sub-pyloric and thence aortic nodes via lymphatics along the right gastro-epiploic vessels.
•Area III drains via lymphatics along the splenic vessels to the suprapancreatic nodes and thence to aortic nodes.

•Area II—the right two-thirds of the inferior one-third of the stomach drain along the right gastro-epiploic vessels to the sub-pyloric nodes and thence to the aortic nodes.

•Area III—the left one-third of the inferior one-third of the stomach drains along the short gastric and splenic vessels lying in the gastro-splenic and lieno-renal ligaments, then, via the supra-pancreatic nodes, to the aortic group.

This extensive lymphatic drainage and the technical impossibility of its complete removal is one of the serious problems in dealing with stomach cancer. Involvement of the nodes along the splenic vessels can be dealt with by removing spleen, gastro-splenic and lieno-renal ligaments and the body and tail of the pancreas. Lymph nodes along the gastro-epiploic vessels are removed by excising the greater omentum. However, involvement of the nodes around the aorta and the head of the pancreas may render the growth incurable.

*The vagal supply to the stomach* (Fig. 53). The anterior and posterior vagi enter the abdomen through the oesophageal hiatus. The anterior nerve lies close to the stomach wall but the posterior, and larger, nerve is at a little distance from it. The anterior vagus supplies branches to the cardia and lesser curve of the stomach and also a large hepatic branch which, in turn, donates a pyloric branch to the upper border of the pylorus and antrum (the nerve of Latarjet). The posterior vagus gives branches to both the anterior and posterior aspects of the body of the stomach but the bulk of the nerve forms the coeliac branch. This runs along the left gastric artery to the coeliac ganglion for distribution to the intestine, as far as the mid-transverse colon, and the pancreas.

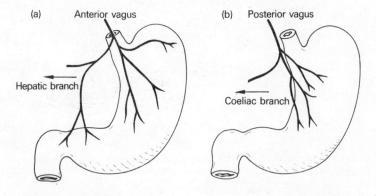

Fig. 53. The vagal supply to the stomach: a. anterior vagus; b. posterior vagus.

The vagus constitutes the motor and secretory nerve supply for the stomach. When divided, in the operation of *vagotomy*, the neurogenic (reflex) gastric secretion is abolished but the stomach is, at the same time, rendered atonic so that it empties only with difficulty. Because of this, vagotomy must always be accompanied by some sort of drainage procedure, either a pyloroplasty (to enlarge the pyloric exit and render the pyloric sphincter incompetent) or by a gastrojejunostomy (to drain the stomach into the proximal small intestine).

CLINICAL FEATURES

1. A posterior gastric ulcer or cancer may erode the pancreas, giving pain referred to the back. Ulceration into the splenic artery—a direct posterior relation—may cause torrential haemorrhage.

2. There may be adhesions across the lesser sac which bring the transverse mesocolon into intimate relationship with the stomach or greater omentum. In these circumstances the middle colic vessels are in danger of damage during mobilization of the stomach for gastrectomy.

3. *Radiology of the stomach* (Fig. 54). A plain erect film of the abdomen reveals a bubble of air below the left diaphragm; this is gas in the stomach fundus. After the subject has swallowed barium sulphate paste, the stomach can be seen and its position, movements and outline studied.

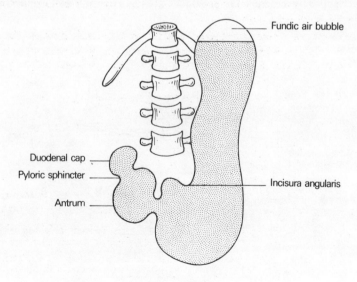

Fig. 54. Tracing of barium meal x-ray of the stomach.

The wide variations in the position and shape of the stomach that we have already mentioned have come to light principally as a result of such investigations.

By tipping the subject head-down, the opaque meal can be made to impinge against the cardia; incompetence of this sphincter mechanism will be demonstrated by seeing barium regurgitate into the oesophagus.

4. *Gastroscopy*—the mucosa of the air-inflated stomach can be inspected in the living subject through the gastroscope. With the modern fiberoptic instrument the whole of the gastric mucosa can be viewed, the duodenum examined, and the common bile duct intubated for retrograde radiological study.

## THE DUODENUM

The duodenum curves in a C around the head of the pancreas and is 10 inches long. At its origin from the pylorus it is completely covered with peritoneum for about 1 inch, but then becomes a retro-peritoneal organ, only partially covered by serous membrane.

*Relations* (Figs. 55, 56)
For descriptive purposes, the duodenum is divided into four sections.

*The first part* (2 in (5 cm)) ascends from the gastro-duodenal junction, overlapped by the liver and gall baldder. Immediately posterior to it lie the

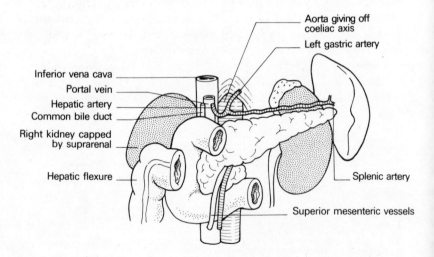

Aorta giving off coeliac axis

Left gastric artery

Inferior vena cava

Portal vein

Hepatic artery

Common bile duct

Right kidney capped by suprarenal

Hepatic flexure

Splenic artery

Superior mesenteric vessels

**Fig. 55.** The relations of the duodenum.

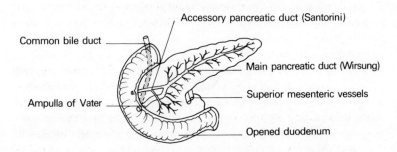

Fig. 56. The duodenum and pancreas dissected to show the pancreatic ducts and their orifices.

portal vein, common bile duct and gastroduodenal artery which separate it from the inferior vena cava.

*The second part* (3 in (7.5 cm)) descends in a curve around the head of the pancreas. It is crossed by the transverse colon and lies on the right kidney and ureter. Half-way along its postero-medial aspect enters the common opening of the bile duct and main pancreatic duct (of Wirsung) onto an eminence called the *duodenal papilla*. This common opening is guarded by the *sphincter of Oddi*. The subsidiary pancreatic duct (of Santorini) opens into the duodenum a little above the papilla.

*The third part* (4 in (10 cm)) runs transversely to the left, crossing the inferior vena cava, the aorta and the third lumbar vertebra. It is itself crossed anteriorly by the root of the mesentery and the superior mesenteric vessels. Its upper border hugs the pancreatic head.

*The fourth part* (1 in (2.5 cm)) ascends upwards and to the left to end at the duodeno-jejunal junction. It is surprisingly easy for the surgeon to confuse this with the ileocaecal junction, a mistake which may be disastrous. He confirms the identity of the duodenal termination by the presence of the *suspensory ligament of Treitz*, which is a well-marked peritoneal fold descending from the right crus of the diaphragm to the duodenal termination, and by visualizing the inferior mesenteric vessels which descend from behind the pancreas immediately to the left of the duodeno-jejunal junction.

### Blood supply of the duodenum

The superior pancreatico-duodenal artery arises from the gastro-duodenal artery; the inferior pancreatico-duodenal artery originates as the first branch of the superior mesenteric artery. These vessels both lie in the curve between the duodenum and the head of the pancreas, supplying both structures.

CLINICAL FEATURES

1.   The first part of the duodenum is overlapped by the liver and gall bladder, either of which may become adherent to, or even ulcerated by, a duodenal ulcer. Moreover, a gallstone may ulcerate from the fundus of the gall bladder into the duodenum. The gallstone may then impact in the lower ileum as it traverses the gut to produce intestinal obstruction *(gallstone ileus)*.

2.   The pancreas, as the duodenum's most intimate relation, is readily invaded by a posterior duodenal ulcer. This should be suspected if the patient's pain radiates into the dorso-lumbar region. Erosion of the gastro-duodenal artery by such an ulcer results in severe haemorrhage.

3.   Extensive dissection of a duodenum, scarred by severe ulceration, may damage the common bile duct which passes behind the first part of the duodenum about 1 inch from the pylorus.

4.   The hepatic flexure of the colon crosses the second part of the duodenum and the latter may be damaged during a right hemicolectomy. Similarly, the right kidney lies directly behind this part of the duodenum, which may be injured in performing a right nephrectomy.

5.   *Radiology of the duodenum.* Within a few minutes of swallowing a barium meal, the first part of the duodenum becomes visible as a triangular shadow termed the *duodenal cap.* Every few seconds the duodenum contracts, emptying this cap, which promptly proceeds to fill again. It is in this region that the great majority of duodenal ulcers occur; an actual ulcer crater may be visualized, filled with barium, or deformity of the cap, produced by scar tissue, may be evident. The rest of the duodenum can also be seen, the shadow being floccular due to the rugose arrangement of the mucosa.

## SMALL INTESTINE

*Length.* The length of the small intestine varies from 3 to 10 metres in different subjects; the average is some 6.1 metres. Resection of up to one-third or even one-half of the small intestine is compatible with a perfectly normal life, and survival has been reported with only 45 cm of small intestine preserved.

*The mesentery* of the small intestine has a 6 in (15 cm) origin from the posterior abdominal wall, which commences at the duodeno-jejunal junction to the left of the 2nd lumbar vertebra, and passes obliquely downwards to the right sacro-iliac joint; it contains the superior mesenteric vessels, the lymph nodes draining the small gut and autonomic nerve fibres.

The upper half of the small intestine is termed the *jejunum*, the remainder is the *ileum*. There is no sharp distinction between the two and this division is a conventional one only. The bowel does, however, change its character from above downwards the following points enabling the surgeon to determine the level of a loop of small intestine at operation.

1. The jejunum has a thicker wall as the circular folds of mucosa *(valvulae conniventes)* are larger and thicker more proximally.

2. The proximal small intestine is of greater diameter than the distal.

3. The jejunum tends to lie at the umbilical region, the ileum in the hypogastrium and pelvis.

4. The mesentery becomes thicker and more fat-laden from above downwards.

5. The mesenteric vessels form only one or two arcades to the jejunum, with long and relatively infrequent terminal branches passing to the gut wall. The ileum is supplied by shorter and more numerous terminal vessels arising from complete series of three, four or even five arcades (Fig. 57).

## LARGE INTESTINE

The large intestine is subdivided, for descriptive purposes, into:
- caecum with the appendix vermiformis,
- ascending colon (5–8 in (12.5–20 cm))
- hepatic flexure,
- transverse colon (18 in (45 cm))
- splenic flexure,
- descending colon (9–12 in (22.5–30 cm))
- sigmoid colon (5–30 in (12.5–75 cm), average 15 in (37.5 cm))
- rectum (5 in (12.5 cm))
- anal canal ($1\frac{1}{2}$ in (4 cm))

The large bowel may vary very considerably in length in different subjects; the average is approximately 5 feet (1.5 m).

The colon (but not the appendix, caecum or rectum), bears characteristic fat-filled peritoneal tags called *appendices epiploicae* scattered over its surface. These are especially numerous in the sigmoid colon.

The colon and caecum (but not the appendix or rectum) are marked

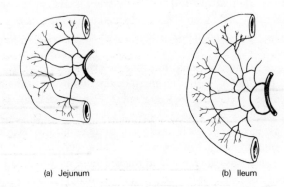

(a) Jejunum                                  (b) Ileum

**Fig. 57.** The simple arterial arcades of the jejunum (a) compared with the complex arcades of the ileum (b).

by the *taeniae coli*. These are three flattened bands commencing at the base of the appendix and running the length of the large intestine to end at the recto-sigmoid junction. They represent the great bulk of the longitudinal muscle of the large bowel; because the taeniae are about a foot shorter than the gut to which they are attached, the colon becomes condensed into its typical sacculated shape. These sacculations may be seen in a plain radiograph of the abdomen when the large bowel is distended and appear as incomplete septa projecting into the gas shadow. The radiograph of distended small intestine, in contrast, characteristically has complete transverse lines across the bowel shadow due to the transverse mucosal folds of the valvulae conniventes.

### Peritoneal attachments

The transverse colon and sigmoid are completely peritonealized (the former being readily identified by its attachment to the greater omentum). The ascending and descending colon have no mesocolon but adhere directly to the posterior abdominal wall (although exceptionally the ascending colon has a mesocolon). The caecum may or may not be completely peritonealized, and the appendix, although usually free within its own mesentery, occasionally lies extraperitoneally behind caecum and ascending colon or adheres to the posterior wall of these structures.

The rectum is extraperitoneal on its posterior aspect in its upper third, posteriorly and laterally in its middle third and completely in its lower third as it sinks below the pelvic peritoneum.

## THE APPENDIX

The appendix arises from the postero-medial aspect of the caecum about one inch below the ileo-caecal valve; its length ranges from $\frac{1}{2}$ in (12 mm) to 9 in (22.5 cm). In the fetus it is a direct out-pouching of the caecum, but differential over-growth of the lateral caecal wall results in its medial displacement.

The position of the appendix is extremely variable—more so than that of any other organ (Fig. 58). Most frequently (75 per cent of cases) the appendix lies behind the caecum. The appendix is usually quite free in this position although occasionally it lies beneath the peritoneal covering of the caecum. If the appendix is very long it may actually extend behind the ascending colon and abut against the right kidney or the duodenum; in these cases its distal portion lies extra-peritoneally.

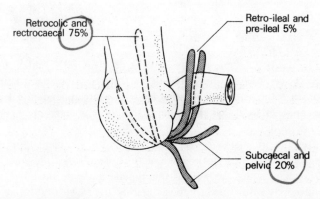

Retrocolic and rectrocaecal 75%

Retro-ileal and pre-ileal 5%

Subcaecal and pelvic 20%

**Fig. 58.** The positions in which the appendix may lie, together with their approximate incidence.

In about 20 per cent of cases, the appendix lies just below the caecum or else hangs down into the pelvis. Less commonly it passes in front of or behind the terminal ileum, or lies in front of the caecum or in the right paracolic gutter.

A long appendix has been known to ulcerate into the duodenum or perforate into the *left* paracolic gutter. It may well be said that "the appendix is the only organ in the body that has no anatomy".

The mesentery of the appendix, containing the appendicular branch of the ileo-colic artery, descends behind the ileum as a triangular fold (Fig. 59). Another peritoneal sheet, the *ileo-caecal fold*, passes to the

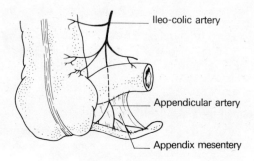

**Fig. 59.** The blood supply of the appendix.

appendix or to the base of the caecum from the front of the ileum. The ileo-caecal fold is termed the *bloodless fold of Treves* although, in fact, it often contains a vessel and, if cut, proves far from bloodless.

CLINICAL FEATURES

1.   The lumen of the appendix is relatively wide in the infant and is frequently completely obliterated in the elderly. Since obstruction of the lumen is the usual precipitating cause of acute appendicitis it is not unnatural, therefore, that appendicitis should be uncommon at the two extremes of life.

2.   The appendicular artery represents the entire vascular supply of the appendix. It runs first in the edge of the appendicular mesentery and then, distally, along the wall of the appendix. Acute infection of the appendix may result in thrombosis of this artery with rapid develop-ment of gangrene and subsequent perforation. This is in contrast to acute cholecystitis, where the rich collateral vascular supply from the liver bed ensures the rarity of gangrene of the gall bladder even if the cystic artery becomes thrombosed.

3.   *Appendicectomy* is usually performed through a muscle-splitting in-cision in the right iliac fossa (see 'abdominal incisions'). The caecum is delivered into the wound and, if the appendix is not immediately visible, it is located by tracing the taeniae coli along the caecum—they fuse at the base of the appendix. When the caecum is extraperitoneal it may be difficult to bring the appendix up into the incision; this is facilitated by first mobilizing the caecum by incising the almost avascular peritoneum along its lateral and inferior borders.

The appendix mesentery, containing the appendicular vessels, is firmly tied and divided, the appendix base tied, the appendix removed and its stump invaginated into the caecum.

## THE RECTUM

The rectum is 5 in (12.5 cm) in length. It commences anterior to the third segment of the sacrum and ends at the level of the apex of the prostate or at the lower quarter of the vagina, where it leads into the anal canal.

The rectum is straight in lower mammals (hence its name) but is curved in man to fit into the sacral hollow. Moreover, it presents a series of three lateral inflexions, capped by the *valves of Houston*, projecting left, right and left from above downwards.

*Relations* (Figs. 60, 61)
The main relations of the rectum are important. They must be visualized in carrying out a rectal examination, they provide the key to the local spread of rectal growths and they are important in operative removal of the rectum.

*Posteriorly* lie sacrum and coccyx and the middle sacral artery, which

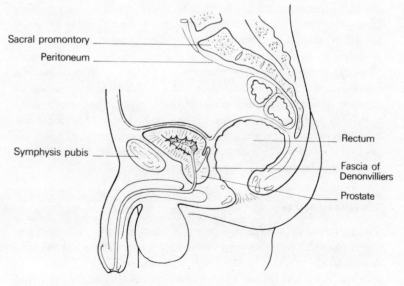

**Fig. 60.** Sagittal section of the rectum and its related viscera in the male.

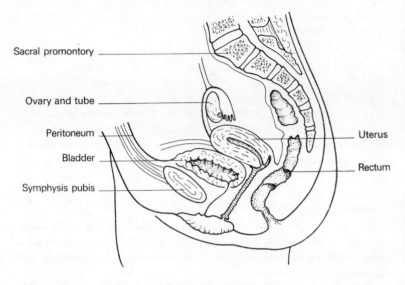

Sacral promontory

Ovary and tube

Peritoneum

Bladder

Symphysis pubis

Uterus

Rectum

**Fig. 61.**  Sagittal section of the rectum and its related viscera in the female.

are separated from it by extra-peritoneal connective tissue containing the rectal vessels and lymphatics. The lower sacral nerves, emerging from the anterior sacral foramina, may be involved by growth spreading posteriorly from the rectum, resulting in severe sciatic pain.

*Anteriorly* the upper two-thirds of the rectum are covered by peritoneum and relate to coils of small intestine which lie in the cul-de-sac of the pouch of Douglas between rectum and the bladder or the uterus. In front of the lower one-third lie prostate, bladder base and seminal vesicles in the male, or the vagina in the female. A layer of fascia (*Denonvilliers*) separates the rectum from the anterior structures and forms the plane of dissection which must be sought after in excision of the rectum.

*Laterally* the rectum is supported by the levator ani.

### The anal canal (Fig. 62)

The anal canal is $1\frac{1}{2}$ in (4 cm) long and is directed downwards and backwards from the rectum to end at the anal orifice. The mid-anal canal represents the junction between endoderm and ectoderm:

1.  The lower half is lined by squamous epithelium and the upper half by columnar epithelium; the latter presents vertical columns of mucosa

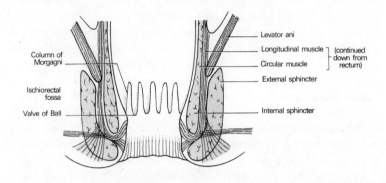

Fig. 62. The sphincters of the anus.

*(the columns of Morgagni)* connected at their distal extremities by valve-like folds *(the valves of Ball)*. A carcinoma of the upper anal canal is thus an adenocarcinoma, whereas that arising from the lower part is a squamous tumour.

2.   The blood supply of the upper half of the anal canal is from the superior haemorrhoidal vessels, whereas that of the lower half is the blood supply of the surrounding anal skin, the inferior haemorrhoidal vessels, which derive from the pudenal, and ultimately the internal iliac vessels. The two venous systems communicate and therefore form one of the anastomoses between the portal and systemic circulations.

3.   The lymphatics above this muco-cutaneous junction drain along the superior haemorrhoidal vessels to the lumbar nodes whereas, below this line, drainage is to the inguinal nodes. A carcinoma of the rectum which invades the lower anal canal may thus metastasize to the groin nodes.

4.   The nerve supply to the upper anal canal is via the autonomic plexuses, the lower part is supplied by the somatic inferior haemorrhoidal nerve. (The lower canal is therefore sensitive to the prick of a hypodermic needle, whereas injection of an internal haemorrhoid with sclerosant fluid, by passing a needle through the mucosa of the upper part of the canal, is painless.)

### The anal sphincter

Forming the walls of the anal canal is a rather complicated muscle arrangement which constitutes a powerful sphincter mechanism (Fig. 62). This comprises:

•*The internal anal sphincter*, of involuntary muscle, which continues above with the circular muscle coat of the rectum.

•*The external anal sphincter*, of voluntary muscle, which surrounds the internal sphincter and which extends further downwards and curves medially to occupy a position below and slightly lateral to the lower rounded edge of the internal sphincter, close to the skin of the anal orifice. The lowermost, or subcutaneous, portion of the external sphincter is traversed by a fan-shaped expansion of the longitudinal muscle fibres of the anal canal which continue above with the longitudinal muscle of the rectal wall. At its upper end the external sphincter fuses with the fibres of levator ani.

In carrying out a digital rectal examination, the ring of muscle on which the flexed finger rests just over an inch from the anal margin is the *ano-rectal ring*. This represents the deep part of the external sphincter where this blends with the internal sphincter and levator ani, and demarcates the junction between anal canal and rectum.

The anal canal is related posteriorly to the fibrous tissue between it and the coccyx (ano-coccygeal body), laterally to the ischio-rectal fossae containing fat, and anteriorly to the perineal body separating it from the bulb of the urethra in the male or the lower vagina in the female.

## Rectal examination

The following structures can be palpated by the finger passed per rectum in the normal patient:

1.  both sexes—the ano-rectal ring (see above), coccyx and sacrum, ischio-rectal fossae, ischial spines;
2.  male—prostate, rarely healthy seminal vesicles;
3.  female—perineal body, cervix, occasionally the ovaries.

Abnormalities which can be detected include:

1.  within the lumen—faecal impaction, foreign bodies;
2.  in the wall—rectal growths, strictures, granulomata, etc., but *not* haemorrhoids unless these are thrombosed;
3.  outside the rectal wall—pelvic bony tumours, abnormalities of the prostate or seminal vesicle, a distended bladder, uterine or ovarian enlargement, collections of fluid or neoplastic masses in the pouch of Douglas.

Do not be deceived by foreign objects placed in the vagina. The commonest are a tampon or a pessary.

During parturition, dilatation of the cervical os can be assessed by rectal examination since it can be felt quite easily through the rectal wall.

CLINICAL FEATURES

1. *Haemorrhoids*—haemorrhoids (piles) are varicosities of the superior haemorrhoidal veins. Initially contained within the anus (1st degree), they gradually enlarge until they prolapse on defaecation (2nd degree) and finally remain prolapsed through the anal orifice (3rd degree).

The superior haemorrhoidal vessels are asymmetrical—there is one branch to the left side of the rectum and two to the right, thus giving the well-known arrangement of piles at 3, 7 and 11 o'clock viewing the patient in the lithotomy position.

Anatomically each pile comprises:

a varicose venous plexus draining into one of the superior haemorrhoidal veins, terminal branches of the corresponding superior haemorrhoidal artery, and a covering of anal canal mucosa and submucosa.

The so-called "thrombosed external pile" is a small tense haematoma at the anal margin caused by rupture of a subcutaneous vein and is much better termed a perianal haematoma.

2. *Perianal abscesses* (Fig. 63) may be localized beneath the anal mucosa (submucous) beneath perianal skin (subcutaneous) or occupy the ischio-rectal fossa (ischio-rectal). Occasionally abscesses lie in the pelvi-rectal space above levator ani, alongside the rectum and deep to the pelvic peritoneum.

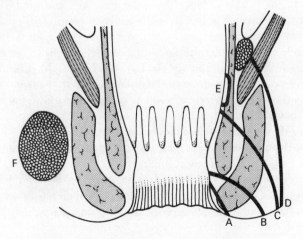

**Fig. 63.** The anatomy of perianal fistulae and abscess.

| | |
|---|---|
| A. Subcutaneous fistula. | D. Ano-rectal fistula. |
| B. Low anal fistula. | E. Submucous fistula. |
| C. High anal fistula. | F. Ischio-rectal abscess. |

3. *Fistulae* (Fig. 63), usually resulting from rupture of such abscesses, are classified anatomically and may be:

•*submucous*—confined to the tissues immediately below the anal mucosa,

•*subcutaneous*—confined to the perianal skin,

•*low-level anal fistula*—passing through the lower part of the superficial sphincter (commonest),

•*high-level anal fistula*—passing through the deeper part of the superficial sphincter,

•*ano-rectal fistula* which has its track passing above the ano-rectal ring and which may or may not open into the rectum.

In laying open fistulae in ano it is essential to preserve the ano-rectal ring if faecal incontinence is to be avoided. The external sphincter, on the other hand, can be divided quite safely without this risk.

4. *Fissure-in-ano*—this is a tear in the anal mucosa and over 90 per cent occur posteriorly in the mid-line. The anatomical basis for this probably lies in the insertion of the superficial component of the external anal sphincter posteriorly into the coccyx; between the two limbs of the V thus formed, the mucosa is relatively unsupported and may therefore be torn by a hard faecal mass at this site.

### Arterial supply of the intestine

The alimentary tract develops from the fore, mid and hind gut; the arterial supply to each is discrete, although anastomosing with its neighbour. The fore-gut comprises stomach and duodenum as far as the entry of the bile duct and is supplied by branches of the coeliac axis which arises from the aorta at T12 vertebral level (see stomach). The mid-gut extends from mid-duodenum to the distal transverse colon and is supplied by the *superior mesenteric artery* (Fig. 64) arising from the aorta at L1. Its branches are:

1. the inferior pancreatico-duodenal artery;
2. jejunal and ileal branches—supplying the bulk of the small intestine;
3. the ileo-colic artery, supplying terminal ileum, caecum and commencement of ascending colon and giving off an *appendicular branch* to the appendix—the most commonly ligated intra-abdominal artery;
4. the right colic artery—supplying the ascending colon;
5. the middle colic artery—supplying the transverse colon.

The hind-gut receives its supply from the *inferior mesenteric artery* (Fig. 64), arising from the aorta at L3 and giving the following branches:

1. the left colic artery—supplying the descending colon;

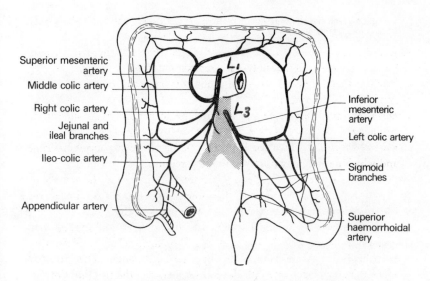

Fig. 64. The superior and inferior mesenteric arteries and their branches.

2. the sigmoid branches—supplying the sigmoid;

3. the superior haemorrhoidal artery—supplying the rectum.

Each branch of the superior and inferior mesenteric artery anastomoses with its neighbour above and below so that there is, in fact, a continuous vascular arcade along the whole length of the gastrointestinal canal.

## The portal system of veins

The portal venous system drains blood to the liver from the abdominal part of the alimentary canal (excluding the anus), the spleen, the pancreas and the gall-bladder and its ducts.

The distal tributaries of this system correspond to, and accompany, the branches of the coeliac and the superior and inferior mesenteric arteries enumerated above; only proximally (Fig. 65) does the arrangement differ.

The *inferior mesenteric vein* ascends above the point of origin of its artery to enter the *splenic vein* behind the pancreas.

The *superior mesenteric vein* joins the splenic vein behind the neck of the pancreas in the transpyloric plane to form the *portal vein*, which ascends behind the first part of the duodenum into the anterior wall of

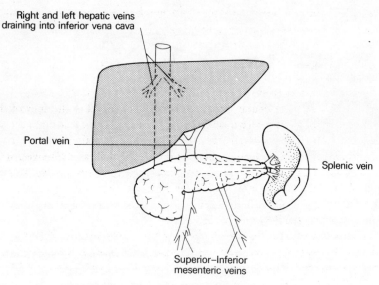

**Fig. 65.**   The composition of the portal system.

the foramen of Winslow and thence to the porta hepatis. Here the portal vein divides into right and left branches and breaks up into capillaries running between the lobules of the liver. These capillaries drain into the radicles of the hepatic vein through which they empty into the inferior vena cava.

## Connections between the portal and systemic venous systems

Normally, portal venous blood traverses the liver as described above and empties into the systemic venous circulation via the hepatic vein and inferior vena cava. This pathway may be blocked by a variety of causes which are classified into:

•pre-hepatic—e.g. thrombosis or congenital obliteration of the portal vein;

•hepatic—e.g. cirrhosis of the liver;

•post-hepatic—e.g. congenital stenosis of the hepatic vein.

If this occurs, the portal venous pressure rises *(portal hypertension)* and collateral pathways open up between the portal and systemic venous systems.

These communications are:

1.   Between the oesophageal branch of the left gastric vein and the

oesophageal veins of the azygos system; these *oesophageal varices* are the cause of the severe haematemeses that may occur in portal hypertension.

2. Between the superior haemorrhoidal branch of the inferior mesenteric vein and the inferior haemorrhoidal veins draining into the interal iliac vein via its internal pudendal tributary.

3. Between the portal tributaries in the mesentery and mesocolon and retroperitoneal veins communicating with the renal, lumbar and phrenic veins.

4. Between the portal branches in the liver and the veins of the abdominal wall via veins passing along the falciform ligament to the umbilicus. This may result in the formation of a cluster of dilated veins which radiate from the navel and which are called the *caput Medusae*.

5. Between the portal branches in the liver and the veins of the diaphragm across the bare area of the liver.

A striking feature of operations upon patients with portal hypertension is the extraordinary dilatation of every available channel between the two systems which renders such procedures tedious and bloody.

## Lymph drainage and the intestine (Fig. 66)

The arrangement of lymph nodes is relatively uniform throughout the small and large intestine. Numerous small nodes lying near, or even on, the bowel wall drain to intermediately placed and rather larger nodes along the vessels in the mesentery or mesocolon and thence to clumps of nodes situated near the origins of the superior and inferior mesenteric arteries. From these, efferent vessels link up to drain into the cisterna chyli.

The lymphatic drainage-field of each segment of bowel corresponds fairly accurately to its blood supply. High ligation of the vessels to the involved segment of bowel with removal of a wide surrounding segment of mesocolon will therefore remove the lymph nodes draining the area. Division of the middle colic vessels and a resection of a generous wedge of transverse mesocolon, for example, would be performed for a growth of transverse colon.

## The structure of the alimentary canal

The alimentary canal is made up of *mucosa* demarcated by the *muscularis mucosae* from the *submucosa*, the *muscle coat* and the *serosa*—the last being absent where the gut is extraperitoneal.

*The Abdomen and Pelvis*

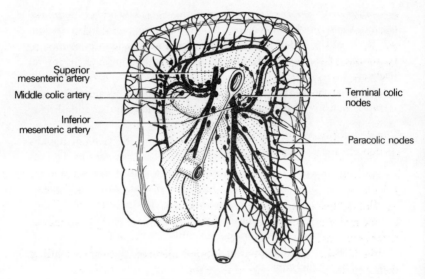

Superior mesenteric artery
Middle colic artery
Inferior mesenteric artery
Terminal colic nodes
Paracolic nodes

**Fig. 66.** Lymph nodes of the large intestine.

The oesophageal mucosa and that of the lower anal canal is squamous; elsewhere it is columnar. At the cardio-oesophageal junction this transition is quite sharp, although occasionally columnar epithelium may line the lower oesophagus.

The gastric mucosa bears simple crypt-like glands projecting down to the muscularis mucosae. The pyloric antrum secretes an alkaline juice containing mucus and the enzyme gastrin. The body of the stomach secretes pepsin and also HCl, the latter from the *oxyntic cells* lying sandwiched deeply between the surface cells. The stomach mucosa also produces intrinsic factor.

The mucosa of the duodenum and small intestine, as well as bearing crypt-like glands, projects into the bowel lumen in villous processes which greatly increase its surface area. The duodenum is distinguished by its crypts extending deep through the muscularis mucosae and opening into an extensive system of acini in the submucosa termed *Brunner's glands*.

The mucosa of the large intestine is lined almost entirely by mucus-secreting goblet cells; there are no villi.

The muscle coat of the alimentary tract is made up of an inner circular layer and an outer longitudinal layer. In the upper two-thirds of the oesophagus and at the anal margin this muscle is voluntary; elsewhere it is involuntary. The stomach wall is reinforced by an innermost oblique coat of muscle and the colon is characterized by the condensation of its longitudinal layer into three *taeniae coli*.

The nerve plexuses of Meissner and Auerbach lie respectively in the submucosal layer and between the circular and longitudinal muscle coats.

## The development of the intestine and its congenital abnormalities (Fig. 67)

The primitive endodermal tube of the gut is divided into:

1.  the fore-gut (supplied by the coeliac axis) extending as far as the entry of the bile duct into the duodenum;
2.  the mid-gut (supplied by the superior mesenteric artery) continuing as far as the distal transverse colon;
3.  the hind-gut (supplied by the inferior mesenteric artery) extending thence to the ectodermal part of the anus.

At an early stage rapid proliferation of the gut wall obliterates its lumen and this is followed by subsequent recanalization.

The fore-gut becomes rotated with the development of the lesser sac so that the original right wall of the stomach comes to form its posterior surface and the left wall its anterior surface. The vagi rotate with the stomach and therefore lie anteriorly and posteriorly to it at the oeso-phageal hiatus.

This rotation swings the duodenum to the right and the mesentery of this organ then blends with the peritoneum of the posterior abdominal wall—this blending process is termed *zygosis*.

The mid-gut enlarges rapidly in the five-week fetus, becomes too large to be contained within the abdomen and herniates into the umbilical cord. The apex of this herniated bowel is continuous with the vitello-intestinal duct and the yolk sac, but this connection, even at this early stage of fetal life, is already reduced to a fibrous strand.

The axis of this herniated loop of gut is formed by the superior mesen-teric artery, which demarcates a cephalic and a caudal limb. The cephalic element develops into the proximal small intestine; the caudal segment differentiates into the terminal 2 feet (62 cm) of ileum, the caecum and the colon as far as the junction of the middle and left thirds of the transverse colon.

A bud which develops on the caudal segment indicates the site of subsequent formation of the caecum; it may well be that this bud delays

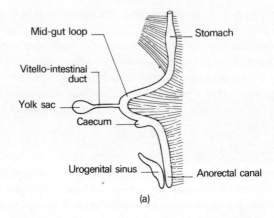

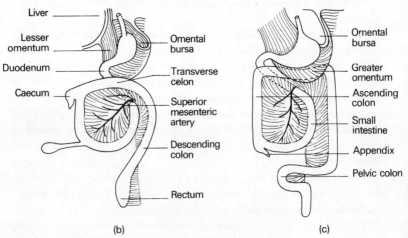

**Fig. 67.** Stages in rotation of the bowel.
(a) The prolapsed mid-gut loop, seen in lateral view.
(b) The mid-gut returns to the abdomen.
(c) The caecum descends to its definitive position. Note the completion of stomach rotation with the formation of the lesser sac (omental bursa).

the return of the caudal limb in favour of the cephalic gut during the subsequent reduction of the herniated bowel.

At 10 weeks this return of the bowel into the abdominal cavity commences. The mid-gut loop first rotates through 90 deg. so that the cephalic limb now lies to the right and the caudal limb to the left.

The cephalic limb returns first, passing upwards and to the left into the space left available by the bulky liver. In doing so, this mid-gut passes *behind* the superior mesenteric artery (which thus comes to cross the third part of the duodenum) and also pushes the hind-gut—the definitive distal colon—over to the left.

When the caudal limb returns, it lies in the only space remaining to it, superficial to, and above, the small intestine with the caecum lying immediately below the liver.

The caecum then descends into its definitive position in the right iliac fossa, dragging the colon with it. The transverse colon thus comes to lie in front of the superior mesenteric vessels and the small intestine.

Finally the mesenteries of the ascending and descending parts of the colon blend with the posterior abdominal wall peritoneum by zygosis.

Numerous anomalies may occur in this highly complex developmental process.

1.  Atresia or stenosis of the bowel may result from failure of recanalization of the lumen. Another cause of this may be damage to the blood supply to the bowel within the fetal umbilical hernia with consequent ischaemic changes.

2.  *Meckel's diverticulum* represents the remains of the embryonic vitello-intestinal duct (communication between the primitive mid-gut and yolk sac) and is therefore always on the anti-mesenteric border of the bowel. As an approximation to the truth it can be said to occur in 2 per cent of subjects, twice as often in males as females, to be situated at 2 feet (62 cm) from the ileo-caecal junction and to be 2 in (5 cm) long. In fact, it may occur anywhere from 6 in (15 cm) to 12 feet (3.70 m) from the terminal ileum and vary from a tiny stump to a 6 in (15 cm) long sac. Occasionally the diverticulum ends in a whip-like solid strand.

As well as a diverticulum—the commonest form—this duct may persist as a fistula or band connecting the intestine to the umbilicus, as a cyst hanging from the anti-mesenteric border of the ileum or as a 'raspberry tumour' at the umbilicus, formed by the red mucosa of a persistent umbilical extremity of the diverticulum pouting at the navel (Fig. 68).

The mucosa lining the diverticulum may contain islands of peptic epithelium with oxyntic (acid-secreting) cells. Peptic ulceration of adjacent intestinal epithelium may then occur with haemorrhage or perforation.

3.  The caecum may fail to descend; the peritoneal fold which normally seals it in the right iliac fossa passes, instead, across the duodenum and

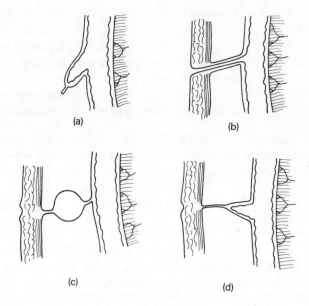

**Fig. 68.**   Abnormalities associated with persistence of the vitello-intestinal tract.

(a) Meckel's diverticulum.

(b) Patent vitello-intestinal duct.

(c) Cyst within a fibrous cord passing from the anti-mesenteric border of the intestine to the umbilicus.

(d) Meckel's diverticulum with terminal filament passing to umbilicus.

causes a neo-natal intestinal obstruction. The mesentery of the small intestine in such a case is left as a narrow pedicle, which allows volvulus of the whole small intestine to occur (*volvulus neonatorum*).

4.   Occasionally reversed rotation occurs, in which the transverse colon comes to lie *behind* the superior mesenteric vessels with the duodenum in front of them; this may again be accompanied by extrinsic duodenal obstruction due to a peritoneal fold.

5.   *Exomphalos* is persistence of the mid-gut herniation at the umbilicus after birth.

# The gastrointestinal adnexae: Liver, gall bladder and its ducts, the pancreas and spleen

## THE LIVER (Fig. 69)

This is the largest organ in the body. It is related by its domed upper surface to the diaphragm, which separates it from pleura, lungs, pericardium and heart. Its postero-inferior (or visceral) surface abuts against the abdominal oesophagus, the stomach, duodenum, hepatic flexure of colon and the right kidney and suprarenal, as well as carrying the gall-bladder.

The liver is divided into a larger right and smaller left lobe separated superiorly by the falciform ligament and postero-inferiorly by an H-shaped arrangement of fossae:

• anteriorly and to the right—the fossa for the gall-bladder;
• posteriorly and to the right—the groove in which the inferior vena cava lies embedded;
• anteriorly and to the left—the fissure containing the ligamentum teres;
• posteriorly and to the left—the fissure for the ligamentum venosum.

The cross-bar of the H is the *porta hepatis*. Two subsidiary lobes are marked out on the visceral aspect of the liver between the limbs of this H—the *quadrate lobe* in front and the *caudate lobe* behind.

The *ligamentum teres* is the obliterated remains of the left umbilical vein which, in utero, brings blood from the placenta back into the fetus. The *ligamentum venosum* is the fibrous remnant of the fetal *ductus venosus* which shunts blood from this left umbilical vein to the inferior vena cava, short-circuiting the liver. It is easy enough to realize, then, that the grooves for the ligamentum teres, ligamentum venosum and inferior vena cava, representing as they do the pathway of a fetal venous trunk, are continuous in the adult.

Lying in the porta hepatis (which is 2 in (5 cm) long) are:

1. The common hepatic duct—anteriorly.
2. The hepatic artery—in the middle.
3. The portal vein—posteriorly.

As well as autonomic nerve fibres (sympathetic from the coeliac axis and parasympathetic from the vagus), lymphatic vessels and lymph nodes.

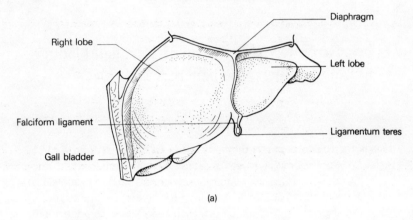

Right lobe

Diaphragm

Left lobe

Falciform ligament

Ligamentum teres

Gall bladder

(a)

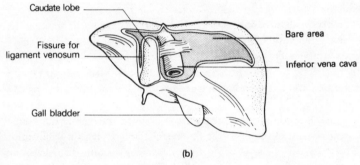

Caudate lobe

Bare area

Fissure for
ligament venosum

Inferior vena cava

Gall bladder

(b)

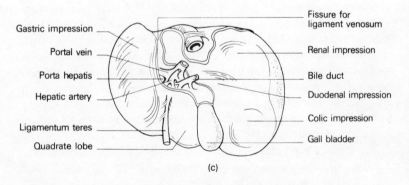

Gastric impression

Fissure for
ligament venosum

Portal vein

Renal impression

Porta hepatis

Bile duct

Hepatic artery

Duodenal impression

Colic impression

Ligamentum teres

Gall bladder

Quadrate lobe

(c)

**Fig. 69.** The liver and its subdivisions.
(a) Anterior aspect. (b) Posterior aspect. (c) Inferior aspect.

## Peritoneal attachments

The liver is enclosed in peritoneum except for a small posterior *bare area*, demarcated by the peritoneum from the diaphragm reflected onto it as the upper and lower layers of the *coronary ligament*. To the right, these fuse to form the *right triangular ligament*.

The *falciform ligament* ascends to the liver from the umbilicus, somewhat to the right of the mid-line, and bears the *ligamentum teres* in its free border. The ligamentum teres passes into its fissure in the inferior surface of the liver while the falciform ligament passes over the dome of the liver and then divaricates. Its right limb joins the upper layer of the coronary ligament and its left limb stretches out as the long narrow *left triangular ligament* which, when traced posteriorly and to the right, joins the lesser omentum in the upper end of the fissure for the ligamentum venosum.

The *lesser omentum* arises from the fissures of the porta hepatis and the ligamentum venosum and passes as a sheet to be attached along the lesser curvature of the stomach.

### Structure

The liver is made up of lobules each with a solitary central vein which is a tributary of the hepatic vein which, in turn, drains into the inferior vena cava. In spaces between the lobules, termed portal canals, lie branches of the hepatic artery (bringing systemic blood) and the portal vein, both of which drain into the central vein by means of sinusoids traversing the lobule.

Branches of the *hepatic duct* also lie in the portal canals and receive fine bile capillaries from the liver lobules.

## Segmental anatomy

The gross anatomical division of the liver into a right and left lobe, demarcated by a line passing from the attachment of the falciform ligament on the anterior surface to the fissures for the ligamentum teres and ligamentum venosum on its posterior surface, is simply a gross anatomical descriptive term with no morphological significance. Studies of the distribution of the hepatic blood vessels and ducts have indicated that the true morphological and physiological division of the liver is into right and left lobes demarcated by a plane which passes through the fossa of the gall bladder and the fossa of the inferior vena cava. Although these two lobes are not differentiated by any visible line on the dome of the liver, each has its own arterial and venous blood supply and separate biliary drainage. This morphological division lies to the right of the gross anatomical plane and in this the quadrate lobe comes to be part of the left morphological lobe of the liver while the caudate lobe divides partly to the left and partly to the right lobe (Fig. 70).

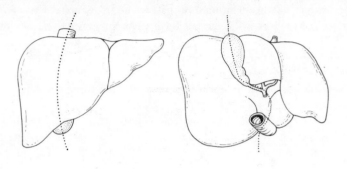

Fig. 70.    The morphological right and left lobes of the liver.

At the hilum of the liver, the hepatic artery, portal vein and bile duct each divide into right and left branches and there is little or no anastomosis between the divisions on the two sides (Fig. 71). From the region of the porta hepatis, the branches pass laterally and spread upwards and downwards throughout the liver substance, defining the morphological left and right lobes.

*The hepatic veins*
These veins are massive and their distribution is somewhat different to that of the portal hepatic arterial and bile duct system already described. There are three major hepatic veins, comprising a right, a central, and a left. These pass upwards and backwards to drain into the inferior vena cava at the superior margin of the liver. Their terminations are somewhat variable, but usually the central hepatic vein enters the left hepatic vein near its termination. In other specimens it may drain directly into the cava. In addition, small hepatic venous tributaries run directly backwards from the substance of the liver to enter the vena cava more distally to the main hepatic veins. Although these are not of great functional

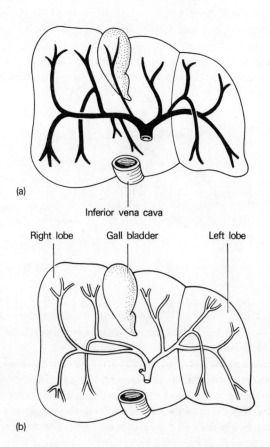

Inferior vena cava

Right lobe     Gall bladder     Left lobe

Fig. 71.

importance they obtrude upon the surgeon during the course of a right
hepatic lobectomy.

The three principal hepatic veins have three zones of drainage corresponding roughly to the right, the middle and left one-thirds of the
liver. The plane defined by the falciform ligament corresponds to the
boundary of the zones drained by the left and middle hepatic veins.
Unfortunately for the surgeon, the middle hepatic vein lies just at the
line of the principle plane of the liver between its right and left morphological lobes and it is this fact which complicates the operation of right
hepatic resection (Fig. 72).

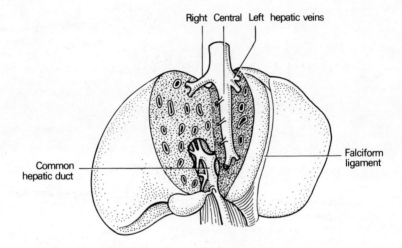

**Fig. 72.** Liver split open to demonstrate the tributaries of the hepatic vein.

## THE BILIARY SYSTEM (Fig. 73)

The *right and left hepatic ducts* fuse in the porta hepatis to form the *common hepatic duct* ($1\frac{1}{2}$ in (4 cm). This joins with the *cystic duct* ($1\frac{1}{2}$ in (4 cm), draining the gall-bladder, to form the *common bile duct* (4 in (10 cm). The common bile duct commences about one inch above the duodenum, then passes behind it to open at a papilla on the medial aspect of the second part of the duodenum. In this course the common duct lies either in a groove in the posterior aspect of the head of the pancreas or is actually buried in its substance.

As a rule the common duct termination joins that of the main pancreatic duct (of Wirsung) in a dilated common vestibule, the *ampulla of Vater*, whose opening in the duodenum is guarded by the *sphincter of Oddi*. Occasionally the bile and pancreatic ducts open separately into the duodenum; another variation is that they may first join together and then open, as a single channel, into the ampulla of Vater.

The common hepatic duct and the supraduodenal part of the common bile duct lie in the free edge of the lesser omentum where they are related as follows (Fig. 47):
• bile duct—anterior to the right;
• hepatic artery—anterior to the left;
• portal vein—posterior;

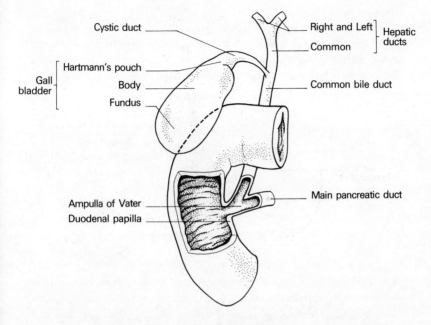

**Fig. 73.** The gall bladder and its duct system. (The anterior wall of the second part of the duodenum has been removed.)

•inferior vena cava—still more posterior, separated from the portal vein by the foramen of Winslow.

## The gall-bladder (Fig. 73)

The gall-bladder normally holds about 50cc of bile and acts as a bile concentrator and reservoir. It lies in a fossa separating the right and quadrate lobes of the liver and is related inferiorly to the duodenum and transverse colon. (An inflamed gall-bladder may occasionally ulcerate into either of these structures.)

For descriptive purposes the organ is divided into *fundus*, *body* and *neck*, the latter opening into the cystic duct. In dilated and pathological gall-bladders there is frequently a pouch present on the ventral aspect just proximal to the neck termed *Hartmann's pouch* in which gall stones may become lodged.

*Blood supply* (Fig. 74)
The gall-bladder is supplied by the *cystic artery* (a branch usually of the right hepatic artery) which lies in the triangle made by the liver, the cystic duct and the common hepatic duct. Other vessels derived from the hepatic artery pass to the gall-bladder from its bed in the liver.

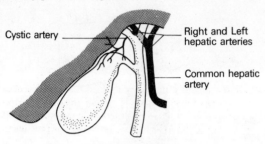

Fig. 74. The arterial supply of the gall bladder.

*Structure*
The gall-bladder wall and the sphincter of Oddi contain muscle, but there are only scattered muscle fibres throughout the remaining biliary duct system. The mucosa is lined throughout by columnar cells and bears mucus-secreting glands.

*Development*
The gall-bladder and ducts are subject to numerous anatomical variations which are best understood by considering their embryological development; a diverticulum grows out from the ventral wall of the duodenum which differentiates into the hepatic ducts and the liver (Fig. 76). Another diverticulum from the side of the hepatic duct bud forms the gall-bladder and cystic duct.

Some variations are shown in a series of diagrams (Fig. 75).

CLINICAL FEATURES
1. Errors in gall-bladder surgery are frequently the result of failure to appreciate the variations in the anatomy of the biliary system; it is important, therefore, before dividing any structures and removing the gall-bladder, to have all the three biliary ducts clearly identified, together with the cystic and hepatic arteries.
2. Haemorrhage during cholecystectomy may be controlled by compressing the hepatic artery (which gives off the cystic branch) between the finger and thumb where it lies in the anterior wall of the foramen of Winslow.

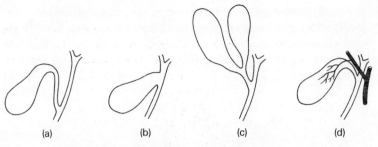

**Fig. 75.** Some variations in biliary anatomy.

(a)   A long cystic duct joining the hepatic duct low down behind the duodenum. (b) absence of the cystic duct—the gall bladder opens directly into the common hepatic duct. (c) A double gall bladder, the result of a rare bifid embryonic diverticulum from the hepatic duct. (d) The right hepatic artery crosses *in front* of the common hepatic duct; this occurs in 25 per cent of cases.

3.   Gangrene of the gall-bladder is rare because even if the cystic artery becomes thrombosed in acute cholecystitis there is a rich secondary blood supply coming in from the liver bed. Gangrene may occur in the unusual event of a gall-bladder on an abnormally long mesentery undergoing torsion, which will destroy both its sources of blood supply.

4.   Stones in the common duct can usually be removed through an incision in the supraduodenal part of the common duct. Occasionally the duodenum and pancreatic head must be mobilized by incising the peritoneum along the lateral duodenal margin to expose the duct in its retro-pancreatic course (Kocher's manoeuvre). Sometimes a stone impacted at the ampulla of Vater must be approached via an incision in the second part of the duodenum. This last approach is also used when it is necessary to divide the sphincter of Oddi or to remove a tumour arising at the termination of the common bile duct.

## THE PANCREAS (See Fig. 55)

The pancreas lies retroperitoneally in roughly the transpyloric plane. For descriptive purposes it is divided into *head*, *neck*, *body* and *tail*.

### Relations

The head lies in the C-curve of the duodenum and sends out the *uncinate process* which hooks posteriorly to the superior mesenteric vessels as these travel from behind the pancreas into the root of the mesentery.

  Posteriorly lie the inferior vena cava, aorta, superior mesenteric vessels, the crura of diaphragm, coeliac plexus and the left kidney. The splenic artery runs along the upper border of the pancreas. The splenic vein runs behind the gland, receives the inferior mesenteric vein and joins the superior mesenteric to form the portal vein behind the pancreatic neck (Fig. 65).

To complete this list of important posterior relationships, the common bile duct lies either in a groove in the right extremity of the gland or embedded in its substance, as it passes to open into the second part of the duodenum.

Anteriorly lies the stomach separated by the lesser sac. To the left, the pancreatic tail lies against the hilum of the spleen.

### The blood supply

Blood is supplied from the splenic and the pancreatico-duodenal arteries; the corresponding veins drain into the portal system.

### The lymphatics

The lymphatics drain into nodes which lie along its upper border, in the groove between its head and the duodenum and along the root of the superior mesenteric vessels.

### Structure

The pancreas macroscopically is lobulated and is contained within a fine capsule; these lobules are made up of alveoli of serous secretory cells draining via their ductules into the principal ducts. Between these alveoli lie the insulin-secreting *islets of Langerhans.*

The *main duct of the pancreas* (Wirsung), (Fig. 56), runs the length of the gland and usually opens at the ampulla of Vater in common with the common bile duct; occasionally it drains separately into the duodenum.

The *accessory duct* (of Santorini) passes from the lower part of the head in front of the main duct, communicates with it, and then opens into the duodenum above it. Occasionally it is absent.

### Development (Fig. 76)

The pancreas develops from a larger dorsal diverticulum from the duodenum and a smaller ventral outpouching from the side of the common bile duct. The ventral pouch swings round posteriorly to fuse with the lower aspect of the dorsal diverticulum, trapping the superior mesenteric vessels between the two parts.

The ducts of the two formative segments of the pancreas then com-

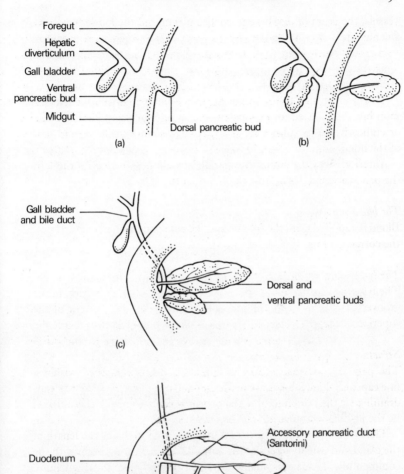

Fig. 76. Development of the intestinal adnexae.

municate, that of the smaller takes over the main pancreatic flow to form the main duct, leaving the original duct of the larger portion of the gland as the accessory duct.

CLINICAL FEATURES

1.   Rarely, the two developing segments of the pancreas completely surround the second part of the duodenum ('annular pancreas') and may produce duodenal obstruction.

2.   Note from the posterior relations of the pancreas that a neoplasm of the head of the pancreas will produce obstructive jaundice by compressing the common bile duct. An extensive growth in the body of the gland may cause portal or inferior vena caval obstruction.

3.   Anterior to the pancreas lies the stomach, separated from it by the lesser sac. This sac may become closed off and distended with fluid either from perforation of a posterior gastric ulcer or from the outpouring of fluid in acute pancreatitis, forming a *pseudo-cyst of the pancreas*—such a collection may almost fill the abdominal cavity.

## THE SPLEEN

The spleen is about the size of the cupped hand. It forms the left lateral extremity of the lesser sac. Passing from it are the *gastro-splenic ligament* to the greater curvature of stomach (carrying the short gastric and left gastro-epiploic vessels) and the *lieno-renal ligament* to the posterior abdominal wall (carrying the splenic vessels and tail of the pancreas).

*Relations* (Fig. 77)

•posteriorly—the left diaphragm, separating it from the pleura, left lung and the 9th, 10th and 11th ribs;

•anteriorly—the stomach;

•inferiorly—the splenic flexure of the colon;

•medially—the left kidney.

The tail of the pancreas abuts against the hilum of the spleen through which vessels and nerves enter and leave this organ.

*Blood supply*

The splenic artery is one of the three main branches of the coeliac axis. The splenic vein is joined by the superior mesenteric to form the portal vein.

*Structure*

The spleen represents the largest reticulo-endothelial accumulation in the body. It has a thin fibrous capsule, to which the peritoneum adheres

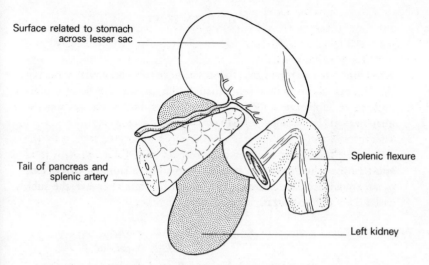

Surface related to stomach
across lesser sac

Tail of pancreas and
splenic artery

Splenic flexure

Left kidney

Fig. 77.   The spleen and its immediate relations.

intimately. The fibrous tissue of the capsule extends into the spleen to form a series of trabeculae between which lies the splenic pulp.

CLINICAL FEATURES

1.   In performing a splenectomy the close relation of the pancreatic tail to the hilum and splenic pedicle must be remembered; it is easily wounded.

2.   Note the close proximity of the lower ribs, left diaphragm, left kidney and the spleen; injuries to the left upper abdomen may damage any combination of these structures. The spleen, with its thin tense capsule, is the commonest intra-abdominal viscus to be ruptured by blunt trauma.

3.   Accessory spleens (one or more) may occur most commonly near the hilum, but also in the tail of pancreas, the mesentery of the spleen, the omentum, small bowel mesentery, ovary and even testis. They occur in about one in ten subjects and, if left behind, may result in persistence of symptoms following splenectomy for congenital acholuric jaundice or thrombocytopenic purpura.

# The urinary tract

## THE KIDNEYS

The kidneys lie retroperitoneally on the posterior abdominal wall; the right kidney is $\frac{1}{2}$ in (12 mm) lower than the left, presumably because of its downward displacement by the bulk of the liver. Each measures approximately $4\frac{1}{2}$ in (11 cm) long, $2\frac{1}{2}$ in (6 cm) wide and $\frac{1}{2}$ in (4 cm) thick.

*Relations* (Figs. 78, 79)
•posteriorly—the diaphragm (separating pleura), quadratus lumborum, psoas, transversus abdominis, the 12th rib and three nerves—the subcostal (D12), ilio-hypogastric and ilio-inguinal (L1).

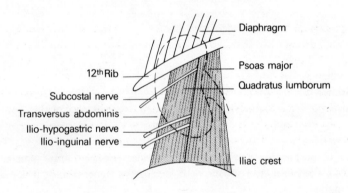

Fig. 78.   The posterior relations of the kidney.

•anteriorly—the right kidney is related to liver, 2nd part of duodenum (which may be opened accidentally in performing a right nephrectomy) and the ascending colon. In front of the left kidney lie the stomach, the pancreas and its vessels, the spleen and the descending colon. The suprarenals sit on each side as a cap on the kidney's upper pole.

The medial aspect of the kidney presents a deep vertical slit, the *hilum*, which transmits, from before backwards, the renal vein, renal artery, pelvis of the ureter and, usually, a subsidiary branch of the renal artery. Lymphatics and nerves also enter the hilum, the latter being sympathetic, mainly vasomotor, fibres.

The *pelvis of the ureter* is subject to considerable anatomical variations

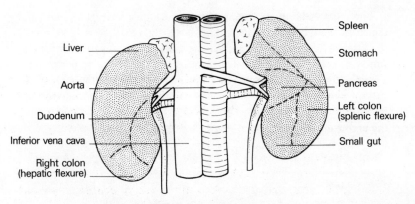

Fig. 79. The anterior relations of the kidney.

(Fig. 80); it may lie completely outside the substance of the kidney (even to the extent of having part of the major calyces extra-renal) or may be almost buried in the renal hilum. All gradations exist between these extremes. If a calculus is lodged in the pelvis of the ureter, its removal is comparatively simple when this is extra-renal and correspondingly difficult when the pelvis is hidden within the substance of the kidney.

Within the kidney, the pelvis of the ureter divides into two or three *major calyces* each of which divides into a number of *minor calyces*. Each of these, in turn, is indented by a papilla of renal tissue and it is here that the collecting tubules of the kidney discharge urine into the ureter.

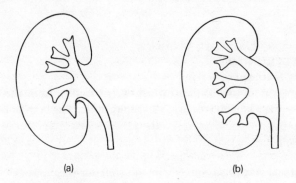

Fig. 80. Variations in the renal pelvis.
(a) The pelvis is buried within the renal parenchyma—pyelo-lithotomy difficult.
(b) The pelvis protrudes generously—pyelo-lithotomy easy.

The kidneys lie in an abundant fatty cushion *(perinephric fat)* contained in the *renal fascia* (Fig. 81). Above, the renal fascia blends with the fascia over the diaphragm, leaving a separate compartment for the suprarenal (which is thus easily separated and left behind in performing a nephrectomy). Medially, the fascia blends with the sheaths of the aorta and inferior vena cava. Laterally it is continuous with the transversalis fascia. Only inferiorly does it remain relatively open—tracking around the ureter into the pelvis.

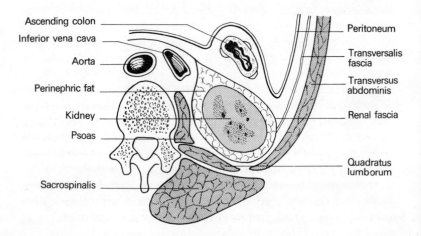

**Fig. 81.** Transverse section demonstrating the fascial compartments of the kidney.

The kidney has, in fact, 3 capsules:

1.  fascial (renal fascia);
2.  fatty (perinephric fat);
3.  true—the fibrous capsule which strips readily from the normal kidney surface but adheres firmly to an organ that has been inflamed.

### Blood supply

The renal artery is derived directly from the aorta.

The renal vein drains directly into the inferior vena cava.

The left renal vein passes *in front* of the aorta immediately below the origin of the superior mesenteric artery. The right renal artery passes *behind* the inferior vena cava.

*Lymph drainage*
Lymphatics drain directly to the para-aortic lymph nodes.

CLINICAL FEATURES
1.   Blood from a ruptured kidney or pus in a perinephric abscess first
distend the renal fascia, then force their way within the fascial compartment downwards into the pelvis. The mid-line attachment of the renal
fascia prevents extravasation to the opposite side.
2.   In hypermotility of the kidney ('floating kidney') this organ can be
moved up and down in its fascial compartment but not from side to side.
To a lesser degree, it is in this plane that the normal kidney moves
during respiration.
3.   *Exposure of the kidney.* An oblique incision is usually favoured midway between the 12th rib and the iliac crest, extending laterally from the
lateral border of erector spinae. Latissimus dorsi and serratus posterior
inferior are divided and the free posterior border of external oblique is
identified, enabling this muscle to be split along its fibres. Internal oblique
and transversus abdominis are then divided, revealing peritoneum anteriorly, which is pushed forward. The renal fascial capsule is then brought
clearly into view and is opened. The subcostal nerve and vessels are
usually encountered in the upper part of the incision and are preserved.
If more room is required, the lateral edge of quadratus lumborum may
be divided and also the 12th rib excised, care being taken to push up, but
not to open, the pleura, which crosses the medial half of the rib.

## THE URETER

The ureter is 10 in (25.5 cm) long and comprises the pelvis of the ureter
(see above) and its abdominal, pelvic and intravesical portions.
    The *abdominal ureter* lies on the medial edge of psoas major (which
separates it from the tips of the transverse processes of L2–L5) and then
crosses into the pelvis at the bifurcation of the common iliac artery in
front of the sacro-iliac joint. Anteriorly, the right ureter is covered at
its origin by the second part of the duodenum and then lies lateral to
the inferior vena cava and behind the posterior peritoneum. It is crossed
by the testicular (or ovarian), right colic, and ileo-colic vessels. The left
ureter is crossed by the testicular (or ovarian) and left colic vessels and
then passes above the pelvic brim, behind the meso-sigmoid and sigmoid
colon.

The *pelvic ureter* runs on the lateral wall of the pelvis in front of the internal iliac artery to just in front of the ischial spine; it then turns forwards and medially to enter the bladder. In the male it lies above the seminal vesicle near its termination and is crossed superficially by the vas deferens (see Fig. 85). In the female, the ureter passes above the lateral fornix of the vagina 1 in (2.5 cm) lateral to the supravaginal portion of the cervix and lies below the broad ligament and uterine vessels.

The *intravesical ureter* passes obliquely through the wall of the bladder for $\frac{3}{4}$ in (2 cm); the vesical muscle and obliquity of this course produce respectively a sphincteric and valve-like arrangement at the termination of this duct.

*Blood supply*

The ureter receives a rich segmental blood supply from all available arteries along its course: the aorta, and the renal, testicular (or ovarian), internal iliac and inferior vesical arteries.

CLINICAL FEATURES

1.   The ureter is readily identified in life by its thick muscular wall which is seen to undergo worm-like (vermicular) writhing movements, particularly if gently stroked or squeezed.

2.   Throughout its abdominal and the upper part of its pelvic course, it adheres to the overlying peritoneum (through which it can be seen in the thin subject), and this fact is used in exposing the ureter—as the parietal peritoneum is dissected up, the ureter comes into view sticking to its posterior aspect.

3.   The ureter is relatively narrowed at three sites:

•at the junction of the pelvis of ureter with its abdominal part,

•at the pelvic brim, and

•at the ureteric orifice (narrowest of all).

A ureteric calculus is likely to lodge at one of these three levels.

4.   In searching for a ureteric stone on a plain radiograph of the abdomen, one must imagine the course of the ureter in relation to the bony skeleton (Fig. 82). It lies along the tips of the transverse processes, crosses in front of the sacro-iliac joint, swings out to the ischial spine and then passes medially to the bladder. An opaque shadow along this line is suspicious of calculus. This course of the ureter is readily studied by examining a radiograph showing a radio-opaque ureteric catheter in situ.

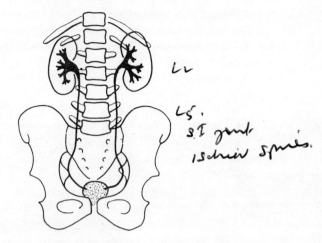

**Fig. 82.** Drawing from an intravenous pyelogram to show the relationship of the ureters to the bony landmarks.

## The embryology and congenital abnormalities of the kidney and ureter (Fig. 83)

The kidney and ureter are mesodermal in origin and develop in an unusual manner of considerable interest to the comparative anatomist.

The *pronephros*, of importance in primitive animals, is transient in man, but the distal part of its duct receives the tubules of the next renal organ to develop, the *mesonephros*, and now becomes the *mesonephric or Wolffian duct*. The mesonephros itself then disappears except for some of its ducts which form the efferent tubules of the testis.

A diverticulum then appears at the lower end of the mesonephric duct which develops into the *metanephric duct*; on top of the latter a cap of tissue differentiates to form the definitive kidney or *metanephros*. The metanephric duct develops into the ureter, pelvis, calyces and collecting tubules, the metanephros into the glomeruli and the proximal part of the renal duct system.

The mesonephric duct now loses its renal connection, atrophies in the female (remaining only as the epoöphoron) but persists in the male, to become the epididymis and vas deferens.

The kidney first develops in the pelvis and then migrates upwards. Its blood supply is first obtained from the common iliac artery but, during migration, a series of vessels form to supply it, only to involute again when the renal artery takes over this duty.

*Developmental abnormalities* (Fig. 84)

1.  It is not uncommon for one or more distally placed arteries to persist *(aberrant renal arteries)* and one may even run to the kidney from the common iliac artery.
2.  Occasionally the kidney will fail to migrate cranially, resulting in a persistent *pelvic kidney*.
3.  The two metanephric masses may fuse in development, forming a *horse-shoe kidney* linked across the mid-line.

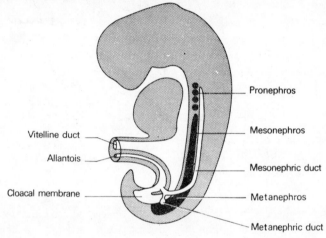

Fig. 83. Development of the pro-, meso- and metanephric systems (after Langman).

4.  In 1 in 2,400 births there is complete failure of development of one kidney *(congenital absence of the kidney)*.
5.  *Congenital polycystic kidneys* (which are nearly always bilateral) are believed to result from failure of metanephric tissue to link up with some of the metanephric duct collecting tubules; blind ducts therefore form which subsequently become distended with fluid. This theory of origin does not explain their occasional association with multiple cysts of the liver, pancreas, lung and ovary.
6.  The mesonephric duct may give off a double metanephric bud so that two ureters may develop on one or both sides. These ureters may fuse into a single duct anywhere along their course or open separately into the bladder (where the *upper* ureter enters *below* the *lower* ureter).

Rarely the extra ureter may open ectopically into vagina or urethra and results in urinary incontinence.

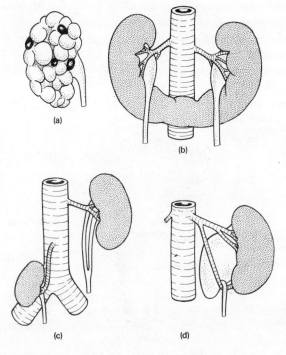

**Fig. 84.** Renal abnormalities.

(a) Polycystic kidney.            (b) Horse-shoe kidney.

(c) Pelvic kidney and double ureter.      (d) Aberrant renal artery and associated
                                               hydronephrosis.

## THE BLADDER (Figs. 60, 61, 85)

The urinary bladder of a normal subject is uncomfortably distended by
half a pint of fluid. When fully distended, the adult bladder projects
from the pelvic cavity into the abdomen, stripping the peritoneum upwards from the anterior abdominal wall. The surgeon utilizes this fact
in carrying out an extraperitoneal incision into the bladder. In children
up to the age of about three years, the pelvis is relatively small and the
bladder is, in fact, intra-abdominal although still extraperitoneal.

*Relations*
• anteriorly—the pubic symphysis.
• superiorly—the bladder is covered by peritoneum with coils of small

intestine and sigmoid colon lying against it. In the female, the body of the uterus flops against its postero-superior aspect.

•posteriorly—in the male the rectum, the termination of the vasa deferentia and the seminal vesicles. In the female, the vagina and the supra-vaginal part of the cervix.

•laterally—the levator ani and obturator internus.

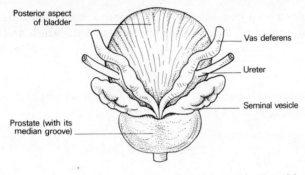

**Posterior aspect of bladder**

**Vas deferens**

**Ureter**

**Seminal vesicle**

**Prostate (with its median groove)**

Fig. 85.   The prostate, seminal vesicles and vasa shown in a posterior view of the bladder.

The neck of the bladder fuses with the prostate in the male; in the female it lies directly on the pelvic fascia surrounding the short urethra.

The muscle coat of bladder is formed by a criss-cross arrangement of bundles; when these hypertrophy in chronic obstruction (due to an enlarged prostate, for example) they account for the typical trabeculated 'open weave' appearance of the bladder wall, readily seen through a cystoscope.

The circular component of the muscle coat condenses as an (involuntary) *internal urethral sphincter* around the internal orifice. This can be destroyed without incontinence providing the external sphincter remains intact (as occurs in prostatectomy).

*Cystoscopy*

The interior of the bladder and its three orifices (the internal meatus and the two ureters) are easily inspected by means of a cystoscope. The ureteric orifices lie 1 in (2.5 cm) apart in the empty bladder, but when this is distended for cystoscopic examination the distance increases to 2 in (5 cm). The submucosa and mucosa of most of the bladder are only loosely adherent to the underlying muscle and are thrown into folds when the bladder is empty, smoothing out during distension of the organ. Over the *trigone*, the triangular area bounded by the ureteric orifices and the internal meatus, the mucosa is adherent and remains smooth even in the empty bladder.

Between the ureters, a raised fold of mucosa can be seen called the inter-ureteric ridge which is produced by an underlying bar of muscle.

*Blood supply*
Blood is supplied from the superior and inferior vesical branches of the internal iliac artery. The vesical veins form a plexus which drains into the internal iliac vein.

*Lymph drainage*
Lymphatics drain alongside the vesical blood vessels to the iliac and then para-aortic nodes.

## THE URETHRA

### The male urethra

The *male urethra* is 8 in (20 cm) long and is divided into the prostatic, membranous and spongy parts.

The *prostatic urethra* $1\frac{1}{4}$ in (82 mm) traverses the prostate and receives the 2 ejaculatory, and the 15 or 20 prostatic ducts.

The *membranous urethra* $\frac{3}{4}$ in (2 cm) pierces the external sphincter urethrae (the voluntary sphincter of the bladder) and the fascial perineal membrane which covers the superficial aspect of the sphincter.

The *spongy urethra* 6 in (15 cm) traverses the corpus spongiosum of the penis. It first passes upwards and forwards to lie below the pubic symphysis and then in its flaccid state bends downwards and forwards.

CLINICAL FEATURES
1.   Where the urethra passes beneath the pubis is a common site for it to be ruptured by a fall astride a sharp object, which crushes it against the edge of the symphysis.
2.   The external orifice is the narrowest part of the urethra and a calculus may lodge there. Immediately within the meatus the urethra dilates into a terminal fossa whose roof bears a mucosal fold (the *lacuna magna*) which may catch the tip of a catheter. Instruments should always be introduced into the urethra beak downwards for this reason.

## The female urethra

The female urethra is $1\frac{1}{2}$ in (4 cm) long; it traverses the sphincter urethrae and lies immediately in front of the vagina. Its external meatus opens 1 inch behind the clitoris. The sphincter urethrae in the female is a tenuous structure and vesical control appears to depend mainly on the intrinsic sphincter of condensed circular muscle fibres of the bladder.

## The mucosa of the urinary tract

The pelvis, ureter, bladder and urethra are lined by a transitional epithelium as far as the entry of the ejaculatory ducts in the prostatic urethra. This is conveniently termed the *uroepithelium* since it has a uniform appearance and is subject to the same pathological processes—for example the development of papillomata. The remainder of the urethra has a columnar lining except at its termination, where the epithelium becomes squamous.

## Radiology of the urinary tract

The renal contours can often be identified on a soft tissue radiograph of the abdomen. Intravenous injection of iodine-containing compounds excreted by the kidney will produce an outline of the calyces and the ureter (intravenous pyelogram). When the injection medium enters the bladder, a cystogram is obtained (Fig. 82).

Further information can be obtained by passing a catheter up the ureter through a cystoscope and injecting radio-opaque fluid to fill the pelvis and calyx system (retrograde pyelogram). Similarly injection of such fluid into the urethra or bladder may be used for the radiographic study of these viscera.

# The male genital organs

## PROSTATE (Fig. 85)

This is a pyramidal-shaped, fibromuscular and glandular organ, $1\frac{1}{4}$ in (3 cm) long, which surrounds the prostatic urethra. It resembles the size and shape of a chestnut.

*Relations* (Fig. 60)

•superiorly—the prostate is continuous with the neck of the bladder.

The urethra enters the upper aspect of the prostate near its anterior border.

•inferiorly—the apex of the prostate rests on the external sphincter of the bladder which lies within the deep perineal pouch.

•anteriorly—lies the pubic symphysis separated by the extraperitoneal fat of the *cave of Retzius* or *retropubic space*. Close against the prostate in this space lies the prostatic plexus of veins. Near the apex of the prostate, the *pubo-prostatic ligament* (a condensation of fibrous tissue) passes forward to the pubis.

•posteriorly—lies the rectum separated by *the fascia of Denonvilliers*.

•laterally—lies levator ani.

The ejaculatory ducts *(vide infra)* enter the upper posterior part of the gland to open into the urethra, dividing off a *median prostatic lobe* lying between these three ducts. A shallow posterior *median groove* (which can be felt on rectal examination) further divides the prostate into left and right *lobes*. Anterior to the urethra, the prostate consists of a narrow *isthmus* only.

*The prostatic capsules*
These are normally two, pathologically three, in number.
1. The true capsule—a thin fibrous sheath which surrounds the gland.
2. The false capsule—condensed extraperitoneal fascia which continues into the fascia surrounding the bladder and with the fascia of Denonvilliers posteriorly. Between layers 1 and 2 lies the prostatic venous plexus.
3. The pathological capsule—when benign 'adenomatous' hypertrophy of the prostate takes place, the normal peripheral part of the gland becomes compressed into a capsule around this enlarging mass (Fig. 86). In performing an enucleation of the prostate, the plane between the adenomatous mass and this compressed peripheral tissue is entered, the 'tumour' enucleated and a condensed rim of prostate tissue, lying deep to the true capsule, left behind. The prostatic venous plexus, lying external to this, is thus undisturbed.

*Blood supply*
The arterial supply is derived from the inferior vesical artery, a branch entering the prostate on each side at its lateral extremity.

The veins form a prostatic plexus which receives the dorsal vein of the penis and drains into the internal iliac vein on each side. Some of the venous drainage passes to the plexus of veins lying in front of the vertebral bodies and within the neural canal. These veins are valveless and constitute the '*valveless vertebral veins of Bateson*'. This communication may explain the readiness with which carcinoma of the prostate spreads to the pelvis and vertebrae.

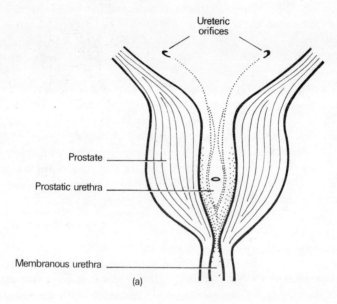

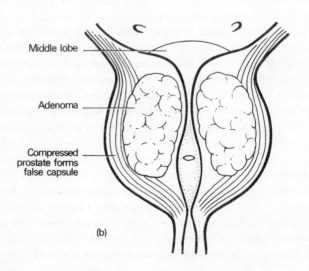

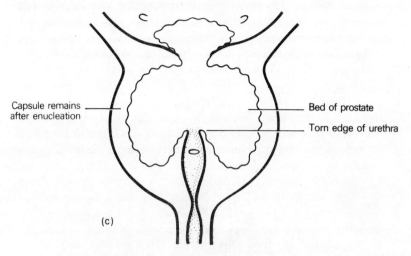

Capsule remains after enucleation

Bed of prostate

Torn edge of urethra

(c)

**Fig. 86.** The surgical anatomy of prostatectomy.
(a) the normal prostate in vertical section.
(b) a prostatic adenoma (benign hypertrophy) compresses the normal prostatic tissue into a false capsule.
(c) 'prostatectomy' removes the adenoma but leaves the capsule.

CLINICAL FEATURES

1. *Approaches to the prostate.* 'Prostatectomy' usually means the enucleation of a hyperplastic mass of glandular tissue from the surrounding prostatic gland which is compressed into a thin rim around it (Fig. 86). The prostate can be approached, for this purpose, either across the bladder *(transvesically)*, or through the prostatic capsule itself *(retropubically)* or, rarely, across the perineum, dissecting in front of the rectum and reaching the prostate through the fascia of Denonvilliers *(the perineal approach)*. In obstructive conditions other than benign hypertrophy, that is to say, carcinoma or fibrosis of the prostate, this plane of cleavage does not exist and the obstructing mass of prostatic tissue may be removed *transurethrally* by means of an operating cystoscope.

2. After the age of 45 years some degree of prostatic hypertrophy is all but invariable; it is as much a sign of ageing as greying of the hair. Usually the lateral lobes are affected and such enlargement is readily detected on rectal examination. The median lobe may also be involved in this process or may be enlarged without the lateral lobes being affected. In such an instance, symptoms of prostatic obstruction may occur (be-

cause of the valve-like effect of this hypertrophied lobe lying over the internal urethral orifice) without prostatic enlargement being detectable on rectal examination.

Anterior to the urethra the prostate consists of a narrow fibro-muscular isthmus containing little, if any, glandular tissue. Benign glandular hypertrophy of the prostate therefore never affects this part of the organ.

3. The fascia of Denonvilliers is important surgically; in excising the rectum it is the plane to be sought after in order to separate off the prostate and urethra without damaging these structures. A carcinoma of the prostate only rarely penetrates this fascial barrier so that ulceration into the rectum is very unusual.


## THE SCROTUM

The scrotum is the pouch in which lie the testes and their coverings. In cryptorchidism, not unnaturally, this pouch is not well developed.

The skin of the scrotum is thin, pigmented, rugose and marked by a longitudinal median raphe. The subcutaneous tissue contains no fat but does contain the involuntary *dartos muscle*.

### CLINICAL FEATURES

The scrotal subcutaneous tissue is continuous with the fasciae of the abdominal wall and perineum and therefore extravasations of urine or blood deep to this plane will gravitate into the scrotum. The scrotum is divided by a septum into right and left compartments but this septum is incomplete superiorly so therefore extravasations of fluid into this sac are always bilateral.

Because of the lax tissues of the scrotum and its dependent position it readily fills with oedema fluid in cardiac or renal failure. Such a condition must be carefully differentiated from extravasation or from a scrotal swelling due to a hernia or hydrocele.


## TESTIS AND EPIDIDYMIS (Fig. 87)

The left testis lies at a lower level than the right within the scrotum; rarely this arrangement is reversed. Each testis is contained by a white fibrous capsule, the *tunica albuginea*, and each is invaginated anteriorly into a double serous covering, the *tunica vaginalis*, just as the intestine is invaginated anteriorly into the peritoneum.

Along the posterior border of the testis, rather to its lateral side, lies the *epididymis*, which is divided into an expanded head, a body, and a pointed tail inferiorly. Medially, there is a distinct groove, the *sinus epididymis*, between it and the testis. The epididymis is covered by the tunica vaginalis except at its posterior margin which is free or, so to say, 'extraperitoneal'.

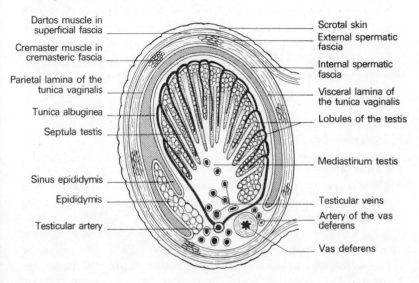

Fig. 87.   Transverse section of the testis.

The testis and epididymis each bear at their upper extremities a small stalked body, termed respectively the *appendix testis* and *appendix epididymis* (hydatid of Morgagni). The appendix testis is a remnant of the upper end of the paramesonephric (Mullerian) duct; the appendix epididymis is a remnant of the mesonephros.

These structures, being stalked, are liable to undergo torsion.

*Blood supply*

The testicular artery arises from the aorta at the level of the renal vessels. It anastomoses with the artery to the vas, supplying the vas deferens and epididymis, which arises from the inferior vesical branch of the internal iliac artery. This cross-connection means that ligation of the testicular artery is not necessarily followed by testicular atrophy.

The pampiniform plexus of veins becomes a single vessel, the testicular vein, in the region of the internal ring. On the right this drains into the inferior vena cava, on the left into the renal vein.

### Lymph drainage

The lymphatic drainage of the testis obeys the usual rule; it accompanies the venous drainage and thus passes to the para-aortic lymph nodes at the level of the renal vessels. Free communication occurs between the lymphatics on either side; there is also a plentiful anastomosis with the para-aortic intrathoracic nodes and, in turn, with the cervical nodes, so that spread of malignant disease from the testis to the root of the neck is not rare.

### Nerve supply

T10 sympathetic fibres via the renal and aortic plexus.

### Structure

The testis is divided into 2–300 lobules each containing one to three *seminiferous tubules*. Each tubule is some 2 feet (62 cm) in length when teased out, and is thus obviously coiled and convoluted to pack away within the testis. The tubules anastomose posteriorly into a plexus termed the *rete testis* from which about a dozen fine *efferent ducts* arise, pierce the tunica albuginea at the upper part of the testis and pass into the head of the epididymis, which is actually formed by these efferent ducts coiled within it. The efferent ducts fuse to form a considerably convoluted single tube which constitutes the body and tail of the epididymis.

### Development of the testis

This is important and is the key to several features which are of clinical interest.

The testis arises from a germinal ridge of mesoderm in the posterior wall of the abdomen just medial to the developing kidney and links up with the epididymis and vas, differentiating from the mesonephric duct. As the testis enlarges, it also undergoes a caudal migration according to the following time-table:

| | |
|---|---|
| 3rd month (of fetal life) | reaches the iliac fossa. |
| 7th month | traverses the inguinal canal. |
| 8th month | reaches the external ring. |
| 9th month | descends into the scrotum. |

A mesenchymal strand, the *gubernaculum testis*, extends from the caudal end of the developing testis along the course of its descent to blend into the scrotal fascia. The exact role of this structure in the descent of the testis is not known; theories are that it acts as a guide (gubernaculum=rudder) or that its swelling dilates the inguinal canal and scrotum.

In the third foetal month, a prolongation of the peritoneal cavity invades the gubernacular mesenchyme and projects into the scrotum as the *processus vaginalis*. The testis slides into the scrotum posterior to this, projects into it and is therefore clothed front and sides with peritoneum. About the time of birth this processus obliterates, leaving the testis covered by the tunica vaginalis. Very rarely, fragments of adjacent developing organs—spleen or suprarenal—are caught up and carried into the scrotum along with the testis.

CLINICAL FEATURES

1. The testis arises at the level of the kidney and drags its vascular, lymphatic and nerve supply from this region. Pain from the kidney is often referred to the scrotum and, conversely, testicular pain may radiate to the loin.

2. When searching for secondary lymphatic spread from a neoplasm of the testis, the upper abdomen must be palpated carefully for enlarged para-aortic nodes: because of cross-communications, these may be present on either side. Mediastinal and cervical nodes may also become involved. It is the beginner's mistake to feel for nodes in the groin; these are only involved if the tumour has ulcerated the scrotal skin and hence invaded scrotal lymphatics which drain to the inguinal nodes.

3. Rarely, a rapidly developing *varicocele* (dilatation of the pampiniform plexus of veins) is a presenting sign of a tumour of the left kidney which, by invading the renal vein, blocks the drainage of the left testicular vein. Most examples of varicocele are idiopathic; why the vast majority are on the left side is unknown, but theories are that the left testicular vein is compressed by a loaded sigmoid colon, obstructed by angulation at its entry into the renal vein or even that it is put into spasm by adrenalin-rich blood entering the renal vein from the suprarenal vein!

4. The testis may *fail to descend* and may rest anywhere along its course—intra-abdominally, within the inguinal canal, at the external ring or high in the scrotum. Failure to descend must be carefully distinguished from *retraction of the testis*; it is common in children for contraction of the cremaster muscle to draw the testis up into the superficial inguinal pouch—a potential space deep to the superficial fascia over the external ring. Gentle pressure from above, or the relaxing effect

of a hot bath, coaxes the testis back into the scrotum in such cases.

Occasionally the testis descends, but into an unusual *(ectopic)* position; most commonly it passes laterally after leaving the external ring to lie superficial to the inguinal ligament, but it may be found in front of the pubis, in the perineum or in the upper thigh. In these cases (unlike the undescended testis), the cord is long and replacement into the scrotum without tension presents no surgical difficulty.

5.　Abnormalities of the obliteration of the processus vaginalis lead to a number of extremely common surgical conditions of which the *indirect inguinal hernia* is the most important.

This variety of hernia may be present at birth or develop in later life; in the latter circumstance it is probable that the processus vaginalis has persisted as a narrow empty sac and that development of the hernia results from some sudden strain due to a cough, straining at micturition or at stool which forces abdominal contents into this peritoneal recess.

In infants, the sac frequently has the testis lying in its wall (congenital inguinal hernia) but this is unusual in older patients.

The closed off tunica vaginalis may become distended with fluid to form a *hydrocele* which may be idiopathic (primary) or secondary to disease in the underlying testis. The anatomical classification of hydroceles is into the following groups (Fig. 88).

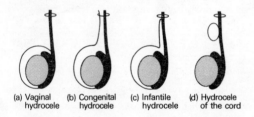

(a) Vaginal hydrocele　　(b) Congenital hydrocele　　(c) Infantile hydrocele　　(d) Hydrocele of the cord

Fig. 88.　Types of hydrocele. (The ring at the upper end of each diagram represents the internal inguinal ring.)

1.　*vaginal*—confined to the scrotum and so called because it distends the tunica vaginalis.
2.　*congenital*—communicating with the peritoneal cavity.
3.　*infantile*—extending upwards to the internal ring.
4.　*hydrocele of the cord*—confined to the cord.

Notice that, from the anatomical point of view, a hydrocele (apart from one of the cord) must *surround* the front and sides of the testis since the tunica vaginalis bears this relationship to it. A cyst of the epididymis in contrast, arises from the efferent ducts of the epididymis and must therefore lie above and behind the testis. This point enables the differential diagnosis between these two common scrotal cysts to be made confidently.

## VAS DEFERENS

This tube is 18 in (45 cm) long (a distance which one may remember is also the length of the thoracic duct, the spinal cord and the femur, and the distance from the incisor teeth to the cardiac end of the stomach).

The vas passes from the tail of the epididymis to traverse scrotum, inguinal canal and so comes to lie upon the side wall of the pelvis. Here, it lies immediately below the peritoneum of the lateral wall, extends almost to the ischial tuberosity then turns medially to the base of the bladder. Here it joins the more laterally placed seminal vesicle to form the *ejaculatory duct* which traverses the prostate to open into the prostatic urethra.

### CLINICAL FEATURES

Infection may track from bladder and urethra along the vas to the epididymis, particularly from the prostatic bed after prostatectomy; division of the vasa at prostatectomy is performed by many surgeons in an attempt to decrease this complication. The vas is identified by its very firm consistency which, in coaching days, was likened to whipcord but which to-day might be compared with fine plastic tubing.

## THE SEMINAL VESICLES

These are coiled sacculated tubes 2 in (5 cm) long which can be unravelled to three times that length. They lie, one on each side, extra-peritoneally at the bladder base, lateral to the termination of the vasa. Each has common drainage with its neighbouring vas via the ejaculatory duct.

### CLINICAL FEATURES

1. The vesicles can be felt on rectal examination if enlarged; this occurs typically in tuberculous infection.
2. The seminal vesicles, as their name implies, act as stores for semen. They receive their nerve supply from the 1st lumbar sympathetic ganglion through the presacral plexus. A bilateral high lumbar sympathectomy results in sterility as ejaculation is prevented.

# The bony and ligamentous pelvis

The pelvis is made up of the innominate bones, the sacrum and the coccyx bound to each other by dense ligaments.

### The os innominatum (Fig. 89)

Examine the bone and revise the following structures.

*The ilium* with its *iliac crest* running between the *anterior* and *posterior superior iliac spines*; below each of these are the corresponding *inferior spines*. Well-defined ridges on its lateral surface are the strong muscle markings of the glutei. Its inner aspect bears the large *auricular surface* which articulates with the sacrum. The *ilio-pectineal line* runs forward from the apex of the auricular surface and demarcates the *true* from the *false pelvis*.

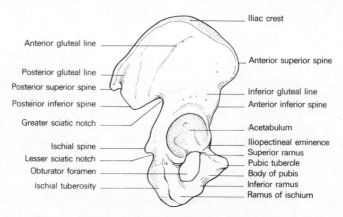

Anterior gluteal line

Posterior gluteal line
Posterior superior spine
Posterior inferior spine

Greater sciatic notch

Ischial spine
Lesser sciatic notch
Obturator foramen
Ischial tuberosity

Iliac crest

Anterior superior spine

Inferior gluteal line
Anterior inferior spine

Acetabulum

Iliopectineal eminence
Superior ramus
Pubic tubercle
Body of pubis
Inferior ramus
Ramus of ischium

Fig. 89.    Lateral view of the os innominatum.

*The pubis* comprises the *body* and the *superior* and *inferior pubic rami*.

*The ischium* has a vertically disposed *body*, bearing the *ischial spine* on its posterior border which demarcates an upper (greater) and lower (lesser), sciatic notch. The inferior pole of the body bears the *ischial tuberosity* then projects forwards almost at right angles into the *ischial ramus* to meet the inferior pubic ramus.

*The obturator foramen* lies bounded by the body and rami of the pubis and the body and ramus of the ischium.

All three bones fuse at the *acetabulum* which forms the socket for the femoral head for which it bears a wide crescentic articular surface.

The pelvis is tilted in the erect position so that the plane of its inlet is at an angle of 60 deg. to the horizontal. (To place a pelvis into this position, hold it against a wall so that the anterior superior spine and the top of the pubic symphysis both touch it.)

### The sacrum (Fig. 90)

The sacrum is made up of five fused vertebrae and is roughly triangular. The anterior border of its upper part is termed the *sacral promontory* and is readily felt at laparotomy.

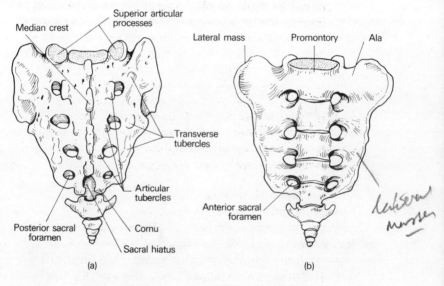

Fig. 90.  The sacrum in: (a) posterior; (b) anterior views.

Its anterior aspect presents a *central mass*, a row of four *anterior sacral foramina* on each side (transmitting the upper four sacral anterior primary rami), and, lateral to these, the *lateral masses* of the sacrum. The superior aspect of the lateral mass on each side forms a fan-shaped surface termed *the ala*.

Note that the central mass is roughly rectangular—the triangular shape of the sacrum is due to the rapid shrinkage in size of the lateral masses of the sacrum from above down.

Posteriorly lies the *sacral canal*, continuing the vertebral canal, bounded by short pedicles, strong laminae and diminutive spinous processes. Perforating through from the sacral canal is a row of four *posterior sacral foramina* on each side. Inferiorly the canal terminates in the *sacral hiatus* which transmits the 5th sacral nerve.

On its lateral aspect, the sacrum presents an *auricular facet* for articulation with the corresponding surface of the ilium.

The 5th lumbar vertebrae may occasionally fuse with the sacrum in whole or in part; alternatively, the 1st sacral segment may be partially or completely separated from the rest of the sacrum. The posterior arch of the sacrum is occasionally bifid.

Note that the dural sheath terminates distally at the second piece of the sacrum. Beyond this the sacral canal is filled with the fatty tissue of the extradural space and the nerve filaments of the filum terminale.

## Coccyx

This is made up of three to five fused vertebrae articulating with the sacrum; occasionally the first segment remains separate. It represents, of course, the tail of more primitive animals.

### The functions of the pelvis

1.  It protects the pelvic viscera.
2.  It supports the weight of the body which is transmitted through the vertebrae, thence through the sacrum, across the sacro-iliac joints to the innominate bones and then to the femora in the standing position or to the ischial tuberosities when sitting. (The sacro-iliac joint is reinforced for this task as will be described below.)
3.  During walking the pelvis swings from side to side by a rotatory movement at the lumbo-sacral articulation which occurs together with similar movements of the lumbar intervertebral joints. Even if the hip joints are fixed, this swing of the pelvis enables the patient to walk reasonably well.
4.  As with all but a few small bones in the hand and foot, the pelvis provides attachments for muscles.
5.  In the female it provides bony support for the birth canal.

## Joints and ligamentous connections of the pelvis

*The symphysis pubis* is the name given to the joint between the pubic bones. Each pubic bone is covered by a layer of hyaline cartilage and is connected across the mid-line by a dense layer of fibro-cartilage. The centre of the latter may break down to form a cleft-like joint space which, however, is not seen before the first decade and which is not lined by a synovial membrane.

The joint is surrounded and strengthened by fibrous ligaments, especially above and below.

*The sacro-iliac joints* are the articulations between the auricular surfaces of the sacrum and ilium on each side and are true synovium-lined and cartilage-covered joints.

The sacrum 'hangs' from these joints supported by the extremely dense *sacro-iliac ligaments* lying posteriorly to the auricular joint surfaces. These support the whole weight of the body, tending to drag the sacrum forward into the pelvis and, not surprisingly, are the strongest ligaments in the body.

Their action is assisted by an interlocking of the grooves between the auricular surfaces of the sacrum and ilium.

*The sacro-tuberous ligament* passes from the ischial tuberosity to the side of the sacrum and coccyx.

*The sacro-spinous ligament* passes from the ischial spine to the side of the sacrum and coccyx.

These two ligaments help to define two important exits from the pelvis:

1. The *greater sciatic foramen*—formed by the sacro-spinous ligament and the greater sciatic notch.
2. The *lesser sciatic foramen*—formed by the sacro-tuberous ligament and the lesser sciatic notch.

*Note*

There is a useful surface landmark in this region, the dimple constantly seen on each side immediately above the buttock, which defines:

1. the posterior superior iliac spine;
2. the centre of the sacro-iliac joint;
3. the level of the second sacral segment;
4. the level of the end of the dural canal of the spinal meninges.

## Differences between the male and female pelvis (Fig. 91)

The pelvis demonstrates a large number of sex differences associated principally with two features: first the heavier build and stronger muscles in the male, accounting for the stronger bone structure and better defined muscle markings in this sex; second, the comparatively wider and shallower pelvic cavity in the female, correlated with its role as the bony part of the birth canal.

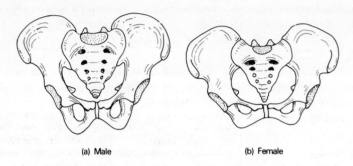

(a) Male          (b) Female

**Fig. 91.**    (a) Male and (b) female pelves compared.

The sex differences are summarized in Table 2.

When looking at a radiograph of the pelvis, the sex is best determined by three features.

1.    The pelvic inlet, heart-shaped in the male, oval in the female.

2.    The angle between the inferior pubic rami, which is narrow in the male, wide in the female. In the former, it corresponds almost exactly to the angle between the index and middle fingers when these are held apart; in the latter the angle equals that between the fully-extended thumb and the index finger.

3.    The soft tissue shadow of the penis and scrotum can usually be seen or, if not, the dense shadow of the lead screen used to shield the testes from harmful radiation.

### Obstetrical pelvic measurements (Fig. 92)

The figures for the measurements of the inlet, mid-cavity and outlet of the true pelvis are readily committed to memory in the form shown (in inches) in Table 3.

The *transverse diameter of the outlet* is assessed clinically by measuring the distance between the ischial tuberosities along a plane passing across the anus; the *antero-posterior outlet diameter* is measured from the pubis to the sacro-coccygeal joint. The most useful measurement clinically is,

Table 2  Comparison of male and female pelves

|   |   | *Male* | *Female* |
|---|---|--------|----------|
| 1 | General structure | Heavy and thick | Light and thin |
| 2 | Joint surfaces | Large | Small |
| 3 | Muscle attachments | Well marked | Rather indistinct |
| 4 | False pelvis | Deep | Shallow |
| 5 | Pelvic inlet | Heart-shaped | Oval |
| 6 | Pelvic canal | 'long segment of a short cone' i.e. long and tapered | 'short segment of a long cone' i.e. short with almost parallel sides |
| 7 | Pelvic outlet | Comparatively small | Comparatively large |
| 8 | 1st piece of sacrum | The superior surface of the body occupies nearly half the width of sacrum | Oval superior surface of the body occupies about one-third the width of sacrum |
| 9 | Sacrum | Long, narrow, with smooth concavity | Short, wide, flat, curving forward in lower part |
| 10 | Sacro-iliac articular facet (auricular surface) | Extends well down the 3rd piece of the sacrum | Extends down only to upper border of 3rd piece |
| 11 | Sub-pubic angle (between inferior pubic rami) | 'The angle between the middle and index finger' | 'The angle between the thumb and index finger' |
| 12 | Inferior pubic ramus | Presents a strong everted surface for attachment of the crus of the penis | This marking is not present |
| 13 | Acetabulum | Large | Small |
| 14 | Ischial Tuberosities | Inturned | Everted |
| 15 | Obturator foramen | Round | Oval |

Table 3

|  | *Transverse* | *Oblique* | *Antero-posterior* |
|---|---|---|---|
| Inlet | 5 | $4\frac{1}{2}$ | 4 inches |
| Mid-pelvis | $4\frac{1}{2}$ | $4\frac{1}{2}$ | $4\frac{1}{2}$ inches |
| Outlett | 4 | $4\frac{1}{2}$ | 5 inches |

however, the *diagonal conjugate*—from the lower border of the pubic symphysis to the promontory of the sacrum. This normally measures 5 in (12.5 cm); from the practical point of view, it is not possible in the normal pelvis to reach the sacral promontory on vaginal examination either readily or without discomfort to the patient.

Another useful clinical guide is:

•the *external conjugate*, from the tip of the 5th lumbar spine to the pubic symphysis, which should not be less than $7\frac{1}{2}$ in (20 cm) (now seldom used);
•the *sub-pubic arch*—the examiner's four knuckles, (i.e. his clenched fist), should rest comfortably between the inferior rami below the pubic symphysis.

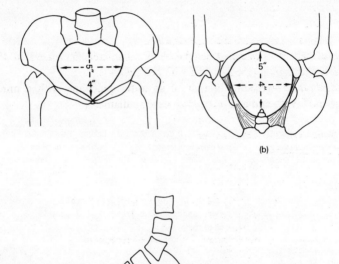

**Fig. 92.** The measurements of the female pel is.
The inlet. (b) The outlet. (c) Lateral view to show the diagonal conjugate.

Note that these measurements are all of the bony pelvis; the 'dynamic pelvis' of the birth-canal, in fact, is narrowed by the pelvic musculature, the rectum and the thickness of the uterine wall. Today accurate radiological techniques enable exact measurements to be made of the bony pelvis.

**Variations of the pelvic shape** (Fig. 93)

The female pelvic shapes may be sub-divided (after Caldwell and Moloy) as follows.

*1.   The normal and its variants*
(a) *Gynaecoid*—normal.
(b) *Android*—the masculine type of pelvis.
(c) *Platypelloid*—shortened in the A. P. diameter, increased in the transverse diameter (the "non-rachitic flat pelvis").
(d) *Anthropoid*—resembling that of an anthropoid ape with a much lengthened A. P. and a shortened transverse diameter.

*2.   Symmetrically contracted pelvis*
That of a small women but with a symmetrical shape.

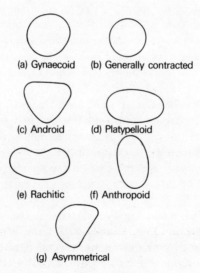

(a) Gynaecoid   (b) Generally contracted

(c) Android   (d) Platypelloid

(e) Rachitic   (f) Anthropoid

(g) Asymmetrical

Fig. 93.   Pelvic variations and abnormalities—shown as diagrammatic outlines of the pelvic inlet.

### 3. The Rachitic flat pelvis

The sacrum is rotated so that the sacral promontory projects forward and the coccyx tips backwards; The A. P. diameter of the inlet is therefore narrowed, but that of the outlet is increased.

### 4. The asymmetrical

Assymetry can be due to a variety of cause such as scoliosis, long-standing hip disease (e.g. congenital dislocation) and the *Naegele pelvis* which is due to the congenital absence of one wing of the sacrum or its destruction by disease.

CLINICAL FEATURES

1. *Fractures of the pelvis* may be isolated lesions due to a localized blow or may be displacements of part of the pelvic ring due to compression injuries. Lateral compression usually results in fractures through both pubic rami on each side, or both rami on one side with dislocation at the symphysis; antero-posterior compression may be followed by dislocation at the symphysis or fractures through the pubic rami accompanied by dislocation at the sacro-iliac joint.

Displacement of part of the pelvic ring must, of course, mean that the ring has been broken in two places.

Falls upon the leg may force the head of the femur through the acetabulum, the so-called central dislocation of the hip. Isolated fractures may be produced by local trauma, especially to the iliac wing, sacrum and pubis.

Associated with pelvic fractures one must always consider soft tissue injuries to bladder, urethra and rectum, which may be penetrated by spicules of bone or torn by wide displacements of the pelvic fragments.

Occasionally in these pelvic displacements the ilio-lumbar branch of the internal iliac artery is ruptured as it crosses above the sacro-iliac joint; this may be followed by a severe or even fatal extra-peritoneal haemorrhage.

2. *Sacral (caudal) anaesthesia.* The sacral hiatus, between the last piece of sacrum and coccyx, can be entered by a needle which pierces skin, fascia and the tough posterior sacro-coccygeal ligament to enter the sacral canal. The hiatus can be defined by palpating the sacral cornua on either side (Fig. 90) immediately above the natal cleft.

Anaesthetic solution injected here will travel extradurally to bathe the spinal roots emerging from the dural sheath.

# The muscles of the pelvic floor and perineum

The canal of the bony and ligamentous pelvis is closed by a diaphragm of muscles and fasciae which the rectum, urethra and, in the female, the vagina, must pierce to reach the exterior.

The muscles are divided into:

1.   The pelvic diaphragm, formed by the levator ani and the coccygeus.
2.   The superficial muscles: *(a)* of the anterior (urogenital) perineum; *(b)* of the posterior (anal) perineum.

*Levator ani* (Fig. 94) is the largest and most important muscle of the pelvic floor. It arises from the back of the pubis, the fascia of the side

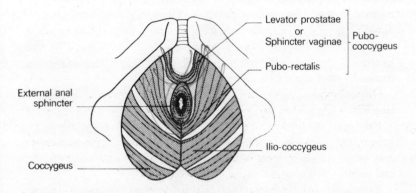

Fig. 94.   Levator ani—inferior aspect.

wall of the pelvis (covering obturator internus) and the spine of the ischium. From this wide origin it sweeps down in a series of loops:

*(a)* to form a sling around the prostate (or vagina), inserting into the perineal body;

*(b)* to form a sling around the rectum and also to insert into and re-inforce the deep part of the anal external sphincter at the ano-rectal ring;

*(c)* to insert into the occyx.

The muscle acts as the principal support of the pelvic floor, has a sphincter action on the rectum and vagina and assists in increasing intra-abdominal pressure during defaecation, micturition and parturition.

Its deep aspect is related to the pelvic viscera and its perineal aspect forms the inner wall of the ischio-rectal fossa (see below).

**The anterior (urogenital) perineum** (Figs. 95, 96)

A line joining the ischial tuberosities passes just in front of the anus. Between this line and the ischio-pubic inferior rami lies the urogenital part of the perineum or the *urogenital triangle*.

Attached to the sides of this triangle is a tough fascial sheet termed the *perineal membrane* which is pierced by the urethra in the male and by the urethra and the vagina in the female. Deep to this membrane is the external *sphincter* of the urethra consisting of voluntary muscle fibres surrounding the membranous urethra; these are competent even when the internal sphincter has been completely destroyed. In the female the superficial sphincter is also pierced by the vagina.

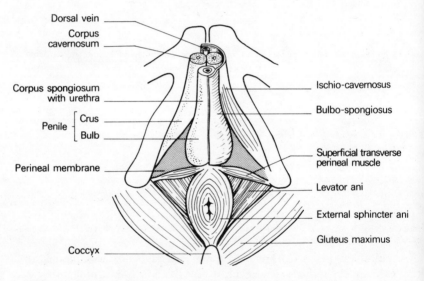

Dorsal vein
Corpus cavernosum
Corpus spongiosum with urethra
Penile { Crus / Bulb }
Perineal membrane
Coccyx

Ischio-cavernosus
Bulbo-spongiosus
Superficial transverse perineal muscle
Levator ani
External sphincter ani
Gluteus maximus

**Fig. 95.** The male perineum—on the right side the muscles of the anterior perineum have been dissected away.

Enclosing the deep aspect of the external sphincter is a second fascial sheath so that this muscle is, in fact, contained within a fascial capsule which is termed the *deep perineal pouch*. This pouch contains, in addition, the deep transverse perineal muscles and, in the male, the two *glands of Cowper* whose ducts pass forward to open into the bulbous urethra. Superficial to the perineal membrane is the *superficial perineal pouch* which contains, in the male:

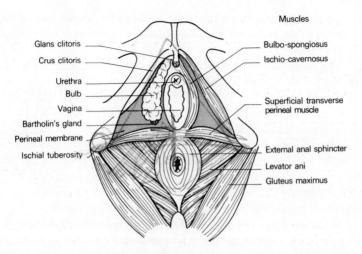

Muscles

Glans clitoris

Crus clitoris

Urethra

Bulb

Vagina

Bartholin's gland

Perineal membrane

Ischial tuberosity

Bulbo-spongiosus

Ischio-cavernosus

Superficial transverse perineal muscle

External anal sphincter

Levator ani

Gluteus maximus

**Fig. 96a.** The female perineum—on the right side the muscles of the anterior perineum have been dissected away.

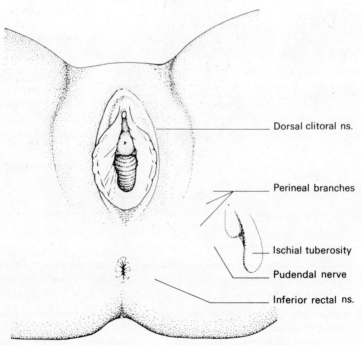

Dorsal clitoral ns.

Perineal branches

Ischial tuberosity

Pudendal nerve

Inferior rectal ns.

**Fig. 96b.** Distribution of the pudendal nerve to the female perineum.

1. The *bulbo-spongiosus* muscle covering the corpus spongiosum which, in turn, surrounds the urethra. (The distal corpus spongiosum expands into the glans penis.)

2. The *ischio-cavernosus* muscle on each side, arising from the ischial ramus and covering the corpus cavernosum.

(The urethra is thus enclosed in a spongy sheath supported by a cavernous tube on each side containing thin walled venous sinuses which become engorged with blood when erection occurs.)

3. The superficial transverse perineal muscle, running transversely from the perineal body to the ischial ramus. It is of no functional importance but it seen during perineal excision of the rectum.

In the female the same muscles are present although much less well developed and the bulbo-spongiosus is pierced by the vagina.

*The perineal body.* This fibro-muscular node lies in the mid-line at the junction of the anterior and posterior perineum.

It is the point of attachment for the anal sphincters, the bulbo-spongiosus, the transverse perineal muscles and fibres of levator ani.

### Posterior (anal) perineum (Figs. 95, 96)

This is the triangle lying between the ischial tuberosities on each side and the coccyx. It comprises, in essentials, the anus with its superficial sphincters, levator ani and, at each side, the ischio-rectal fossa.

*The ischio-rectal fossa* (Fig. 97) is of considerable surgical importance because of its great tendency to become infected. Its boundaries are:

•Laterally—the fascia over obturator internus (i.e. the side wall of the

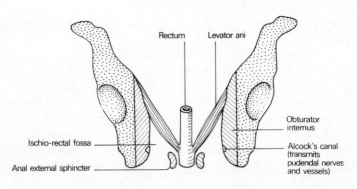

Fig. 97. The ischio-rectal fossa.

pelvis); contained in this wall within a fascial tunnel termed *Alcock's canal* are the pudendal vessels and nerve which give off respectively the inferior haemorrhoidal vessels and nerve to the anus, then pass forward to supply the perineal tissues;

•Medially—the fascia over levator ani and the external anal sphincter;
•Posteriorly—the sacro-tuberous ligament covered posteriorly by gluteus maximus;
•Anteriorly—the urogenital perineum;
•Floor—skin and subcutaneous fat.

### CLINICAL FEATURES

1.   The content of the fossa is coarsely lobulated fat. It is important to note that the ischio-rectal fossae communicate with each other behind the anus—infection in one passes readily to the other.

Infection of this space may occur from boils or abrasions of the perianal skin, from lesions within the rectum and anal canal, from pelvic infection bursting through levator ani or, rarely, via the bloodstream. The fossa contains no important structures and can therefore be fearlessly incised when infected.

2.   The pudendal nerves can be blocked in Alcock's canal on either side to give useful regional anaesthesia in forceps delivery (see Fig. 96 and page 272).

# The female genital organs

## THE VULVA

The *vulva* (or *pudendum*) is the term applied to the female external genitalia.

The *labia majora* are the prominent hair-bearing folds extending back from the *mons pubis* to meet posteriorly in the mid-line of the perineum. They are equivalent to the male scrotum.

The *labia minora* lie between the labia majora as lips of soft skin which meet posteriorly in a sharp fold, the *fourchette*. Anteriorly they split to enclose the *clitoris*, forming an anterior *prepuce* and posterior *frenulum*.

The *vestibule* is the area enclosed by the labia minora and contains the *urethral orifice* (which lies immediately behind the clitoris) and the *vaginal orifice*.

The vaginal orifice is guarded in the virgin by a thin mucosal fold, the *hymen*, which is perforated to allow the egress of the menses, and may have an annular, semilunar, septate or cribriform appearance. Rarely it is imperforate and menstrual blood distends the vagina (haematocolpos). At first coitus the hymen tears, usually posteriorly or postero-laterally, and after childbirth nothing is left of it but a few tags termed *carunculae myrtiformes*.

*Bartholin's glands* (the greater vestibular glands) are a pair of lobulated, pea-sized, mucus-secreting glands lying deep to the posterior parts of the labia majora. They are impalpable when healthy but become obvious when inflamed or distended. Each drains by a duct one inch long which opens into the groove between the hymen and the posterior part of the labium minus.

Anteriorly, each gland is overlapped by the *bulb of the vestibule*—a mass of cavenous erectile tissue equivalent to the bulbus spongiosum of the male. This tissue passes forwards, under cover of bulbo-spongiosus, around the sides of the vagina to the roots of the clitoris.

CLINICAL FEATURES

At childbirth the introitus may be enlarged by making an incision in the perineum *(episiotomy)*. This starts at the fourchette and extends medio-laterally on the right side for 3.75 cm. The skin, vaginal epithelium, subcutaneous fat, perineal body and superficial transverse perineal muscle are incised. After delivery the episiotomy is carefully sutured in layers.

## THE VAGINA (Fig. 98)

The vagina surrounds the cervix of the uterus, then passes downwards and forwards through the pelvic floor to open into the vestibule.

The cervix projects into the anterior part of the vault of the vagina so that the continuous vaginal gutter surrounding the cervix is shallow anteriorly (where the vaginal wall is 3 in (7.5 cm) in length) and is deep posteriorly (where the wall is 4 in (10 cm) long). This continuous gutter is, for convenience of description, divided into the anterior, posterior and lateral *fornices*.

*Relations*
•Anteriorly—the base of the bladder and the urethra (which is embedded in the anterior vaginal wall).
•Posteriorly—from below upwards, the anal canal (separated by the perineal body), the rectum and then the peritoneum of the pouch of

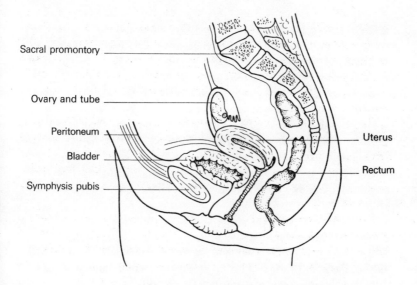

Fig. 98.  Sagittal section of the uterus and its relations.

Douglas which covers the *upper quarter of the posterior vaginal wall.*
•Laterally—levator ani, pelvic fascia and the *ureters, which lie close to lateral fornices.*

The amateur abortionist (or inexperienced gynaecologist)without a knowledge of anatomy, fails to realize that the uterus passes upwards and forwards from the vagina; he pushes the instrument or I.U.C.D. (intra-uterine contraceptive device), which he intends to enter the cervix directly backwards through the posterior fornix. To make matters worse, this is the only part of the vagina which is intraperitoneal; the peritoneal cavity is thus entered and peritonitis follows.

*Blood supply*
Arterial supply is from the internal iliac artery via its vaginal, uterine, internal pudendal and middle rectal branches.

A venous plexus drains via the vaginal vein into the internal iliac vein.

*Lymphatic drainage* (see Fig. 102)
•Upper third to the external and internal iliac nodes.
•Middle third to the internal iliac nodes.
•Lower third to the superficial inguinal nodes.

*Structure*

A stratified squamous epithelium lines the vagina and the vaginal cervix; it contains no glands and is lubricated partly by cervical mucus and partly by desquamated vaginal epithelial cells. In the nulliparous women the vaginal wall is rugose, but it becomes smoother after childbirth.

Beneath the epithelial coat is a thin areolar layer separating it from the muscular wall which is made up of a criss-cross arrangement of involuntary muscle fibres. This muscle layer is ensheathed in a fascial capsule which blends with adjacent pelvic connective tissues, so that the vagina is firmly supported in place.

In old age the vagina shrinks in length and diameter. The cervix projects far less into it so that the fornices all but disappear.

## THE UTERUS (Figs. 98, 99)

The uterus is a pear-shaped organ, 3 in (7.5 cm) in length, made up of the *fundus*, *body* and *cervix*. The *Fallopian tubes* enter into each superolateral angle (the *cornu*) above which lies the *fundus*.

The *body* of the uterus narrows to a waist termed the *isthmus*, continuing into the *cervix* which is embraced about its middle by the vagina; this attachment delimits a supravaginal and vaginal part of the cervix.

The *isthmus* is 1.5 mm wide, the anatomical internal os marks its junction with the uterine body but its mucosa is histologically similar to the endometrium. The isthmus is that part of the uterus which becomes the lower segment in pregnancy.

The *cavity* of the uterine body is triangular in coronal section, but in the sagittal plane it is no more than a slit. This cavity communicates via the *internal os* with the *cervical canal* which, in turn, opens into the vagina by the *external os*.

The nulliparous external os is circular but after childbirth it becomes a transverse slit with an anterior and a posterior lip.

The non-pregnant cervix has the firm consistency of the nose; the pregnant cervix has the soft consistency of the lips.

In fetal life the cervix is considerably larger than the body; in childhood (the *infantile uterus*) the cervix is still twice the size of the body but, during puberty, the uterus enlarges to its adult size and proportions by relative overgrowth of the body. The adult uterus is bent forward on itself at about the level of the internal os to form an angle of 170 deg; this is termed *anteflexion of the uterus*. Moreover, the axis of the cervix forms an angle of 90 deg with the axis of the vagina—*anteversion of the uetrus*. The uretus thus lies in an almost horizontal plane.

In *retroversion of the uterus*, the axis of the cervix is directed upwards and backwards. Normally on vaginal examination the lowermost part of the cervix to be felt is its anterior lip; in retroversion either the os or the posterior lip becomes the presenting part.

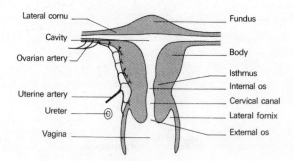

**Fig. 99.** Coronal section of the uterus and vagina. Note the important relationships of ureter and uterine artery.

In *retroflexion* the axis of the body of the uterus passes upwards and backwards in relation to the axis of the cervix.

Frequently these two conditions co-exist. They may be mobile and symptomless—as a result of distension of the bladder or purely as a development anomaly. Less commonly they are fixed; the result of adhesions, endometriosis or the pressure of a tumour in front of the uterus (Fig. 100).

*Relations*
•Anteriorly—the body is related to the utero-vesical pouch of peritoneum and lies either on the superior surface of the bladder or on coils of intestine. The supravaginal cervix is related directly to bladder, separated only by connective tissue. The infravaginal cervix has the anterior fornix immediately in front of it.
•Posteriorly—lies the pouch of Douglas, with coils of intestine within it.
•Laterally—the broad ligament and its contents (see below); *the ureter lies 12 mm lateral to the supravaginal cervix.*

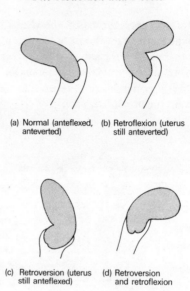

(a) Normal (anteflexed, anteverted)  (b) Retroflexion (uterus still anteverted)

(c) Retroversion (uterus still anteflexed)  (d) Retroversion and retroflexion

Fig. 100. Variations in uterine position and their terminology.

CLINICAL FEATURES

The most important single practical relationship in this region is that of the ureter to the supravaginal cervix. At this point, the ureter lies just above the level of the lateral fornix, below the uterine vessels as these pass across within the broad ligament (Fig. 101). In performing a hysterectomy, the ureter may be accidentally divided in clamping the uterine vessels, especially when the pelvic anatomy has been distorted by a previous operation, a mass of fibroids, infection or malignant infiltration.

The ureter is readily compressed by lateral extension of growth from the uterus; bilateral hydronephrosis with uraemia is a frequent mode of termination in this disease.

The close relationship of ureter to the lateral fornix is best appreciated by realizing that a ureteric stone at this site can be palpated on vaginal examination. (This is the answer to the examination question: 'When can a stone in the ureter be felt?')

*Blood supply*

The *uterine artery* (from the internal iliac) runs in the base of the broad ligament and crosses above and at right angles to the ureter to reach the uterus at the level of the internal os. The artery then ascends in a tortuous manner alongside the uterus, supplying the corpus, and then anastomoses with the *ovarian artery*. The uterine artery also gives off a descending branch to the cervix and branches to the upper vagina. The veins accompany the arteries and drain into the internal iliac veins, but they also communicate via the pelvic plexus with the veins of the vagina and bladder.

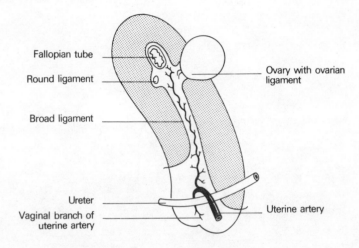

Fallopian tube

Round ligament

Broad ligament

Ovary with ovarian ligament

Ureter

Vaginal branch of uterine artery

Uterine artery

Fig. 101.  Lateral view of the uterus (schematic) to show:
  1  Composition of the broad ligament.
  2  The relations of ureter and uterine artery.
  3  The peritoneal covering of the uterus (stippled).

*Lymph drainage* (Fig. 102)

1.  The fundas (together with the ovary and Fallopian tube) drains along the ovarian vessels to the aortic nodes, apart from some lymphatics which pass along the round ligament to the inguinal nodes.

2.  The body drains via the broad ligament to glands lying alongside the external iliac vessels.

3.  The cervix drains in three directions—laterally, in the broad ligament, to the external iliac nodes; postero-laterally along the uterine vessels to the internal iliac nodes, and posteriorly along the recto-uterine folds to the sacral nodes.

Always examine the inguinal glands in a suspected carcinoma of the corpus uteri—they may be involved by lymphatic spread along the round ligament.

*Structure*

The uterus is covered with peritoneum except where this is reflected off at two sites; anteriorly onto the bladder at the uterine isthmus and laterally at the broad ligaments. The muscle wall is thick and made up of a criss-cross of involuntary fibres mixed with fibro-elastic connective tissue.

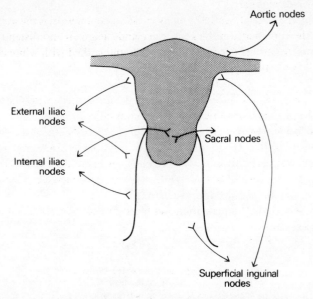

**Fig. 102.**    Lymph drainage of the uterus and vagina.

The *mucosa* is applied directly to muscle with no submucosa intervening. The mucosa of the body of the uterus is the *endometrium*, made up of a single layer of cuboidal ciliated cells forming simple tubular glands which dip down to the underlying muscular wall. Below this epithelium is a stroma of connective tissue containing blood-vessels and round cells.

The cervical canal epithelium is made up of tall columnar cells which form a series of complicated branching glands; these secrete an alkaline mucus which forms a protective '*cervical plug*' filling the canal.

The vaginal aspect of the cervix is covered with a stratified squamous epithelium continuous with that of the vagina.

The mucosa of the corpus undergoes extensive changes during the menstrual cycle which may be briefly summarized thus:

1. first 4 days—desquamation of its superficial two-thirds with bleeding;
2. subsequent 2–3 days—rapid reconstitution of the raw mucosal surface by growth from the remaining epithelial cells in the depths of the glands;
3. by the 14th day the endometrium has reformed; this is the end of the *proliferative phase*;
4. from the 14th day until the menstrual flow commences is the *secretory phase*; the endometrium thickens, the glands lengthen and distend with fluid and the stroma becomes oedematous and stuffed with white cells.

At the end of this phase three layers can be defined:
1. A compact superficial zone.
2. A spongy middle zone—with dilated glands and oedematous stroma.
3. A basal zone of inactive non-secreting tubules.

With degeneration of the corpus luteum there is shrinkage of the endometrium, the arteries retract and coil, producing ischaemia of the middle and superficial zones which then desquamate. It is probable that spasm of the vessels in the basal layer (which remains non-desquamated) prevents the woman bleeding to death.

Only very slight desquamation and bleeding takes place in the mucosa of the cervical canal.


## THE FALLOPIAN TUBES (Fig. 103)

The Fallopian, or uterine, tubes are about 4 in (10 cm) long; they lie in the free edge of the broad ligaments and open into the cornu of the uterus. Each comprises four parts.

1. The *infundibulum*—the bugle shaped extremity extending beyond the broad ligament and opening into the peritoneal cavity by the *ostium*. Its mouth is fimbriated and overlies the ovary, to which one long fimbria actually adheres *(fimbria ovarica)*.
2. The *ampulla*—wide, thin-walled and tortuous.
3. The *isthmus*—narrow, straight and thick-walled.
4. The *interstitial part*—which pierces the uterine wall.


*Structure*

Apart from the interstitial part, the tube is clothed in peritoneum. Beneath this is a muscle layer of outer longitudinal and inner circular fibres.

The mucosa is formed of columnar, mainly ciliated cells and lies in longitudinal ridges each of which is thrown into numerous folds.

The ova are propelled to the uterus along this tube, partly by peristalsis and partly by cilial action.

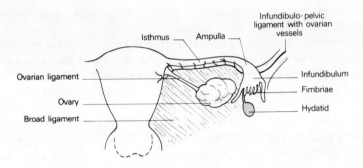

Fig. 103.   The Fallopian tube, ovary and broad ligament.

### CLINICAL FEATURES

1.   Note that the genital canal in the female is the only direct communication into the peritoneum from the exterior and is a potential pathway for infection, for example, in gonorrhoea.

2.   The fertilized ovum may implant ectopically, i.e. in a site other than the endometrium of the corpus uteri. When this occurs in the Fallopian tube it is called, according to the exact site, fimbrial, ampullary, isthmic or interstitial, of which the ampullary is commonest and interstitial rarest.

As the ectopic embryo enlarges, it may abort into the peritoneal cavity (where rarely it continues to grow as a secondary abdominal pregnancy) or else ruptures the tube. This second fate is particularly likely to occur in the narrow and relatively non-distensible isthmus; rupture is usually into the peritoneal cavity but may rarely occur into the broad ligament.

## THE OVARY (Fig. 103)

The ovary is an almond-shaped organ, $1\frac{1}{2}$ in (4 cm) long, attached to the back of the broad ligament by the *mesovarium*. The ovary has two other attachments, the *infundibulo-pelvic ligament*, along which pass the ovarian vessels and lymphatics from the side wall of the pelvis, and the *ovarian ligament*, which passes to the cornu of the uterus.

*Relations*

The ovary is usually described as lying on the side wall of the pelvis opposite the *ovarian fossa*, which is a depression bounded by the external iliac vessels in front and the ureter and internal iliac vessels behind and which contains the obturator nerve. However, the ovary is extremely variable in its position and is frequently found prolapsed into the pouch of Douglas in perfectly normal women.

The ovary, like the testis, develops from the genital ridge and then descends into the pelvis. In the same way as the testis, it therefore drags its blood supply and lymphatic drainage downwards with it from the posterior abdominal wall.

*Blood supply*

Blood supply, lymph drainage and nerve supply is from the ovarian artery which arises from the aorta at the level of the renal arteries.

The ovarian vein drains, on the right side, to the inferior vena cava, on the left, to the left renal vein.

*Lymphatics* pass to the aortic glands at the level of the renal vessels.

*Nerve supply* is from the aortic plexus (T10).

All these structures pass to the ovary in the infundibulo-pelvic ligament.

*Structure*

The ovary has no peritoneal covering; the serosa ends at the mesovarian attachment. It consists of a connective tissue stroma containing *Graafian follicles* at various stages of development, *corpora lutea* and *corpora albicantia* (hyalinized, regressing corpora lutea, which take several months to absorb completely).

The surface of the ovary in young children is covered with a so-called '*germinal epithelium*' of cubical cells. It is now known, however, that the primordial follicles develop in the ovary in early fetal life and do not differentiate from these cells. In adult life, in fact, the epithelial covering of the ovary disappears, leaving only a fibrous capsule termed the *tunica albuginea*.

After the menopause the ovary becomes small and shrivelled; in old age the follicles disappear completely.

## The endopelvic fascia and the pelvic ligaments (Fig. 104)

The *pelvic fascia* is the term applied to the connective tissue floor of the pelvis covering levator ani and obturator internus. The *endopelvic* fascia is the extraperitoneal cellular tissue of the uterus (the *parametrium*), vagina, bladder and rectum. Within this endopelvic fascia are three important condensations of connective tissue which sling the pelvic viscera from the pelvic walls:

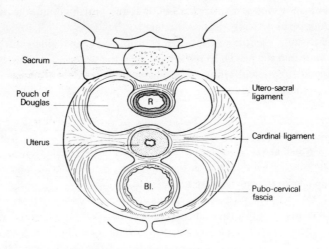

Fig. 104. The pelvic ligaments seen from above.

1. The *cardinal ligaments* (transverse cervical, or Mackenrodt's ligaments) which pass laterally from the cervix and upper vagina to the side walls of the pelvis along the lines of attachment of levator ani. They are composed of white fibrous connective tissue with some involuntary muscle fibres and are pierced in their upper part by the ureters.

2. The *utero-sacral ligaments*, which pass backwards from the postero-lateral aspect of the cervix at the level of the isthmus and from the lateral vaginal fornices deep to the utero-sacral folds of peritoneum in the lateral boundaries of the pouch of Douglas. They are attached to the periosteum in front of the sacro-iliac joints and the lateral part of the third piece of the sacrum.

3. The *pubo-cervical fascia*, which extends forward from the cardinal ligament to the pubis on either side of the bladder, to which it acts as a sling.

These three ligaments act as supports to the cervix of the uterus and the vault of the vagina, in conjunction with the important elastic muscular foundation provided by levator ani. In prolapse these ligaments lengthen (in procidentia they may be 6 inches long) and any repair operation must include their reconstitution.

Two other pairs of ligaments take attachments from the uterus.

1. The *broad ligament*, which is a fold of peritoneum connecting the lateral margin of the uterus with the side wall of the pelvis on each side. The uterus and its broad ligaments therefore form a partition across the pelvic floor dividing off an anterior compartment, containing bladder (the *utero-vesical pouch*), from a posterior compartment, containing rectum (the *pouch of Douglas* or *recto-uterine pouch*).

The broad ligament contains or carries (Figs. 101, 103).

- the Fallopian (uterine) tube in its free edge;
- the ovary, attached by the mesovarium to its posterior aspect;
- the round ligament;
- the ovarian ligament, crossing from the ovary to the uterine cornu (see ovary);
- the uterine vessels and branches of the ovarian vessels;
- lymphatics and nerve fibres.

The ureter passes forwards to the bladder deep to this ligament and lateral to the lateral fornix of the vagina.

2. The *round ligament*—a fibromuscular cord. It passes from the lateral angle of the uterus in the anterior layer of the broad ligament to the internal inguinal ring; thence it traverses the inguinal canal to the labium majus. Taken together with the ovarian ligament, it is equivalent to the male gubernaculum testis and can be thought of as the pathway along which the female gonad might have, but in fact did not, descend to the labium majus (the female homologue of the scrotum).

## Vaginal examination

The relations of the vagina to the other pelvic organs must be constantly borne in mind when carrying out a vaginal examination.

- *Inspection* (by means of a speculum) enables the vaginal walls and cervix to be examined and a biopsy or cytological smear to be taken. Inspection of the introitus while straining detects prolapse and the presence of stress incontinence.
- *Anteriorly*—the urethra, bladder and symphysis pubis are felt.
- *Posteriorly*—the rectum (invasion of the vagina by a rectal neoplasm must always be sought after in this disease). Collections of fluid, malig-

nant deposits, prolapsed uterine tubes and ovaries or coils of distended bowel may be felt in the pouch of Douglas.

•*Laterally*—the ovary and tube, and the side wall of pelvis. Rarely a stone in the ureter may be felt through the lateral fornix. The strength of the perineal muscles can be assessed by asking the patient to tighten up her perineum.

•*Apex*—the cervix is felt projecting back from the anterior wall of the vagina. In the normal anteverted uterus the anterior lip of the cervix presents; in retroversion either the cervical os or the posterior lip are first to be felt.

Pathological cervical conditions, for example, neoplasm, can be felt as can the softening of the cervix in pregnancy and its dilatation during labour.

*Bimanual examination* assesses the size and position of the uterus, en-largements of ovary or uterine tubes and the presence of other pelvic masses.

The obstetrician can assess the size of the pelvis both in the transverse and antero-posterior diameter. Particularly important is the distance from the lower border of the symphysis pubis to the sacral promontory which is termed the *diagonal conjugate*. If the pelvis is of normal size the examiner's fingers should fail to reach the promontory of the sacrum. If it is readily palpable, pelvic narrowing is present (see 'obstetrical measurements', page 136).

## Embryology of the Fallopian tubes, uterus and vagina (Fig. 105)

The *Mullerian* or *paramesonephric ducts* develop, one on each side, ad-jacent to the *mesonephric (Wolffian) ducts* in the posterior abdominal wall—they are mesodermal in origin. All these four tubes lie close to-gether caudally, projecting into the anterior (urogenital) compartment of the cloaca.

One system disappears in the male, the other in the female, each leaving behind congenital remnants of some interest to the clinician.

In the male, the Mullerian system disappears, apart from the appendix testis and the prostatic utricle; in the female, the mesonephric system (which in the male develop into the vas deferens and epididymal ducts) persists as remnants in the broad ligament termed the epöophoron, paröophoron and ducts of Gartner.

The Mullerian ducts in the female form the Fallopian tubes cranially. More caudally they come together and fuse in the mid-line (dragging, as they do so, a peritoneal fold from the side wall of the pelvis which becomes the broad ligament). The median structure so formed differentiates into the epithelium of the uterine body (endometrium), cervical canal and upper one-third of the vagina which are first solid and later become canalized. The rest of the vaginal epithelium develops by canalisation of the solid sino-vaginal node at the back of the urogenital sinus. The muscle of the Fallopian tubes, uterine body, cervix and vagina develops from surrounding mesoderm, so that remnants of the Wolffian duct system of the female are found in the myometrium, cervix and vaginal wall.

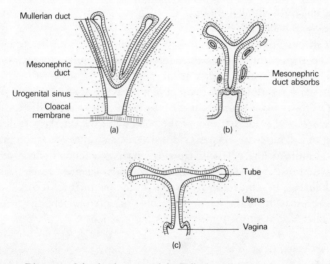

**Fig. 105.** Diagrams of the development of the Fallopian tubes, uterus and vagina from the Mullerian (paramesonephric) ducts and the urogenial sinus (after Hollinshead).

Development abnormalities of this system can easily be deduced. All stages of division of the original double tube may persist from a bicornuate uterus to a complete reduplication of the uterus and vagina. Alternatively there may be absence, hypoplasia or atresia of the duct system on one or both sides.

Failure of canalization of the originally solid caudal end of the duct results, after puberty, in the accumulation of menstrual blood above the obstruction. First the vagina may distend with blood, then the uterus and then the tubes (haematocolpos, haematometra and haematosalpinx respectively).

# The posterior abdominal wall

The bed of the posterior abdominal wall is made up of three bony and four muscular structures.

The bones are:
• the bodies of the lumbar vertebrae,
• the sacrum,
• the wings of the ilium.

The muscles are:
• the diaphragm—posterior part,
• the quadratus lumborum,
• the psoas major,
• the iliacus.

The diaphragm has been considered in the section on thorax.

The psoas must be dealt with in more detail because of the involvement of its sheath in the formation of a psoas abscess.

*The Psoas major* arises from the transverse processes of all the lumbar vertebrae and from the sides of the bodies and the intervening discs of T12 to L5 vertebrae. It passes downwards and laterally at the margin of the brim of the pelvis, narrowing down to a tendon which crosses the front of the hip joint beneath the inguinal ligament to be inserted, with iliacus, into the lesser trochanter of the femur (Fig. 106).

The psoas flexes the hip on the trunk, or, alternatively, the trunk on the hips (e.g. in sitting up from the lying position).

CLINICAL FEATURES

1. The femoral artery lies on the psoas tendon, and it is this firm posterior relation of the femoral artery at the groin which enables it here to be identified and compressed easily by the finger.

2. The psoas is enclosed in the *psoas sheath* which is a compartment of the lumbar fascia. Pus from a tuberculous infection of the lumbar vertebrae is limited in its anterior spread by the anterior longitudinal vertebral ligament, and therefore passes laterally into this sheath *(psoas abscess)*, which may also be entered by pus tracking down from the posterior mediastinum in disease of the thoracic vertebrae. Pus may then spread under the inguinal ligament into the femoral triangle where it produces a soft swelling (Fig. 106). Occasionally, in completely neglected cases, pus tracks along the femoral vessels, along the subsartorial canal and eventually appears in the popliteal fossa.

*The retroperitoneal organs* are: the pancreas, kidneys and ureters (which have already been considered), the suprarenals, the aorta and inferior vena cava and their main branches, the para-aortic lymph nodes and the lumbar sympathetic chain.

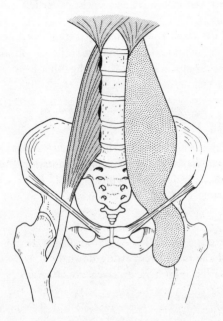

**Fig. 106.** Psoas sheath and psoas abscess.
On the right is a normal psoas sheath; on the left is shown distended with pus, which tracks under the inguinal ligament to present in the groin.

## The suprarenals (Fig. 79)

The suprarenal glands cap the upper poles of the kidneys and lie against the diaphragm. The left is related anteriorly to the stomach across the lesser sac, the right lies behind the right lobe of the liver and tucks medially behind the inferior vena cava.

Each gland, although weighing only 3 to 4 grammes, has three arteries supplying it:
1. a direct branch from the aorta;
2. a branch from the phrenic artery;
3. a branch from the renal artery.

The single main vein drains from the hilum of the gland into the nearest available vessel—the inferior vena cava on the right, the renal vein on the left. The stubby right suprarenal vein, coming directly from the i.v.c., presents the most dangerous feature in performing an adrenalectomy—the tiro should always choose the easier left side and leave the right to his chief.

The suprarenal gland comprises a cortex and medulla. The latter is derived from the neural crest (ectoderm) whose cells also give rise to the sympathetic ganglia. The cortex, on the other hand, is derived from the mesoderm.

### Abdominal aorta (Fig. 107)

The aorta enters the abdomen via the aortic hiatus in the diaphragm at the level of the 12th thoracic vertebra and ends at L4. It lies throughout this course against the vertebral bodies.

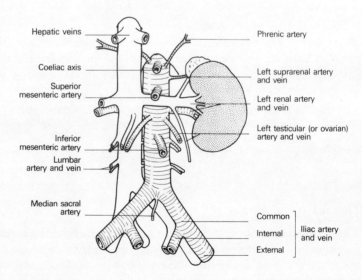

Hepatic veins

Coeliac axis

Superior mesenteric artery

Inferior mesenteric artery

Lumbar artery and vein

Median sacral artery

Phrenic artery

Left suprarenal artery and vein

Left renal artery and vein

Left testicular (or ovarian) artery and vein

Common
Internal
External
} Iliac artery and vein

Fig. 107. The abdominal aorta, the inferior vena cava and their main branches.

Anteriorly, from above down, it is related to the pancreas (separating it from the stomach), the third part of the duodenum and coils of small intestine. A large tumour of pancreas or stomach, or a mass of enlarged

para-aortic glands or a large ovarian cyst may transmit the pulsations of
the aorta and be mistaken for an aneurysm.

The branches of the aorta are:

*(a)* 3 anterior unpaired branches passing to viscera:

1.  the *coeliac axis*—giving off the $\begin{cases} \text{hepatic artery} \\ \text{splenic artery} \\ \text{left gastric artery} \end{cases}$

2.  the *superior mesenteric artery*
3.  the *inferior mesenteric artery*

*(b)* 3 lateral paired branches passing to viscera:

1.  the *suprarenal artery*
2.  the *renal artery*
3.  the *testicular* or *ovarian artery*

*(c)* 5 lateral paired branches to the parietes:

1.  the *inferior phrenic artery*
2.  4 *lumbar* branches

*(d)* terminal branches:

1.  the *common iliacs*
2.  the *median sacral artery*

*The common iliac arteries* pass, one on each side, downwards and outwards to bifurcate into the internal and external iliacs in front of the
sacro-iliac joint, at the level of the sacral promontory. They give no
other branches.

At the bifurcation, the common iliac artery is crossed superficially by
the ureter—a convenient site to identify this latter structure in pelvic
operations.

*The external iliac artery* runs along the brim of the pelvis on the medial
side of psoas major. The artery passes below the inguinal ligament to
form the femoral artery giving off, immediately before its termination,
the *inferior epigastric artery* which demarcates the medial edge of the
internal inguinal ring (Fig. 43).

*The internal iliac artery* passes backwards and downwards into the
pelvis, sandwiched between the ureter anteriorly and the internal iliac
vein posteriorly. At the upper border of the greater sciatic notch it
divides into an anterior and posterior division which give off branches
to supply the pelvic organs, perineum, buttock and sacral canal.

## Inferior vena cava (Fig. 107)

The i.v.c. commences at L5 by the junction of the common iliac veins
*behind* the right common iliac artery (unlike the usual arrangement of

a vein being superficial to its corresponding artery). It lies to the right of the aorta as it ascends until separated from it by the right crus of the diaphragm when the aorta pierces this muscle. The i.v.c. itself passes through the diaphragm at T8, traverses the pericardium and drains into the right atrium.

As the i.v.c. ascends, it is related anteriorly to coils of small intestine, the third part of the duodenum, the head of the pancreas with the common bile duct, and the first part of duodenum. It then passes behind the foramen of Winslow, in front of which lies the portal vein separating it from the common bile duct and hepatic artery. Finally the i.v.c. lies in a deep groove in the liver before piercing the diaphragm. Within the liver it receives the right and left hepatic veins. Occasionally these veins fuse into a single trunk which opens directly into the i.v.c.; on other occasions a third hepatic vein may open into the i.v.c., draining either the left side of the right lobe of the liver or the main part of the left lobe. These variations are now of importance because of the possibility of carrying out resection of one or other lobe of the liver.

## Lumbar sympathetic chain

The lumbar part of the sympathetic trunk commences deep to the medial arcuate ligament of the diaphragm as a continuation of the thoracic sympathetic chain. On each side it lies against the bodies of the lumbar vertebrae overlapped, on the right side, by the inferior vena cava and on the left by the aorta.

The lumbar arteries lie deep to the chain but lumbar veins may cross superficial to it and are of importance because they may be damaged in performing a sympathectomy.

Below, the lumbar trunk passes deep to the iliac vessels to continue as the sacral trunk in front of the sacrum.

Usually the lumbar trunk carries four ganglia although sometimes these are condensed to three. All four send grey rami communicantes to the lumbar spinal nerves; in addition, the upper two ganglia receive white rami.

Branches from the chain pass to plexuses around the abdominal aorta and its branches, which also receive fibres from the splanchnic nerves and the vagus. Other branches pass in front of the common iliac vessels as the hypogastric plexus ('presacral nerves') to supply the pelvic viscera via plexuses of nerves distributed along the branches of the internal iliac artery.

The *parasympathetic* supply to the pelvic viscera arises from the anterior primary rami of S2, 3 and 4 and is distributed with the pelvic plexuses.

### CLINICAL FEATURES
*Lumbar sympathectomy* is carried out via an extra-peritoneal approach. A paramedian or transverse mid-abdominal incision is used, the peritoneum exposed and peeled medially from the posterior abdominal wall. The ureter, which adheres to the peritoneum like a fly to fly-paper, is seen and carefully preserved. Psoas major comes into view with the genito-femoral nerve upon it, then the lumbar vertebrae, against which the sympathetic chain can be felt.

Usually the second, third and fourth ganglia are excised with the intermediate chain; this effects an adequate sympathectomy of the lower limb, the skin of which then becomes warm, pink and dry.

# PART 3
# THE UPPER LIMB

# The Female Breast

The female breast overlies the second to the sixth rib; two-thirds of it rests on pectoralis major, one-third on serratus anterior, while its lower medial edge just overlaps the upper part of the rectus sheath.

## Structure

The breast is made up of 15–20 lobules of glandular tissue embedded in fat; the latter accounts for its smooth contour and most of its bulk. These lobules are separated by fibrous septa running from the sub-cutaneous tissues to the fascia of the chest wall (the *ligaments of Cooper*).

Each lobule drains by its lactiferous duct onto the *nipple*, which is surrounded by the pigmented *areola*. This area is lubricated by the *areolar glands of Montgomery*; these are sebaceous glands which may form sebaceous cysts which may, in turn, become infected.

The male breast is rudimentary, comprising small ducts without alveoli and supported by fibrous tissue and fat. Insignificant it may be, but it is still prone to the major diseases that affect the female organ.

## Blood supply

1.   From the axillary artery via its lateral thoracic and acromio-thoracic branches.
2.   From the internal mammary artery via its perforating branches; these pierce the first to the fourth intercostal spaces, then traverse pectoralis major to reach the breast along its medial edge. The first and second perforators are the largest of these branches.
3.   From the intercostal arteries via their lateral perforating branches; a relatively unimportant source.

The venous drainage is to the corresponding veins.

## Lymphatic drainage

This is of considerable importance in the spread of breast tumours. The description given here is at variance with some standard text-books but it is based on recent work on this subject.

The lymph drainage of the breast, as with any other organ, follows the pathway of its blood supply and therefore travels:
1.   along tributaries of the axillary vessels to axillary lymph nodes;

2. along the tributaries of the internal mammary vessels, piercing pectoralis major to traverse each intercostal space to lymph nodes along the internal mammary chain; these also receive lymphatics penetrating along the lateral perforating branches of the intercostal vessels.

Although the lymph vessels lying between the lobules of the breast freely communicate, there is a tendency for the lateral part of the breast to drain towards the axilla and the medial part to the internal mammary chain. Fig. 108).

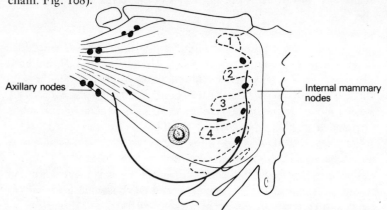

**Fig. 108.** Diagram of the principal pathways of lymphatic drainage of the breast. These follow the venous drainage of the breast—to the axilla and to the internal mammary chain.

A subareolar plexus of lymphatics below the nipple (the plexus of Sappey) and another deep plexus on the pectoral fascia have, in the past, been considered to be the central points to which, respectively, the superficial and deep parts of the breast drain before communicating with main efferent lymphatics. These plexuses appear, however, to be relatively unimportant, the vessels, in the main, passing directly to the regional lymph nodes.

The axillary lymph nodes (some 20 to 30 in number) drain not only the lymphatics of the breast, but also those of the pectoral region, upper abdominal wall and the upper limb, and are arranged in five groups (Fig. 109):

1. *Anterior*—lying deep to pectoralis major along the lower border of pectoralis minor.
2. *Posterior*—along the subscapular vessels.
3. *Lateral*—along the axillary vein.
4. *Central*—in the axillary fat.

5.   *Apical* (through which all the other axillary nodes drain)—immediately behind the clavicle at the apex of the axilla above pectoralis minor and along the medial side of the axillary vein. From the apical nodes emerges the *subclavian lymph trunk*. On the right, this either drains directly into the subclavian vein or else joins the right jugular trunk; on the left it usually drains directly into the thoracic duct.

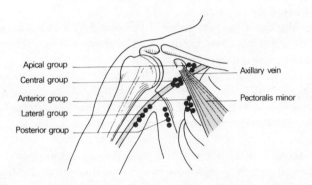

Fig. 109.   The lymph nodes of the axilla.

   Lymphatic spread of a growth of the breast may occur further afield when these normal pathways have become interrupted by malignant deposits, surgery or radiotherapy. Secondaries may then be found in the lymphatics of the opposite breast or opposite axilla, the groin lymph nodes (via lymph vessels in the trunk wall), the cervical nodes (as a result of retrograde extension from the blocked thoracic duct or jugular trunk), or in peritoneal lymphatics, spreading there in a retrograde manner from the lower internal mammary nodes. This in addition, of course, to spread via the blood stream.

## Development

The breasts develop as an invagination of chest wall ectoderm which forms a series of branching ducts. Shortly before birth this site of invagination everts to form the nipple. At puberty, alveoli sprout from the ducts and considerable fatty infiltration of the breast tissue takes place. With pregnancy there is a tremendous development of the alveoli which, in lactation, secrete the fatty droplets of milk. At the menopause the gland tissue atrophies.

CLINICAL FEATURES

1. Developmental abnormalities are not uncommon. The nipple may fail to evert and it is important to find out from the patient whether or not an inverted nipple is a recent event or has been present since birth. Supernumerary nipples or even breasts may occur along a vertical 'milk line'—a reminder of the line of mammary glands in more primitive mammals; on the other hand the breast on one or both sides may be small or even absent (amazia).

2. An abscess of the breast should be opened by a radial incision to avoid cutting across a number of lactiferous ducts. Such an abscess may rupture from one fascial compartment into its neighbours, and it is important at operation to break down any loculi which thus form in order to provide ample drainage.

3. Dimpling of the skin over a carcinoma of the breast results from malignant infiltration and fibrous contraction of Cooper's ligaments—as these pass from breast to skin, their shortening results in tethering of the skin to the underlying tumour. This may also occur, however, in chronic infection, after trauma and, very rarely, in cystic mastitis, so that skin fixation to a breast lump is not necessarily diagnostic of malignancy.

4. Retraction of the nipple, if of recent origin, is suggestive of involvement of the milk ducts in the fibrous contraction of a scirrhous tumour.

5. The excision of a breast carcinoma by *radical mastectomy* involves the removal of a wide area of skin around the tumour, all the breast tissue, the pectoralis major (through which lymphatics pass to the internal mammary chain), the pectoralis minor (which lies as a gateway to the axilla), and the whole axillary contents of fatty tissue and contained lymph nodes. This excision also removes the bulk of the lymphatics from the arm which pass along the anterior and medial aspects of the axillary vein. A few lymph vessels from the upper limb pass above the axillary vein and are therefore saved.

Many surgeons today perform less extensive surgery for breast cancer; either a *simple mastectomy*, in which the breast alone is removed, or an *extended simple mastectomy*, which combines this with clearance of the axillary fat and its contained nodes, or a *Patcy mastectomy*, in which pectoralis minor is divided to facilitate the axillary dissection.

Oedema of the arm after mastectomy usually only occurs if further damage is done to this precarious lymph drainage by infection, malignant infiltration or heavy irradiation, or if additional strain is put on the evacuation of fluid from the limb by ligation or thrombosis of the axillary vein.

# Surface Anatomy and Surface Markings
# of the Upper Limb

Much of the anatomy of the limbs can be revised on oneself; otherwise choose a thin colleague.

## Bones and Joints

The subcutaneous border of the *clavicle* can be palpated throughout its length; the supraclavicular sensory nerves crossing it can be rolled against the bone.

The *acromion process* forms a sharp bony edge at the lateral extremity of the *scapular spine*. It lies immediately above the smooth bulge of the *deltoid muscle* which itself covers the *greater tuberosity of the humerus*. Less easily identified is the *coracoid process* of the scapula, lying immediately below the clavicle at the junction of the middle and outer thirds, covered by the anterior fibres of the deltoid.

The medial border of the scapula can be both seen and felt. Abduction of the arm is a complex affair made up of abduction at the shoulder joint, elevation at the sternoclavicular joint and rotation of the scapula; the last two are readily confirmed on self-palpation.

With the shoulder abducted, the *head of the humerus* can be felt in the axilla; note its movement with rotation of the arm.

At the elbow, the three bony landmarks are the *olecranon process* and the *medial and lateral epicondyles*. A supracondylar fracture lies above these points, which therefore remain in their triangular relationship to each other; in dislocation of the elbow, however, the olecranon comes more or less in line with the epicondyles (Fig. 110).

Note a hollow in the posterolateral aspect of the extended elbow distal to the lateral epicondyle; this lies over the *head of the radius* which can be felt to rotate during pronation and supination.

The posterior border of the *ulna* is completely subcutaneous and crossed by no named vessel or nerve; it can therefore be exposed surgically from end to end without danger.

At the wrist, the *styloid processes* of the radius and ulna can be felt; the former extends more distally. The *dorsal tubercle* of Lister is palpable on the posterior aspect of the distal end of the radius.

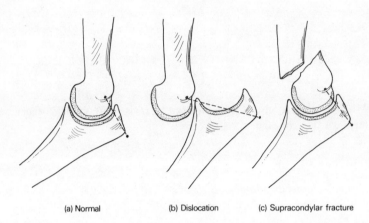

(a) Normal    (b) Dislocation    (c) Supracondylar fracture

**Fig. 110.** The relationship of the medial and lateral epicondyles to the olecranon process (a) is disturbed in a dislocation of the elbow (b) but maintained in a supracondylar fracture (c).

In the palm of the hand, palpate the *pisiform* at the base of the hypothenar eminence. Flexor carpi ulnaris is inserted into it and when this tendon is relaxed by flexing the wrist the pisiform can be moved a little from side to side. The *hook of the hamate* can be felt by deep palpation just distal to the pisiform. The *scaphoid* is felt at the base of the thenar eminence and also within the anatomical snuff-box where there is characteristic tenderness when this bone is fractured.

## Muscles and tendons

The anterior fold of the axilla is formed by the *pectoralis major*, and its posterior fold by the *teres major* and *latissimus dorsi*. The digitations of *serratus anterior* can be seen in a muscular subject on the medial axillary wall.

In the upper arm the *deltoid* forms the smooth contour of the shoulder. The *biceps* and *brachialis* constitute the bulk of the anterior aspect of the arm, and the *triceps* its posterior aspect.

When the forearm is flexed against resistance, the *brachio-radialis* presents prominently along its radial border.

At the wrist (Figs. 111–3) it is convenient to commence at the radial pulse. The tendon medial to this is that of the *flexor corpi radialis*, then *palmaris longus* (which may be absent), then the cluster of tendons of *flexor digitorum sublimis*. The tendon of *flexor carpi ulnaris* lies most medially, inserting into the pisiform; the ulnar pulse can be felt just lateral to this tendon.

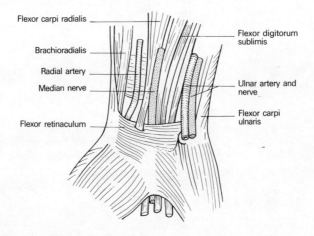

Flexor carpi radialis

Brachioradialis

Radial artery

Median nerve

Flexor retinaculum

Flexor digitorum sublimis

Ulnar artery and nerve

Flexor carpi ulnaris

**Fig. 111.** The structures on the anterior aspect of the wrist. (Palmaris longus, which overlies the median nerve and is inserted into the flexor retinaculum, has been removed).

On the dorsal aspect of the wrist (Figs. 112, 113) the anatomical snuff-box is formed by the tendons of *abductor pollicis longus* and of *extensor pollicis brevis* laterally and that of *extensor pollicis longus* medially—the latter can be traced to the base of the terminal phalanx of the extended thumb. The tendons of *extensor digitorum* are seen in the extended hand passing to be inserted into the bases of the proximal phalanges of the fingers.

## Vessels

Feel the pulsations of the *subclavian artery* against the first rib, the *brachial artery* against the humerus, the *radial* and *ulnar* arteries at the wrist and the *radial artery* again in the anatomical snuff-box.

The brachial artery bifurcates into its radial and ulnar branches at the level of the neck of the radius and the line of the radial artery then corresponds to the slight groove which can be seen along the ulnar border of the tensed brachio-radialis.

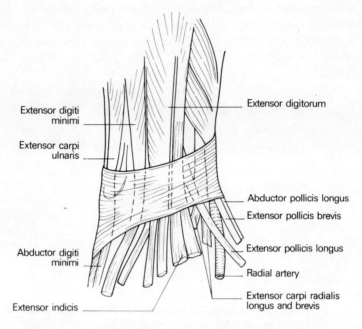

Extensor digiti minimi

Extensor carpi ulnaris

Extensor digitorum

Abductor pollicis longus

Extensor pollicis brevis

Extensor pollicis longus

Radial artery

Extensor carpi radialis longus and brevis

Abductor digiti minimi

Extensor indicis

Fig. 112. The structures on the posterior aspect of the right wrist.

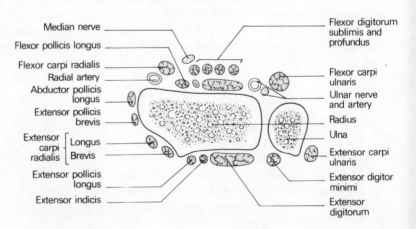

Median nerve

Flexor pollicis longus

Flexor carpi radialis

Radial artery

Abductor pollicis longus

Extensor pollicis brevis

Extensor carpi radialis — Longus / Brevis

Extensor pollicis longus

Extensor indicis

Flexor digitorum sublimis and profundus

Flexor carpi ulnaris

Ulnar nerve and artery

Radius

Ulna

Extensor carpi ulnaris

Extensor digitor minimi

Extensor digitorum

Fig. 113. Schematic section immediately above the wrist joint.

The veins of the upper limb (Fig. 114) comprise the deep venae comitantes, which accompany all the main arteries, usually in pairs, and the much more important superficial veins—more important both in size and in practical value because of their use for venepuncture and transfusion.

These superficial veins can be seen as a dorsal venous network on the back of the hand which drains into a lateral *cephalic* and medial *basilic* vein.

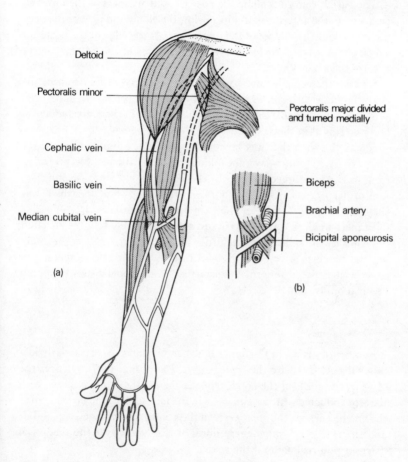

Fig. 114. (a) The superficial veins of the upper limb. (b) Detail of the bicipital aponeurosis, lying between the median cubital vein and the brachial artery.

The cephalic vein at its origin lies fairly constantly in the superficial fascia just posterior to the radial styloid; even if not visible it can be cut down upon confidently at this site. It then runs up the anterior aspect of the forearm to lie in a groove along the lateral border of the biceps and then, after piercing the deep fascia, in the groove between pectoralis major and the deltoid, where again it can readily be exposed for an emergency cut-down. It finally penetrates the clavi-pectoral facia to enter the axillary vein.

The basilic vein runs along the postero-medial aspect of the forearm, passes on to the anterior aspect just below the elbow and pierces the deep fascia at about the middle of the upper arm. At the edge of the posterior axillary fold it is joined by the venae comitantes of the brachial artery to form the axillary vein.

Linking the cephalic and basilic veins just distal to the front of the elbow is the *median cutital* vein, usually the most prominent superficial vein in the body and visible or palpable when all others are hidden in fat or collapsed in shock.

It was this vein that was favoured for the operation of bleeding, or phlebotomy, in former days; the underlying brachial artery was protected from the barber-surgeon's knife by the *bicipital aponeurosis*, a condensation of deep fascia passing across from the biceps tendon, which was therefore termed the 'grâce à Dieu' (praise be to God) fascia.

In more modern times one tries to avoid using this vein for injection of Pentothal and other irritating drugs because of the slight risk of entering the brachial artery and also because of the danger of piercing a superficially-placed abnormal ulnar artery in occasional instances of high brachial bifurcation.

## Nerves

A number of nerves in the upper limb can be palpated, particularly in a thin subject; these are the *supraclavicular nerves*, as they pass over the clavicle, the cords of the *brachial plexus* against the humeral head (with the arm abducted), the *median nerve* in the mid-upper arm, crossing over the brachial artery, the *ulnar nerve* in the groove of the medial epicondyle and the *superficial radial nerve* fibres as they pass over the tendon of extensor pollicis longus at the wrist.

The median nerve lies first lateral then medial to the brachial artery, crossing it at the mid upper arm, usually superficially but occasionally deeply. This close relationship is of historical interest; Nelson had his median nerve accidentally incorporated in the ligature around the artery when his arm was amputated above the elbow.

Useful surface markings of other, impalpable, nerves may be listed as follows.

1.  The *circumflex nerve* is related closely to the surgical neck of the humerus a handsbreadth below the acromion process.

2.  The *radial nerve* crosses the posterior aspect of the humeral shaft at its mid-point.

3.  The *posterior interosseous branch* of the radial nerve is located by Henry's method as it winds round the radius. Place three fingers along the radius, the uppermost lying just distal to the radial head; the third finger then lies over this nerve.

4.  The *median nerve* (Figs. 111 and 113) in the forearm lies, as its names suggests, in the median plane; its area of distribution in the hand is thus anaesthetized if local anaesthetic be injected exactly in the mid-line at the wrist.

5.  The *ulnar nerve* at the wrist lies immediately medial to the ulnar pulse (Figs. 111 and 113).

# The Bones and Joints of the Upper Limb

### The scapula (Fig. 115)

This triangular bone bears three prominent features: the *glenoid fossa* laterally (which is the scapula's contribution to the shoulder joint), the *spine* on its posterior aspect, projecting laterally as the *acromion process*, and the *coracoid process* on its anterior aspect.

Because of its strong muscular coverings the scapula is well protected and rarely fractured.

### The clavicle (Fig. 115)

This long bone has a number of unusual features.

1.  It has no medullary cavity.

2.  It is the first to ossify in the fetus (5th–6th week).

3.  Although a long bone, it develops in membrane and not in cartilage.

4.  It is the most commonly fractured bone in the body.

The clavicle is made up of a medial two-thirds which is circular in section and convex anteriorly, and a lateral one-third which is flattened in section and convex posteriorly.

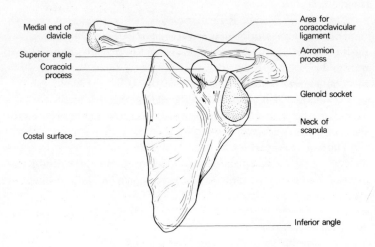

**Fig. 115.** The left scapula and clavicle.

Medially it articulates with the manubrium at the manubrio-sternal joint (this joint containing an articular disc), and is also attached to the first costal cartilage by the *costo-clavicular ligament*.

Laterally it articulates with the acromion (the joint containing an incomplete articular disc) and, in addition, is attached to the coracoid process by the tough *coraco-clavicular ligament*.

The third parts of the subclavian vessels and the trunks of the brachial plexus pass behind the medial third of the shaft of the clavicle separated only by the thin subclavius muscle. Rarely these vessels are torn by the fragments of a fractured clavicle; this was the cause of death of Sir Robert Peel following a fall from his horse.

The sternal end of the clavicle has important posterior relations; behind the sterno-clavicular joints lie the common carotid artery on the left and the bifurcation of the innominate (brachiocephalic) artery on the right. The internal jugular vein lies a little more laterally on either side. These vessels are separated from bone by the strap muscles—the sterno-hyoid and sterno-thyroid.

CLINICAL FEATURES

The clavicle has three functions.

1. To transmit forces from the upper limb to the axial skeleton.
2. To act as a strut holding the arm free from the trunk, to hang supported principally by trapezius.
3. To provide attachment for muscles.

The weakest point along the clavicle is the junction of middle and outer third. Transmission of forces to the axial skeleton in falls on the shoulder or hand may prove greater than the strength of the bone at this site and this indirect force is the usual cause of fracture.

When fracture occurs, the trapezius is unable to support the weight of the arm so that the characteristic picture of the patient with a fractured clavicle is that of a man supporting his sagging upper limb with his opposite hand. The lateral fragment is not only depressed but also drawn medially by the shoulder adductors, principally the teres major, latissimus dorsi and pectoralis major (Fig. 116).

## The humerus (Fig. 117)

The upper end of the humerus consists of a *head* (one-third of a sphere) facing medially, upwards and backwards, separated from the *greater* and *lesser tuberosities* by the *anatomical neck*. The tuberosities, in turn, are separated by the *bicipital groove* along which emerges the long head of biceps from the shoulder joint.

Where the upper end and the shaft of the humerus meet there is the narrow *surgical neck* against which lie the circumflex nerve and vessels. The shaft itself is circular in section above and flattened in its lower part. The posterior aspect of the shaft bears the faint *spiral groove*, demarcating the origins of the medial and lateral heads of the triceps between which wind the radial nerve and the profunda vessels.

The lower end of the humerus bears the rounded *capitulum* laterally, for articulation with the radial head, and the spool-shaped *trochlea* medially, articulating with the trochlear notch of the ulna.

The *medial* and *lateral epicondyles*, on either side, are extracapsular; the medial is the larger of the two, extends more distally and bears a groove on its posterior aspect for the ulnar nerve.

Three important nerves thus come into close contact with the humerus —the circumflex, the radial and the ulnar; they may be damaged respectively in fractures of the humeral neck, midshaft, and lower end (Fig. 117).

It is an important practical point to note that the lower end of the

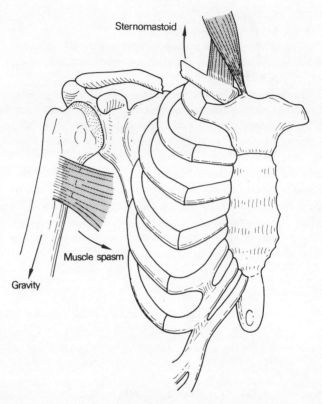

**Fig. 116.** The deformity of a fractured clavicle—downward displacement and adduction of the outer fragment by gravity and muscle spasm respectively; slight elevation of the inner fragment by the sternomastoid.

humerus is angulated forward 45 deg. on the shaft. This is easily confirmed by examining a lateral radiograph of the elbow when it will be seen that a vertical line continued downwards along the front of the shaft bisects the capitulum. Any decrease of this angulation indicates backward displacement of the distal end of the humerus and is good radiographic evidence of a supracondylar fracture.

### The radius and ulna (Fig. 118)

The radius consists of the *head*, *neck*, *shaft* (with its *radial tuberosity*) and expanded distal end. The ulna comprises *olecranon*, *trochlear fossa*, *coro-*

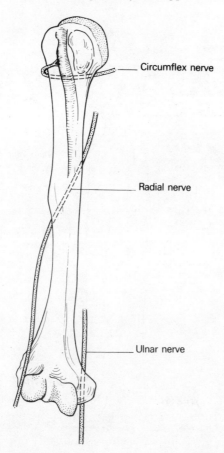

**Fig. 117.**   The humerus with its three major related nerves—circumflex, radial and ulnar—
all of which are in danger of injury in humeral fractures.

*noid process* (with its *radial notch* for articulation with the radial head),
*shaft* and small distal *head*, which articulates with the medial side of the
distal end of the radius at the inferior radio-ulnar joint.

In pronation and supination, the head of the radius rotates against the
radial notch of the ulna, the shaft of the radius swings round the relatively
fixed ulnar shaft (the two bones being connected by a fibrous interosseous
ligament) and the distal end of the radius rotates against the head of the
ulna. This axis of rotation passes from the radial head proximally to the
ulnar head distally.

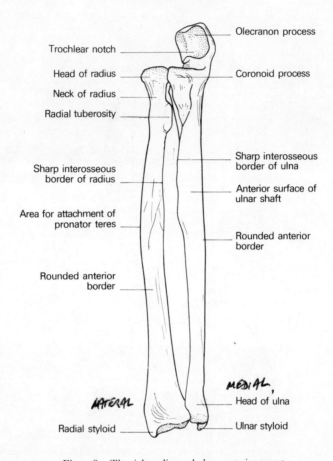

**Fig. 118.** The right radius and ulna—anterior aspect.

CLINICAL FEATURES

1.    The pronator teres is inserted midway along the radial shaft. If the radius is fractured proximal to this, the proximal fragment is supinated (by the action of the biceps) and the distal fragment is pronated by pronator teres. The fracture must therefore be splinted with the forearm supinated so that the distal fragment is aligned with the supinated proximal end. If the fracture is distal to the midshaft, the actions of biceps and the pronator muscles more or less balance and the fracture is therefore immobilized with the forearm in the neutral position (Fig. 119).

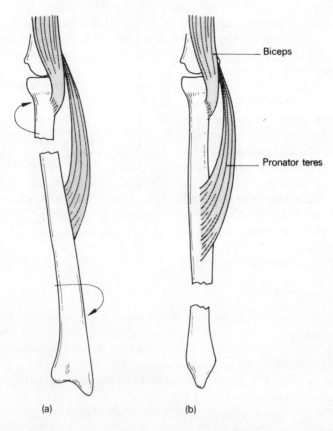

Biceps

Pronator teres

(a)                    (b)

**Fig. 119.**    The important role of pronator teres in radial fractures. (a) In proximal fractures, above the insertion of pronator teres, the distal fragment is pronated. Such a fracture must be splinted in the supinated position. (b) When the fracture is distal to pronator teres insertion, the action of this muscle on the proximal fragment is cancelled by the supinator action of biceps. This fracture is therefore held reduced in the neutral position, mid-way between pronation and supination.

2.    The force of a fall on the hand produced different effects in different age groups; in a child it may cause a posterior displacement of the distal radial epiphysis, in the young adult the shafts of the radius and ulna may fracture whereas, in the elderly, the most likely result will be a *Colles'* *fracture*. In this latter injury, the radius fractures about one inch proximal to the wrist joint; the distal fragment is displaced posteriorly and usually

becomes impacted. The shortening which results brings the styloid processes of the radius and ulna more or less in line with each other.

Another forearm injury resulting from a fall on the outstretched hand is fracture of the head of the radius, due to its being crushed against the capitulum of the humerus.

3.  The olecranon process may be fractured by direct violence but more often it is avulsed by forcible contraction of the triceps, which is inserted into its upper aspect. In these circumstances the bone ends are widely displaced and operative repair, to reconstruct the integrity of the elbow joint, becomes essential.

4.  A subcutaneous bursa is constantly present over the olecranon and is likely to become inflamed when exposed to repeated trauma. Students and coal miners share this hazard so that olecranon bursitis goes by the nicknames of 'student's elbow' and 'miner's elbow'. Although I have seen many miners with this lesion I have yet, however, to see a medical student thus disabled.

### The bones of the hand (Fig. 120)

The *carpus* is made up of two rows each containing four bones. In the proximal row, from the lateral to the medial side, are the *scaphoid*, *lunate*, and *triquetral*, the last bearing the *pisiform* on its anterior surface, into which bone the flexor carpi ulnaris is inserted.

In the distal row, from the lateral to the medial side, are the *trapezius*, *trapezoid*, *capitate* and *hamate*.

The carpus as a whole is arched transversely, the palmar aspect being concave. This is maintained by:

1.  the shapes of the individual bones, which are broader posteriorly than anteriorly (except for the lunate, which is broader anteriorly);
2.  the tough *flexor retinaculum* passing from the scaphoid and the ridge of the trapezium laterally to the pisiform and the hook of the hamate medially (Fig. 121).

CLINICAL FEATURES
1.  A fall on the hand may dislocate the rest of the carpal arch backwards from the lunate which, as commented on above, is wide-based anteriorly (perilunate dislocation of the carpus). The dislocated carpus may then reduce spontaneously, only to push the lunate forward and tilt it over so that its distal articular surface faces forward (dislocation of the lunate).

2. The scaphoid may be fractured by a fall on the palm with the hand abducted, in which position the scaphoid lies directly facing the radius.

The blood supply of the scaphoid in one third of cases enters distally along its waist so that, if the fracture is proximal, the blood supply to this

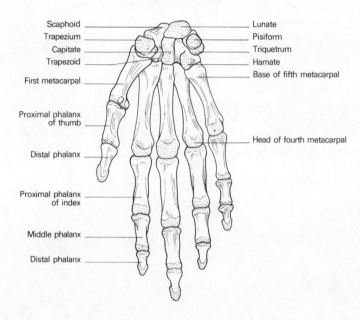

Scaphoid ——————
Trapezium ——————
Capitate ——————
Trapezoid ——————
First metacarpal ——————
Proximal phalanx of thumb ——————
Distal phalanx ——————
Proximal phalanx of index ——————
Middle phalanx ——————
Distal phalanx ——————

—————— Lunate
—————— Pisiform
—————— Triquetrum
—————— Hamate
—————— Base of fifth metacarpal
—————— Head of fourth metacarpal

Fig. 120.   The right carpus, metacarpus and phalanges.

small proximal fragment may be completely cut off with resultant aseptic necrosis of this portion of bone (Fig. 122).

3. '*The carpal tunnel syndrome.*' The flexor retinaculum forms the roof of a tunnel the floor and walls of which are made up of the concavity of the carpus. Packed within this tunnel are the long flexor tendons of the fingers and thumb together with the median nerve. Any lesion diminishing the size of the compartment, for example an old fracture or arthritic change, may result in compression of the median nerve, resulting in paraesthesiae, numbness and motor weakness in its distribution. Since the superficial palmar branch of the nerve is given off proximal to the retinaculum, there is usually no sensory impairment in the palm.

It is interesting that this syndrome also often occurs without any very

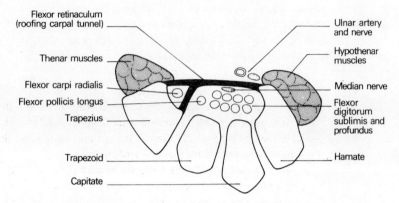

Fig. 121.  Transverse section through the carpus, showing the attachments of the flexor retinaculum.

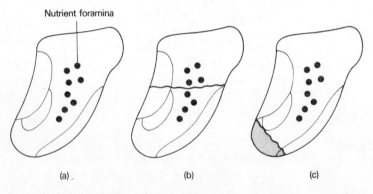

Fig. 122.  Blood supply of the scaphoid. (a) Blood vessels enter the bone principally in its distal half. (b) A fracture through the waist of the scaphoid—vessels to the proximal fragment are preserved. (c) A fracture near the proximal pole of the scaphoid— in this case there are no vessels supplying the proximal fragment and aseptic necrosis of bone is therefore inevitable.

obvious cause, although symptoms are relieved by dividing the retinaculum longitudinally.

## The shoulder (Figs. 123, 124)

The shoulder is a ball and socket joint between the relatively large *head of humerus* and relatively small and shallow *glenoid fossa*, although the

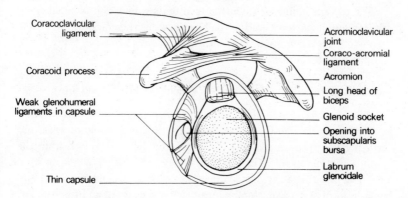

Coracoclavicular ligament

Coracoid process

Weak glenohumeral ligaments in capsule

Thin capsule

Acromioclavicular joint

Coraco-acromial ligament

Acromion

Long head of biceps

Glenoid socket

Opening into subscapularis bursa

Labrum glenoidale

**Fig. 123.** The left shoulder joint—its ligaments are shown after removal of the humerus.

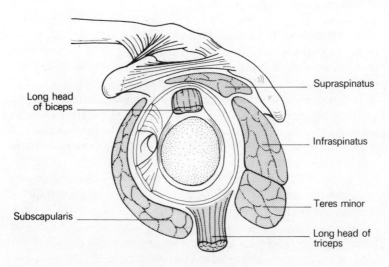

Long head of biceps

Subscapularis

Supraspinatus

Infraspinatus

Teres minor

Long head of triceps

**Fig. 124.** The shoulder joint—the same view as in Fig. 123 but now with the addition of the surrounding muscles.

latter is deepened somewhat by the cartilaginous *labrum glenoidale*.

The joint *capsule* is lax and is attached around the epiphyseal lines of both the glenoid and the humeral head. However, it does extend down onto the diaphysis on the medial aspect of the neck of the humerus, so that an osteomyelitis of the upper end of the humeral shaft may involve the joint by direct spread.

The capsule is lined by synovial membrane which is prolonged along the tendon of the long head of the biceps as this traverses the joint. The synovium also communicates with the *subscapular bursa* beneath the tendon of subscapularis.

The stability of the shoulder joint depends almost entirely on the strength of the surrounding muscles which may be grouped into:

1. The closely related short muscles of the 'rotator cuff' (see below).
2. The long head of biceps, arising from the supraglenoid tubercle and crossing over the head of the humerus, thus lying actually within the joint, although enclosed in a tube of synovium.
3. The more distantly related long muscles of the shoulder; the deltoid, long head of triceps, pectoralis major, latissimus dorsi and teres major.

*Movements of the shoulder girdle*

The movements of the shoulder joint itself cannot be divorced from those of the whole shoulder girdle. Even if the shoulder joint is fused, a wide range of movement is still possible by elevation, depression, rotation and protraction of the scapula, leverage occurring at the sterno-clavicular joint.

Abduction of the shoulder is initiated by the supraspinatus; the deltoid can then abduct to 90 deg. Further movement to 180 deg. (elevation) is brought about by rotation of the scapula upwards by the trapezius and serratus anterior. Shoulder and shoulder girdle movements combine into one smooth action.

The '*rotator cuff*' (Fig. 125) is the name given to the sheath of tendons of the short muscles of the shoulder which covers and blends with all but the inferior aspect of that joint. The muscles are the supraspinatus, infraspinatus and teres minor, which are inserted from above down into the humeral greater tuberosity, and the subscapularis, which is inserted into the lesser tuberosity. All originate from the scapula.

Of these muscles, the supraspinatus is of the greatest practical importance. It passes over the apex of the shoulder beneath the acromion process, from which it is separated by the *subacromial bursa*. This bursa is continued beneath the deltoid as the *subdeltoid bursa* forming, together, the largest bursa in the body.

The supraspinatus initiates the abduction of the humerus on the scapula; if the tendon is torn as a result of injury, active initiation of abduction becomes impossible and the patient has to develop the trick movement of tilting his body towards the injured side so that gravity passively swings the arm from his trunk. Once this occurs, the deltoid and the scapular rotators can then come into play.

Inflammation of the supraspinatus tendon ('supraspinatus tendinitis') is characterized by a painful arc of shoulder movement between 60° and 120°; in this range, the tendon impinges against the overlying acromion.

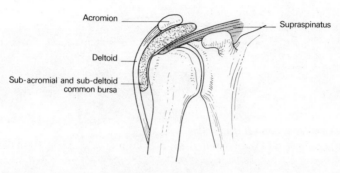

Fig. 125. Supraspinatus and the subacromial-subdeltoid bursa. Note that the supraspinatus tendon lies close against the acromion—if this tendon is inflamed, there is a painful arc of movement as the shoulder is abducted from 60° to 120°, because, in this range, the inflamed tendon impinges against the acromion.

*Principal muscles acting on the shoulder joint*

| | | | |
|---|---|---|---|
| *Abductors*— | supraspinatus<br>deltoid | *Adductors*— | pectoralis major<br>latissimus dorsi |
| *Flexors*— | pectoralis major<br>coraco-brachialis<br>deltoid (anterior fibres) | *Extensors*— | teres major<br>latissimus dorsi<br>deltoid (posterior fibres) |

| *Medial rotators* | *Lateral rotators* |
|---|---|
| pectoralis major | infraspinatus |
| latissimus dorsi | teres minor |
| teres major | deltoid (posterior fibres) |
| deltoid (anterior fibres) | |
| subscapularis | |

CLINICAL FEATURES

*Dislocation of the shoulder*

The wide range of movement possible at the shoulder is achieved only at the cost of stability, and for this reason it is the most commonly dislocated major joint. Its inferior aspect is completely unprotected by muscles and it is here that, in violent abduction, the humeral head may slip away from the glenoid to lie in the subglenoid region whence it usually passes anteriorly into a subcoracoid position (Fig. 126).

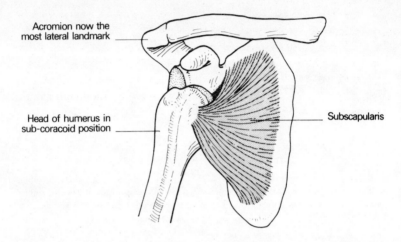

Acromion now the
most lateral landmark

Head of humerus in
sub-coracoid position

Subscapularis

**Fig. 126.** The deformity of shoulder dislocation. The dislocated head of the humerus is held adducted by the shoulder girdle muscles and internally rotated by subscapularis.

The circumflex nerve, lying in relation to the surgical neck of the  humerus, may be torn in this injury.

The head of the humerus is drawn medially by the powerful adductors of the shoulder; its greater tuberosity therefore no longer remains the most lateral bony projection of the shoulder region, being replaced for this honour by the acromion process. The normal bulge of the deltoid over the greater tuberosity is lost; instead there is the characteristic flattening of this muscle.

In reducing the dislocation by *Kocher's method* the elbow is flexed and the forearm rotated outwards; this stretches the subscapularis which is holding the humeral head internally rotated. The elbow is then swung medially across the trunk, thus levering the head of the humerus laterally so that it slips back into place.

In the *Hippocratic method*, the foot is used as a fulcrum in the axilla, traction and adduction being applied to the forearm; in this way the humeral head is levered outwards into its normal position.

## The elbow joint (Figs. 127, 128)

The elbow joint, although a single synovial cavity, is made up of three distinct articulations, which are:

1. the *humero-ulnar*, between the trochlea of the humerus and the trochlear notch of the ulna (a hinge-joint);

2. the *humero-radial*, between the capitulum and the upper concave surface of the radial head (a ball and socket joint);

3. the *superior radio-ulnar*, between the head of the radius and the radial notch of the ulna, the head being held in place by the tough annular ligament (a pivot joint).

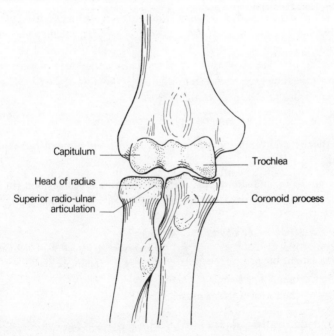

Capitulum

Trochlea

Head of radius

Superior radio-ulnar articulation

Coronoid process

**Fig. 127.** The bony components of the elbow joint. Note the three sets of articular surfaces.

The *capsule* of the elbow joint is closely applied around this complex articular arrangement; the non-articular medial and lateral epicondyles are extra-capsular. The capsule is thin and loose anteriorly and posteriorly

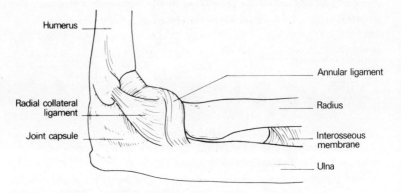

Fig. 128.  The joint capsule of the right elbow—lateral aspect.

to allow flexion and extension, whereas it is strongly thickened on either side to form the *medial* and *lateral collateral* ligaments. The lateral ligament is attached distally to the *annular ligament* around the radial head. In order to allow rotation of the radius, the lower margin of the annular ligament is free and, beneath it, the synovium of the elbow bulges downwards onto the neck of the radius.

Two sets of movements take place at the elbow:

1. flexion and extension at the humero-ulnar and humero-radial joints and

2. pronation and supination at the proximal radio-ulnar joint (in conjunction with associated movements of the distal radio-ulnar joint).

*Muscles acting on the elbow*

*Flexors*—biceps
                brachialis
                brachio-radialis
                the forearm flexor muscles

*Extensors*—triceps
                anconeus

*Pronators*—pronator teres
                pronator quadratus
                flexor carpi radialis

*Supinators*—biceps
                supinator
                extensor pollicis
                    longus
                abductor pollicis
                    longus

The supinator action of the biceps is due to its insertion onto the *posterior* aspect of the tuberosity of the radius. When the biceps contracts, not only is the forearm flexed, but the radius 'unwinds' as its tuberosity is rotated anteriorly, i.e. the forearm supinates (Fig. 129).

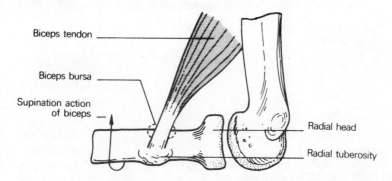

Biceps tendon

Biceps bursa

Supination action of biceps

Radial head

Radial tuberosity

Fig. 129. The supination action of biceps.

CLINICAL FEATURES

1. The elbow joint is safely approached by a vertical posterior incision dividing the triceps expansion.
2. Because the capsule is relatively weak anteriorly and posteriorly it will be distended at these sites by an effusion; particularly posteriorly, since the anterior aspect is covered by muscles and dense deep fascia. Aspiration of such an effusion is readily performed posteriorly on one or other side of the olecranon.
3. The annular ligament is funnel-shaped in adults, but its sides are vertical in young children. A sudden jerk on the arm of a child under the age of 8 years may subluxate the radial head through this ligament ('*pulled elbow*'). Reduction is easily affected by firm supination of the elbow which 'screws' the radial head back into place.
4. *Posterior dislocation of the elbow* may occur as a result of the indirect violence of a fall on the hand. Occasionally the coronoid process of the ulna is fractured in this injury, being snapped off against the trochlea of the humerus. Characteristically, the triangular relationship between the olecranon and the two humeral epicondyles is lost (Fig. 110).

Reduction is affected by traction, to overcome the protective spasm of the muscles acting on the joint, together with flexion of the elbow, which levers the humero-ulnar joint back into place.

## The wrist joint (Fig. 130)

The articular disc of the inferior radio-ulnar joint covers the head of the ulna and is attached to the base of the ulnar styloid process. This disc, together with the distal end of the radius, form the proximal face of the wrist joint, the distal surface being the proximal articular surfaces of the scaphoid, lunate and triquetral.

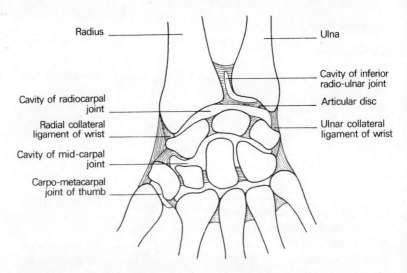

Radius — 

Ulna

Cavity of inferior radio-ulnar joint

Cavity of radiocarpal joint

Articular disc

Radial collateral ligament of wrist

Ulnar collateral ligament of wrist

Cavity of mid-carpal joint

Carpo-metacarpal joint of thumb

Fig. 130.   The wrist, carpal and carpo-metacarpal joints in section.

The wrist is a condyloid joint, that is to say, it allows flexion, extension, abduction, adduction and circumduction, the last being a combination of the previous four. Flexion and extension are increased by associated sliding movements of the inter-carpal joints; although the range of flexion at the wrist is actually less than that of extension, these associated movements make it apparently greater.

Because of the greater distal projection of the radial styloid, the range of abduction at the wrist is considerably less than that of adduction.

*Muscles acting on the wrist*
*Flexors*—all the long muscles crossing the anterior aspect of the wrist
    joint

*Extensors*—all the long muscles crossing the posterior aspect of the joint

*Adductors*—flexor carpi ulnaris acting in concord with extensor carpi ulnaris

*Abductors*—flexor carpi radialis and extensores carpi radialis longus et brevis together with the long abductor and short extensor of the thumb.

## The joints of the hand (Fig. 130)

The joints between the individual carpal bones allow gliding movements to occur which increase the range of extesion and, more particularly, flexion permitted at the wrist joint.

The *carpo-metacarpal joint* of the thumb is saddle-shaped and permits flexion and extension (in a plane parallel to the palm of the hand), abduction and adduction (in a plane at right-angles to the palm) and opposition, in which the thumb is brought across in contact with the 5th finger. This joint's range contrasts with the limited movements of the other carpo-metacarpal joints which allow a few degrees of gliding movement of the 2nd and 3rd metacarpals and a small range of flexion and extension of the 4th and 5th metacarpals.

The opposite state of affairs holds at the *metacarpo-phalangeal joints*; only a 60-degree range of flexion and extension is possible at the m/p joint of the thumb whereas a 90-degree range of flexion and extension, together with abduction, adduction and circumduction, are possible at the four other m/p joints, which are condyloid in shape.

Note that, when the m/p joints of the fingers are flexed, abduction and adduction become impossible. This is because each metacarpal head, although rounded at its distal extremity, is flattened anteriorly; when the base of the proximal phalanx moves onto this flattened surface side movements become impossible. Moreover, the collateral ligaments on either side of the m/p joints become taut in flexion and thus prevent abduction and adduction.

The m/p joints of the fingers, but not the thumb, are linked by the tough *deep transverse ligaments* which prevent any spreading of the palm when a firm grip is taken.

All the *inter-phalangeal joints* have pulley-shaped opposing surfaces and are therefore hinge-joints allowing flexion and extension only. At all the m/p and i/p joints the ligamentous arrangements are the same.

1.    Posteriorly—the joint capsule is replaced by the expansion of the extensor tendon of the digit concerned.

2. Anteriorly—the capsule is formed by a dense plate of fibro-cartilage. This *palmar ligament* is the response to the friction of the adjacent flexor tendons.

3. On either side the joints are reinforced by the *collateral ligaments*, which are lax in extension and taut in flexion of the joint.

### The muscles acting on the hand

The long flexors of the fingers are:

1. flexor digitorum profundus, inserted into the base of the four distal phalanges, and

2. flexor digitorum sublimis inserted into the sides of the four middle phalanges.

The profundus tendon pierces that of sublimis over the proximal phalanx.

The profundus flexes the distal phalanx, sublimis the middle phalanx; acting together they flex the fingers and the wrist (Fig. 131).

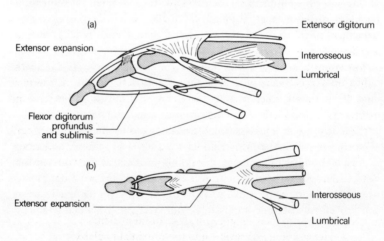

**Fig. 131.** The tendons of a finger. (a) Lateral view. (b) Posterior view.

The long extensors of the fingers are:

● extensor digitorum longus, reinforced by

● extensor indicis ⎫ which join the appropriate tendons of
   extensor digiti minimi ⎬ extensor digitorum longus on their medial
          ⎭ sides

The extensor tendons are inserted into the base of the *proximal* phalanx of each finger and therefore can *only* act on the metacarpo-phalangeal joints. From this insertion, the tendon spreads distally as the extensor expansion, which is attached by a central slip to the middle phalanx and by two lateral slips to the distal phalanx (Fig. 131).

Into this extensor expansion are inserted the intrinsic muscles of the fingers:

(1) The dorsal and palmar interossei, arising from the sides and fronts of the metacarpals respectively.

(2) The lumbricals, arising from the flexor tendons.

These muscles, arising from the palmar aspect of the hand and insert-ing along the dorsal aspect of the fingers, *flex the m/p joints and extend* the i/p joints; the only muscles, in fact, capable of carrying out this latter movement.

The interossei, together with abductor digiti minimi, are responsible for abduction and adduction of the fingers. A weak abduction movement accompanies the action of extensor digitorum and the long flexors adduct the fingers in the movement of full flexion. However, if these movements of extension and flexion are eliminated by laying the hand flat on the table, abduction and adduction become purely the actions of the intrinsic muscles. A card gripped between the fingers in this position of the hand is kept there entirely by intrinsic muscle action.

The 5th finger receives two further intrinsic muscles, opponens digiti minimi and flexor digiti minimi, from the hypothenar eminence.

The eight muscles acting on the thumb may be divided into the long (proceeding from the forearm), and the short or intrinsic muscles.

Long—flexor pollicis longus—inserted into the distal phalanx
       extensor pollicis longus—into the distal phalanx
       extensor pollicis brevis—into the proximal phalanx
       abductor pollicis longus—into the metacarpal
Short—adductor pollicis ⎫
       flexor pollicis brevis ⎬ into the base of the proximal phalanx
       abductor pollicis brevis ⎭
       opponens pollicis—along the metacarpal

The flexors and extensors of the wrist play an important synergic role in movements of the hand. Notice how weak the grip becomes when the wrist is fully flexed; it must be held firmly in the extended or neutral position by balanced muscle action in order to allow the long flexors of the fingers and thumb to work at their full stretch and therefore at their maximum efficiency.

# The Arteries of the Upper Limb

The *axillary artery* commences at the lateral border of the first rib, as a continuation of the subclavian, and ends at the lower border of the axilla (i.e. the lower border of teres major) to become the brachial artery. It is divided into three parts by pectoralis minor and, apart from its distal extremity, it lies covered by pectoralis major.

Above pectoralis minor, the brachial plexus lies above and behind the artery, but, distal to this, the cords of the plexus take up their positions around the artery according to their names, i.e. lateral, medial and posterior.

The branches of the axillary artery supply the chest wall and shoulder; conveniently the 1st, 2nd and 3rd parts give off one, two and three branches respectively:

- •1st part — 1, superior thoracic artery
- •2nd part— 1, acromio-thoracic trunk
    2, lateral thoracic artery
- •3rd part— 1, subscapular artery
    2, anterior circumflex humeral artery
    3, posterior circumflex humeral artery.

All but the circumflex humeral vessels are encountered in the axillary dissection of a radical mastectomy.

*The brachial artery* continues on from the axillary and ends at the level of the neck of the radius by dividing into the radial and ulnar arteries. It is superficial (immediately below the deep fascia), along its whole course except where it is crossed, at the level of the middle of the humerus, by the median nerve which passes superficially from its lateral to medial side; occasionally the nerve crosses deep to the artery. Fairly frequently the artery divides into its two terminal branches in the upper arm.

The named branches of the artery are:
the *profunda* (accompanying the radial nerve),
*ulnar collateral* (accompanying the ulnar nerve),
*nutrient* (to the humerus) and
*supratrochlear*.

*The radial artery* (Fig. 132) commences at the level of the radial neck by lying on the tendon of biceps. In its upper half it lies overlapped by brachio-radialis, the surface marking of the artery being the groove which can be seen on the medial side of this tensed muscle in the muscular subject. Distally in the forearm the artery lies superficially between

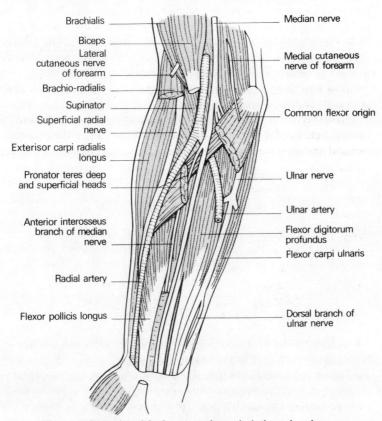

Brachialis

Biceps

Lateral cutaneous nerve of forearm

Brachio-radialis

Supinator

Superficial radial nerve

Exterisor carpi radialis longus

Pronator teres deep and superficial heads

Anterior interosseus branch of median nerve

Radial artery

Flexor pollicis longus

Median nerve

Medial cutaneous nerve of forearm

Common flexor origin

Ulnar nerve

Ulnar artery

Flexor digitorum profundus

Flexor carpi ulnaris

Dorsal branch of ulnar nerve

Fig. 132.　Dissection of the forearm to show principal vessels and nerves.

brachio-radialis and flexor carpi radialis, and it is between these two tendons that it is palpated at the wrist (Fig. 111).

In the middle third of the forearm the radial nerve lies along the lateral side of the artery; the nerve may here be incorporated in a carelessly-placed ligature.

Distal to the radial pulse, the artery gives off a branch to assist in forming the superficial palmar arch. It then passes below the tendons of the anatomical snuff-box (in which it can be felt), pierces the first dorsal interosseous muscle and adductor pollicis, between the first and second metacarpals, and goes on to form the *deep palmar arch* with the deep branch of the ulnar artery.

*The ulnar artery* (Fig. 132) is the larger of the two terminal branches of the brachial artery. From its commencement it passes beneath the muscles arising from the common flexor origin, lies upon flexor digitorum profundus and is overlapped by flexor carpi ulnaris. The median nerve crosses superficially to the ulnar artery, separated from it by only part of one muscle, the deep head of pronator teres.

In the distal half of the forearm the artery becomes superficial between the tendons of flexor carpi ulnaris and flexor digitorum sublimis; it then crosses the flexor retinaculum to form the *superficial palmar arch* with the superficial branch of the radial artery.

The ulnar nerve accompanies the artery on its medial side in the distal two-thirds of its course in the forearm and across the flexor retinaculum (Fig. 111).

*Note*

There is a rich anastomosis of artéries around all major joints. Apart from remembering this fact, the clinical student need not commit to memory the numerous named branches involved.

# The Brachial Plexus

The brachial plexus is of great practical importance to the surgeon. It may be damaged in open, closed or obstetrical injuries, be pressed upon by a cervical rib or be involved in tumour. It is encountered, and hence put in danger, in operations upon the root of the neck.

The plexus is formed as follows (Fig. 133):

- **5** *roots*—derived from the anterior primary rami of C5, 6, 7, 8 and T1 link up into:
- **3** *trunks*—formed by the union of C5 and 6 (upper)
  - C7 alone (middle)
  - C8 and T1 (lower)

which split into:

- **6** *divisions*—formed by each trunk dividing into an anterior and posterior division

which link up again into:

•3 *cords*—a lateral, from the fused anterior divisions of the upper and
middle trunks.

a medial, from the anterior division of the lower trunk.

a posterior, from the union of all three posterior divisions.

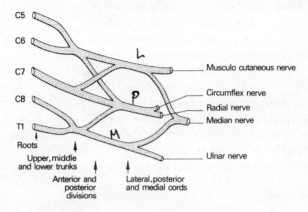

Fig. 133.   Scheme of the brachial plexus.

*The roots* lie between the anterior and middle scalene muscles.
*The trunks* traverse the posterior triangle of the neck.
*The divisions* lie behind the clavicle.
*The cords* lie in the axilla.

The cords continue distally to form the main nerve trunks of the upper
limb thus:

1.   the lateral cord continues as the *musculo-cutaneous nerve*;
2.   the medial cord—as the *ulnar nerve*;
3.   the posterior cord—as the *radial nerve* and the *circumflex nerve*;
4.   a cross communication between the lateral and medial cords forms
the *median nerve*.

For reference purposes, the derivatives of the various components of
the brachial plexus are given below (Fig. 134).

*From the roots*—nerve to rhomboids,
                 nerve to subclavius,
                 nerve to serratus anterior (C5, 6, 7).

*From the trunk*—suprascapular nerve—from the upper trunk (supplies
                 supraspinatus and infraspinatus).

*From the lateral cord*—musculo-cutaneous nerve,
                        lateral pectoral nerve,
                        lateral root of median nerve.

*From the medial cord*—medial pectoral nerve,
medial cutaneous nerves of arm and forearm,
ulnar nerve,
medial root of median nerve.

*From the posterior cord*—subscapular nerves,
nerve to latissimus dorsi,
circumflex nerve,
radial nerve.

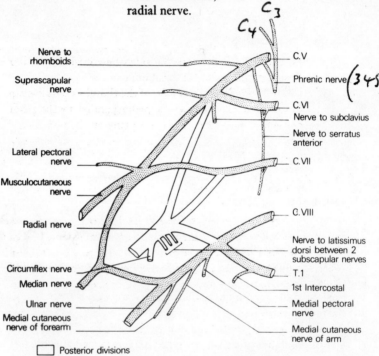

**Fig. 134.** The derivatives of the brachial plexus.

Note that the posterior cord supplies the skin and muscles of the posterior aspect of the limb whereas the anteriorly placed lateral and medial cords supply the anterior compartment structures.

## The segmental cutaneous supply of the upper limb (Fig. 135)

In spite of this complex interlacing of the nerve roots in the brachial plexus, the skin of the upper limb, as with the skin of the rest of the body,

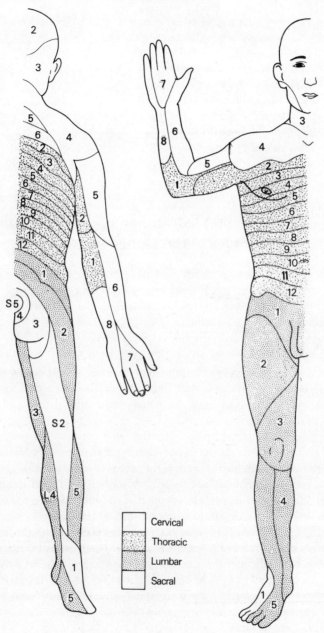

**Fig. 135.** The segmental cutaneous innervation of the body.

has a perfectly regular segmental nerve supply. This is derived from C4 to T2 which is arranged approximately as follows:

- •C4—supplies skin over the shoulder tip
- •C5—radial side of upper arm
- •C6—radial side of forearm
- •C7—the skin of the hand
- •C8—ulnar side of forearm
- •T1—ulnar side of upper arm
- •T2—skin of the axilla.

# The Course and Distribution of the Principal Nerves of the Upper Limb

The nerves of the upper limb are derived from the brachial plexus.

 *The circumflex nerve* (C5, 6) arises from the posterior cord of the plexus and winds round the surgical neck of the humerus in company with the circumflex humeral vessels (Fig. 117, 136). Its *branches* are:
muscular—to deltoid and teres minor.
cutaneous—to a palm-sized area of skin over the deltoid.

The circumflex nerve may be injured in fractures of the humeral neck or in dislocations of the shoulder. This will be followed by weakness of shoulder abduction, wasting of the deltoid and a small patch of anaesthesia over this muscle.

*The radial nerve* (C5, 6, 7, 8, T1) is the main branch of the posterior cord. Lying first behind the axillary artery, it then passes backwards between the long and medial heads of the triceps to lie in the spiral groove on the back of the humerus between the medial and lateral heads of triceps (Fig. 136). The profunda branch of the brachial artery accompanies the nerve in this part of its course (Fig. 117).

At the lower third of the humerus, the radial nerve pierces the lateral intermuscular septum to re-enter the anterior compartment of the arm between brachialis and brachio-radialis (a convenient site for surgical exposure Fig. 132). At the level of the lateral epicondyle its important *posterior interosseous nerve* is given off, which winds round the radius within the supinator muscle then sprays out to be distributed to the extensor muscles of the forearm.

The radial nerve itself continues as the superficial radial nerve, lying

Post view

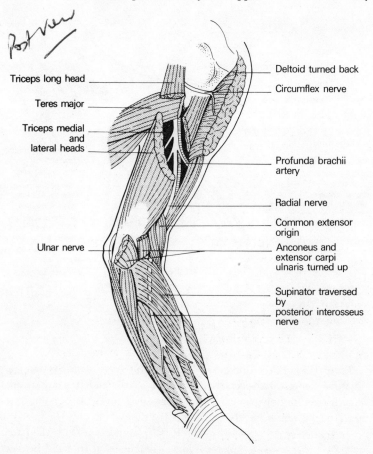

Triceps long head

Teres major

Triceps medial and lateral heads

Ulnar nerve

Deltoid turned back

Circumflex nerve

Profunda brachii artery

Radial nerve

Common extensor origin

Anconeus and extensor carpi ulnaris turned up

Supinator traversed by posterior interosseus nerve

**Fig. 136.** The distribution of the radial nerve.

deep to brachio-radialis (Fig. 132). Above the wrist, it emerges from beneath this muscle to end by dividing into cutaneous nerves to the posterior aspects of the radial $3\frac{1}{2}$ digits.

*Branches.* The radial nerve is the nerve of supply to the extensor aspect of the upper limb. The main trunk itself innervates:
triceps, anconeus, brachio-radialis and extensor carpi radialis longus; it also gives a twig to the lateral part of brachialis.

The posterior interosseous branch supplies:
all the remaining extensor muscles of the forearm, the supinator and abductor pollicis longus.

Cutaneous branches are distributed to the back of the arm, forearm

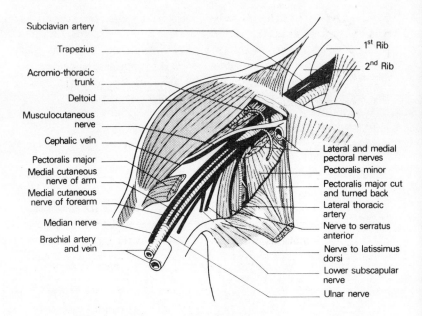

**Fig. 137.** Dissection of the upper arm to show the course of the major nerves.

and radial side of the dorsum of the hand. So great is the overlap from adjacent nerves, however, that division of the radial nerve results, surprisingly, in only a small area of anaesthesia over the dorsum of the hand, in the web between the thumb and index finger.

   *The musculocutaneous nerve* (C5, 6, 7) continues on from the lateral cord of the plexus. It pierces coraco-brachialis then runs between biceps and brachialis (supplying all these three muscles) to innervate, by its terminal cutaneous branch, the skin of the lateral forearm.

   *The ulnar nerve* (C7, 8, T1) (Fig. 137) is formed from the medial cord of the plexus. It lies medial to the axillary and brachial artery as far as the middle of the humerus, then pierces the medial intermuscular septum (in company with the ulnar collateral artery) to descend on the anterior face of triceps. It passes behind the medial epicondyle (where it can readily be rolled against the bone), to enter the forearm (Fig. 117). Here it descends beneath flexor carpi ulnaris until this muscle thins out into its tendon, leaving the nerve to lie superficially on its radial side. In the distal two-thirds of the forearm the nerve is accompanied by the ulnar artery which lies on the nerve's radial side. About 2 in (5 cm) above the wrist, a

dorsal cutaneous branch passes deep to flexor carpi ulnaris to supply the
dorsal aspects of the ulnar $1\frac{1}{2}$ fingers.

The ulnar nerve crosses the flexor retinaculum superficially (Fig. 132)
to break up into a superficial terminal branch, supplying the ulnar $1\frac{1}{2}$
fingers, and a deep terminal branch which supplies the hypothenar
muscles and the intrinsic muscles of the hand.

Its *branches* are:

•muscular—  to flexor carpi ulnaris,
  medial half of flexor digitorum profundus,
  the hypothenar muscles,
  the interossei, 3rd and 4th lumbricals and the adductor
  pollicis.

  (i.e. it supplies all the intrinsic muscles of the hand apart
  from those of the thenar eminence and the 1st and 2nd
  lumbricals, which are innervated by the median nerve).

•cutaneous—to the ulnar side of both aspects of the hand and both
  surfaces of the ulnar $1\frac{1}{2}$ fingers.

*The median nerve* (C6, 7, 8, T1) (Fig. 137) arises by the junction of a
branch from the medial and another from the lateral cord of the plexus,
which unite anterior to the third part of the axillary artery. Continuing
along the lateral aspect of the brachial artery, the nerve then crosses
superficially (occasionally deep) to the artery at the mid-humerus to lie
on its medial side. The nerve enters the forearm between the heads of
pronator teres, the deeper of which separates it from the ulnar artery
(Fig. 132). Here the nerve gives off its *anterior interosseous branch* and then
lies on the deep aspect of flexor digitorum sublimis, to which it adheres.

At the wrist, the median nerve becomes superficial on the ulnar side
of flexor carpi radialis, exactly in the mid-line (Fig. 111). It then passes
deep to the flexor retinaculum, giving off an important branch to the
thenar muscles beyond the distal skin crease, twigs to the radial two
lumbricals and cutaneous branches to the palmar aspects of the radial
$3\frac{1}{2}$ digits.

Its *branches* are:

•muscular—to all the muscles of the flexor aspect of the forearm, apart
from the flexor carpi ulnaris and the ulnar half of flexor digitorum profundus; the thenar eminence muscles and the radial two lumbricals.

•cutaneous—to the skin of the radial side of the palm; the palmar, and
a variable degree of the dorsal, aspect of the radial $3\frac{1}{2}$ digits.

Note that there is considerable variation in the exact cutaneous distribution of the nerves in the hand; for example, the ulnar nerve may encroach on median territory and supply the whole of the 4th and 5th digits (Fig. 138).

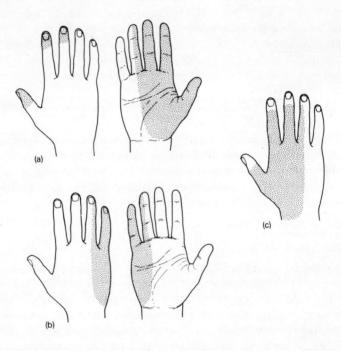

**Fig. 138.** The usual cutaneous distribution of the median (a), ulnar (b) and radial (c) nerves in the hand (considerable variations and overlap occur).

# The Anatomy of Upper Limb Deformities

Many deformities of the upper limb, particularly those resulting from nerve injuries, are readily interpreted anatomically.

*Brachial plexus injuries* may occur from traction on the arm during birth. The force of downward traction falls upon roots C5 and 6, resulting in paralysis of the deltoid and short muscles of the shoulder, and of brachialis and biceps which flex and supinate at the elbow. The arm therefore hangs limply by the side with the forearm pronated and the palm facing backwards, like a porter hinting for a tip. (*Erb—Duchenne paralysis.*) In adults this lesion is seen in violent falls on the side of the head and shoulder forcing the two apart and thus putting a tearing strain on the upper roots of the plexus.

Upward traction on the arm (e.g. in a forcible breech delivery) may tear the lowest root, T1, which is the segmental supply of the intrinsic hand muscles. The hand assumes a clawed appearance because of the unopposed action of the long flexors and extensors of the fingers; the extensors, inserting into the bases of the proximal phalanges, extend the m/p joints while the flexor profundus and sublimis, inserting into the distal and middle phalanges, flex the interphalangeal joints *(Klumpke's paralysis).* There is often an associated Horner's syndrome due to traction on the cervical sympathetic chain.

A mass of malignant supraclavicular lymph nodes or the direct invasion of a pulmonary carcinoma (Pancoast's syndrome) may produce a similar neurological picture by involvement of the lowest root of the plexus.

Not infrequently the lower trunk of the plexus (C8, T1) is pressed upon by a cervical rib, or by the fibrous strand running from the extremity of such a rib, resulting in paraesthesiae along the ulnar border of the arm and weakness and wasting of the small muscles of the hand.

*The radial nerve* may be injured in the axilla by the pressure of a crutch ('crutch palsy') or may be compressed when a drunkard falls into an intoxicated sleep with the arm hanging over the back of a chair ('Saturday night palsy'). Fractures of the humeral shaft may damage the main radial nerve, whereas its posterior interosseous branch, to the extensor muscles of the forearm, may be injured in fractures or dislocations of the radial head. An ill-placed incision to expose the head of the radius taken more than three fingers' breadth below the head will divide the nerve as it lies in the supinator muscle.

Damage to the main trunk of the radial nerve results in a *wrist drop* due to paralysis of all the wrist extensors (Fig. 139). Damage to the posterior interosseous nerve, however, leaves extensor carpi radialis longus intact, as it is supplied from the radial nerve above its division; this muscle alone is sufficiently powerful to maintain extension of the wrist.

The disability produced by a wrist drop is inability to grip firmly, since, unless the flexor muscles are stretched by extending the wrist, they act at a mechanical disadvantage. Try yourself to grip strongly with the wrist flexed and realize how, by operative fusion of the wrist joint in extension, the weakness produced by a radial nerve paralysis would be overcome.

Because of nerve overlap, division of the radial nerve produces only

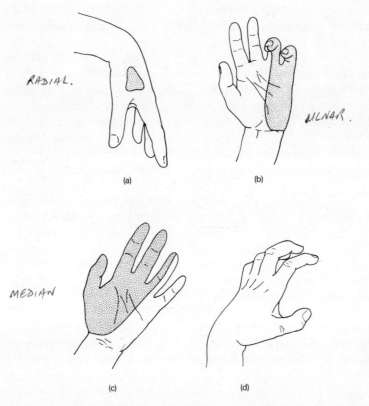

**Fig. 139.** Deformities of the hand. (a) Radial palsy—wrist drop. (b) Ulnar nerve palsy—'Main en griffe' or claw hand. (c) Median nerve palsy—'Monkey's hand'. (d) Volkmann's contracture—another claw hand deformity. (The shaded areas represent the usual distribution of anaesthesia.)

a small area of anaesthesia of the dorsum of the hand between the 1st and 2nd metacarpals.

*The ulnar nerve,* in its vulnerable position behind the medial epicondyle of the humerus, may be damaged in fractures or dislocations of the elbow; it is also frequently divided in lacerations of the wrist. In the latter case, all the intrinsic muscles of the fingers (apart from the radial two lumbricals) are paralysed so that the hand assumes the clawed position already described under Klumpke's palsy (Fig. 139). The clawing is slightly less intense in the 2nd and 3rd digits because of their intact lumbricals, supplied by the median nerve. In late cases, wasting of the interossei is readily seen on inspecting the dorsum of the hand. Sensory loss over the ulnar $1\frac{1}{2}$ fingers is present.

If the nerve is injured at the elbow, the flexor digitorum profundus to the 4th and 5th fingers is paralysed so that the clawing of these two fingers is less intense than in division at the wrist. Paralysis of the flexor carpi ulnaris results in a tendency to radial deviation of the wrist.

Division of the ulnar nerve leaves a surprisingly efficient hand. The long flexors enable a good grip to be taken, the thumb, apart from loss of adductor pollicis, is intact and sensation over the palm of the hand is largely maintained. Indeed, it may be difficult to determine clinically with certainty that the nerve is injured; a reliable test is loss of ability to adduct and abduct the fingers with the hand laid flat, palm downwards on the table; this eliminates 'trick' movements of adduction and abduction of the fingers brought about as part of their flexion and extension respectively.

*The median nerve* is occasionally damaged in supracondylar fractures but is in greatest danger in lacerations of the wrist.

If divided at the wrist, only the thenar muscles (excluding adductor pollicis) and the radial two lumbricals are paralysed and wasting of the thenar muscles occurs. It might be thought that such a lesion is relatively trivial since the only motor defect is loss of accurate opposition movement of the thumb to other fingers. In point of fact this injury is a serious disability because of the loss of sensation over the thumb, adjacent $2\frac{1}{2}$ fingers and the radial two-thirds of the palm of the hand, which prevents the accurate the delicate adjustments the hand makes in response to tactile stimuli (Fig. 139).

If the median nerve is divided at the elbow there is serious muscle impairment. Pronation of the forearm is lost and is replaced by a trick movement of rotation of the upper arm. Wrist flexion is weak and accompanied by ulnar deviation, since this now depends on the flexor carpi ulnaris and the ulnar half of flexor digitorum profundus.

*Volkmann's contracture* of the hand follows ischaemia and subsequent fibrosis and contraction of the long flexor and extensor muscles of the forearm (Fig. 139).

The deformities are readily explained as follows.

1. Since the flexors of the wrist are bulkier than the extensors, their fibrous contraction is greater and the wrist is therefore flexed.
2. The long extensors of the fingers are inserted into the proximal

phalanges; their contracture extends the m/p joints.

3.   The long flexors are inserted into the distal and middle phalanges and therefore flex the i/p joints.

There is therefore flexion at the wrist, extension at the m/p and flexion at the i/p joints.

If the wrist is passively further flexed by the examiner, the tight flexor tendons are somewhat relaxed and therefore the fingers become a little less clawed.

*Dupuytren's contracture* results from a fibrous contraction of the palmar fascia, particularly of the 4th and 5th fingers.

The palmar fascia is merely part of the deep fascial sheath of the upper limb; it passes from the palm along either side of each finger, blends with the fibrous flexor sheath of the fingers and is attached to the sides of the proximal and middle phalanges. Contracture of this fascia results in a longitudinal thickening in the palm together with flexion of the m/p and proximal i/p joints. However, the distal i/p joints are not involved and, in fact, in an advanced case, are actually *extended* by the distal phalanx being pushed backwards against the palm of the hand.

# The Spaces of the Hand

The spaces of the hand are of practical significance because they may become infected and, in consequence, become distended with pus. The important spaces are:

1.   the superficial pulp spaces of the fingers.
2.   the synovial tendon sheaths of the 2nd, 3rd and 4th fingers.
3.   the ulnar bursa.
4.   the radial bursa.
5.   the mid-palmar space.
6.   the thenar space.

### The superficial pulp space of the fingers (Fig. 140)

The tips of the fingers and thumb are composed entirely of subcutaneous fat broken up and packed between fibrous septa, which pass from the skin down to the periosteum of the terminal phalanx. The tight packing

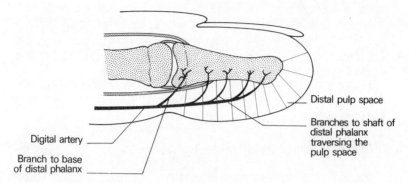

Fig. 140. The distal pulp space of the finger—note the distribution of the arterial supply to the distal phalanx.

of this compartment is responsible for the severe pain of a 'septic finger' —there is little room for the expansion of inflamed and oedematous tissues.

The blood-vessels to the shaft of the distal phalanx must traverse this space and may become thrombosed in a severe pulp infection with resulting necrosis of the diaphysis of the bone. The base of the distal phalanx receives its blood supply more proximally from a branch of the digital artery in the middle segment of the finger and therefore survives. At each of the skin creases of the fingers, the skin is bound down to the underlying flexor sheath so that the pulp over each phalanx is in a separate compartment cut off from its neighbours. Infection may, however, track from one space to another along the neurovascular digital bundles.

Over the palm of the hand there is very little subcutaneous tissue, the skin adhering to the underlying palmar fascia; in contrast, the skin of the dorsum of the fingers and hand is loose and fluid can therefore readily collect beneath it. Unless this is remembered the marked dorsal oedema which may accompany sepsis of the palmar aspect of the fingers or hand may result in the primary site of the infection being overlooked.

## The ulnar and radial bursae and the synovial tendon sheaths of the fingers (Fig. 141)

The flexor tendons traverse a fibro-osseous tunnel in each digit. This tunnel is made up posteriorly by the metacarpal head, the phalanges and the fronts of the intervening joints. The anterior fibrous part consists of condensed deep fascia attached to the sharp anterolateral margin of each phalanx and termed the *fibrous flexor sheath*. This is particularly tough over the phalanges but loose over the front of each joint; it therefore holds the flexor tendons in place without 'bow-stringing' during flexion of the fingers, but does not impede movement of the joints.

Distally the fibrous sheath ends at the insertion of the profundus tendon (or flexor pollicis longus tendon in the case of the thumb) at the base of the distal phalanx.

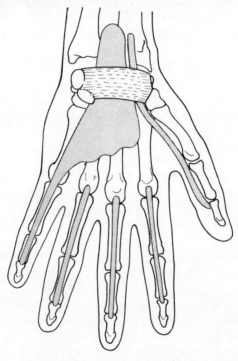

Fig. 141. The synovial sheaths of the flexor tendons of the hand—the radial and ulnar bursae track proximally deep to the flexor retinaculum and provide a potential pathway of infection into the forearm. In many cases these bursae communicate.

These fibrous sheaths are lined by synovial membrane which is reflected around each tendon. The tendons of the 2nd, 3rd and 4th fingers have synovial sheaths which are closed off proximally at the metacarpal head, but the synovial sheaths of the thumb and little finger extend proximally into the palm.

That of the long flexor tendon of the thumb extends through the palm, deep to the flexor retinaculum, to about 1 in (2.5 cm) proximal to the wrist and is termed the *radial bursa*. The synovial sheath of the 5th finger continues as the *ulnar bursa*; an expanded synovial sheath which encloses all the finger tendons in the palm and which also extends proximally below the flexor retinaculum for 1 in (2.5 cm) above the wrist. In about 50 per cent of cases the radial and ulnar bursae communicate. These synovial sheaths may become infected either directly, for example, following the entry of a splinter, or may be secondarily involved from a neglected pulp-space

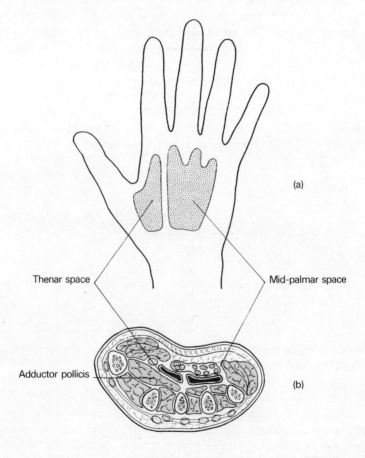

**Fig. 142.** The mid-palmar and thenar spaces. (a) Projected on to the surface of the hand. (b) In transverse section. (In both diagrams these spaces are cross-hatched.)

infection. Infection of the 2nd, 3rd and 4th sheaths are confined to the finger concerned, but sepsis in the 1st and 5th sheaths may spread proximally into the palm through the radial and ulnar bursa respectively, and may pass from one bursa to the other via the frequent cross communication between the two.

Since these bursae both extend proximally beyond the wrist, infection may, on occasion, spread into the forearm.

Two spaces deep in the palm of the hand may rarely become distended with pus; these are the mid-palmar and thenar spaces (Fig. 142).

*The mid-palmar* space lies behind the flexor tendons and ulnar bursa in the palm and in front of the 3rd, 4th and 5th metacarpals with their attached interossei. The 1st and 2nd metacarpals are curtained off from this space by the adductor pollicis, which arises from the shaft of the 3rd metacarpal and passes as a triangular sheet to the base of the proximal phalanx of the thumb.

*The thenar space* is the fascial capsule of the adductor pollicis.

Infection of these two spaces sometimes results from penetrating wounds or may be due to secondary involvement from a long neglected tendon sheath infection. Nowadays they are fortunately extremely rare.

# PART FOUR

# THE LOWER LIMB

# The Anatomy and surface Markings
# of the Lower Limb

## Bones and joints

The tip of the *anterior superior spine of the ilium* is easily felt and may be visible in the thin subject. The *greater tuberosity* of the femur lies a hands-breadth below the iliac crest; it is best palpated with the hip abducted so that the overlying hip abductors (tensor fasciae latae and gluteus medius and minimus) are relaxed. In the very thin, wasted patient the greater trochanter may be seen as a prominent bulge and its overlying skin is a common site for a pressure sore to form in such a case.

The *ischial tuberosity* is covered by gluteus maximus when one stands. In the sitting position, however, the muscle slips away laterally so that weight is taken directly on the bone. To palpate this bony point, there-fore, feel for it uncovered by gluteus maximus in the flexed position of the hip.

At the knee, the *patella* forms a prominent landmark. When quadriceps femoris is relaxed this bone is freely mobile from side to side; note that this is so when you stand erect. The *condyles* of the femur and tibia, the *head of the fibula* and the joint line of the knee are all readily palpable; less so is the *adductor tubercle* of the femur, best identified by running the fingers down the medial side of the thigh until they are halted by it, the first bony prominence so to be encountered.

The *tibia* can be felt throughout its course along its anterior sub-cutaneous border. The *fibula* is subcutaneous for its terminal 3 in (7.5 cm) above the *lateral malleolus*, which extends more distally than the stumpier *medial malleolus* of the tibia.

Immediately in front of the malleoli can be felt a block of bone which is the *head of the talus*.

The *tuberosity of the navicular* stands out as a bony prominence 1 inch in front of the medial malleolus; it is the principal point of insertion of tibialis posterior. The base of the 5th metatarsal is easily felt on the lateral side of the foot and is the site of insertion of peroneus brevis.

If the *os calcis* is carefully palpated, the *peroneal tubercle* can be felt 1 in (2.5 cm) below the tip of the lateral malleolus and the *sustentaculum tali* 1 in (2.5 cm) below the medial malleolus; these represent pulleys respectively for peroneus longus and for flexor hallucis longus.

## Bursae of the lower limb

A number of these bony prominences are associated with overlying bursae which may become distended and inflamed; the one over the ischial tuberosity may enlarge with too much sitting ('weaver's bottom') That in front of the patella is affected by prolonged kneeling forwards, as in scrubbing floors or hewing coal ('housemaid's knee', the 'beat knee' of north-country miners or prepatellar bursitis), whereas the bursa over the ligamentum patellae is involved by years of kneeling in a more erect position—as in praying ('clergyman's knee' or infrapatellar bursitis). Young women who affect fashionable but tight shoes are prone to bursitis over the insertion of the tendo-Achillis into the os calcis and may also develop bursae over the navicular tuberosity and dorsal aspects of the phalanges.

A 'bunion' is a thickened bursa on the inner aspect of the first metatarsal head, usually associated with hallux valgus deformity.


## Mensuration in the lower limb

Measurement is an important part of the clinical examination of the lower limb. Unfortunately, students find difficulty in carrying this out accurately and still greater difficulty in explaining the results they obtain, yet this is nothing more or less than a simple exercise in applied anatomy.

First note the difference between *real* and *apparent* shortening of the lower limbs. Real shortening is due to actual loss of bone length, for example where a femoral fracture has united with a good deal of over-riding of the two fragments. Apparent shortening is due to a *fixed* deformity of the limb (Fig. 143). Stand up and flex your knee and hip on one side, imagine these are both ankylosed at 90° and note that, although there is no loss of tissue in this leg, it is apparently some 2 feet (62 cm) shorter than its partner.

If there is a fixed pelvic tilt or fixed joint deformity in one limb, then, there may be this *apparent* difference between the lengths of the two legs. By experimenting on yourself you will find that *adduction* apparently *shortens* the leg, whereas it is apparently *lengthened* in abduction.

To measure the real length of the limbs (Fig. 144), overcome any disparity due to fixed deformity by putting both legs into exactly the same position; where there is no joint fixation, this means that the patient lies with his pelvis 'square', his legs abducted symmetrically and both lying flat on the couch. If, however, one hip is in 60° of fixed flexion, for ex-

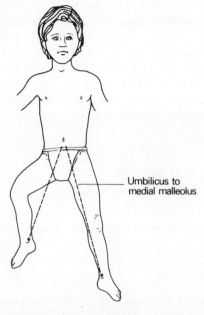

Umbilicus to
medial malleolus

**Fig. 143.** Apparent shortening—one limb may be apparently shorter than the other because of fixed deformity; the legs in this illustration are actually equal in length but the right is *apparently* considerably shorter because of a gross flexion contracture at the hip. Apparent shortening is measured by comparing the distance from the umbilicus to the medial malleolus on each side.

ample, the other hip must first be put into this identical position. The length of each limb is then measured from the anterior superior iliac spine to the medial malleolus. In order to obtain identical points on each side, slide the finger upwards along Poupart's ligament and mark the bony point first encountered by the finger. Similarly, slide the finger upwards from just distal to the malleolus to determine the apex of this landmark on each side.

To determine apparent shortening, the patient lies with his legs parallel (as they would be when he stands erect) and the distance from umbilicus to each medial malleolus is measured (Fig. 143).

Now suppose we find 4 in (10 cm) of apparent shortening and 2 in (5 cm) of real shortening of the limb; we interpret this as meaning that 2 in (5 cm) of the shortening is due to true loss of limb length and another 2 in (5 cm) is due to fixed postural deformity.

If the apparent shortening is *less* than the real, this can only mean that the hip has ankylosed in the abducted, and hence apparently elongated, position.

Note this important point: one reason why the orthopaedic surgeon immobilizes a tuberculous hip in the abducted position is that, when the hip becomes ankylosed, shortening due to actual destruction at the hip (i.e. true shortening) will be compensated, to a considerable extent, by the apparent lengthening produced by the fixed abduction.

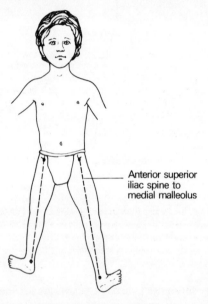

Anterior superior
iliac spine to
medial malleolus

**Fig. 144.**   Measuring real shortening—the patient lies with the pelvis 'square' and the legs placed symmetrically. Measurement is made from the anterior superior spine to the medial malleolus on each side.

Having established that there is real shortening present, the examiner must then determine whether this is at the hip, the femur or the tibia, or at a combination of these sites.

*At the hip*—place the thumb on the anterior superior spine and the index finger on the greater trochanter on each side; a glance is sufficient to tell if there is any difference between the two sides.

Examiners may still ask about Nelaton's line and Bryant's triangle (Fig. 145).

*Nelaton's line* joins the anterior superior iliac spine to the ischial tuberosity and should normally lie above the greater trochanter; if the line passes through or below the trochanter, there is shortening at the head or neck of the femur.

*Bryant's triangle* might better be called 'Bryant's T' because it is not necessary to construct all of its three sides. With the patient supine, a perpendicular is dropped from each anterior superior spine and the distance between this line and the greater trochanter compared on each side. (The third side of the triangle, joining the trochanter to the anterior spine, need never be completed.)

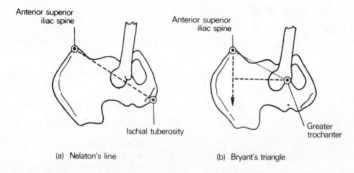

Anterior superior
iliac spine

Anterior superior
iliac spine

Ischial tuberosity

Greater
trochanter

(a)  Nelaton's line          (b)  Bryant's triangle

**Fig. 145.**  (a) Nelaton's line joins the anterior superior iliac spine to the ischial tuberosity—normally this passes above the greater trochanter. (b) Bryant's triangle—drop a vertical from each anterior superior spine; compare the perpendicular distance from this line to the greater trochanter on either side. (There is no need to complete the third side of the triangle.)

*At the femur*—measure the distance from the anterior superior spine (if hip disease has been excluded) or from the greater trochanter to the line of the knee joint (not to the patella, whose height can be varied by contraction of the quadriceps).

*At the tibia*—compare the distance from the line of the knee joint to the medial malleolus on each side.

## Muscles and tendons

*Quadriceps femoris* forms the prominent muscle mass on the anterior aspect of the thigh; its insertion into the medial aspect of the patella can be seen to extend more distally than on the lateral side. In the well-developed subject, *sartorius* can be defined when the hip is flexed and externally rotated against resistance. It extends from the anterior superior iliac spine to the medial side of the upper end of the tibia and, as the lateral border of the *femoral triangle*, it is an important landmark.

*Gluteus maximus* forms the bulk of the buttock and can be felt to contract in extension of the hip.

*Gluteus medius and minimus* and the *adductors* can be felt to tighten respectively in resisted abduction and adduction of the hip.

Define the tendons around the knee with this joint comfortably flexed in the sitting position:

- laterally—the *biceps* tendon passes to the head of the fibula
  - the *ilio-tibial tract* lies about $\frac{1}{2}$ in (12 mm) in front of this tendon and passes to the lateral condyle of the tibia;
- medially—the bulge which one feels is the *semimembranosus insertion* on which two tendons, *semitendinosus* laterally and *gracilis* medially are readily palpable.

Between the tendons of biceps and semitendinosus can be felt the heads of origin of *gastrocnemius*. This muscle, with soleus, forms the bulk of the posterior bulge of the calf; the two end distally in the *tendo Achillis*.

At the front of the ankle (Fig. 146) the tendon of *tibialis anterior* lies most medially, passing to its insertion at the base of the first metatarsal and the medial cuneiform. More laterally, the tendons of *extensor hallucis longus* and *extensor digitorum longus* are readily visible in the dorsi-flexed foot. *Peroneus longus and brevis* tendons pass behind the lateral malleolus. Behind the medial malleolus, from the medial to the lateral side, pass the tendons of *tibialis posterior* and *flexor digitorum longus*, the *posterior tibial artery* with its venae comites, the *tibial nerve* and, finally, *flexor hallucis longus* (Fig. 147).

## Vessels

The *femoral artery* (Fig. 148) can be felt pulsating at the mid-inguinal point, half-way between the anterior superior iliac spine and the pubic symphysis. The upper two-thirds of a line joining this point to the adductor tubercle, with the hip somewhat flexed and externally rotated, accurately defines the surface markings of this vessel. A finger on the femoral pulse lies directly over the head of the femur, immediately lateral to the femoral vein and a finger's-breadth medial to the femoral nerve.

The pulse of the *popliteal artery* is often not easy to detect. It is most readily felt with the patient prone, his knee flexed and his muscles relaxed by resting the leg on the examiner's arm. The pulse is sought by firm pressure downwards against the popliteal fossa of the femur.

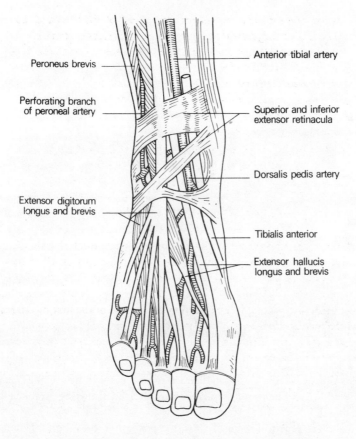

Peroneus brevis

Perforating branch
of peroneal artery

Extensor digitorum
longus and brevis

Anterior tibial artery

Superior and inferior
extensor retinacula

Dorsalis pedis artery

Tibialis anterior

Extensor hallucis
longus and brevis

**Fig. 146.** The structures passing over the dorsum of the ankle.

The pulse of *dorsalis pedis* (Fig. 146) is felt between the tendons of extensor hallucis longus and extensor digitorum on the dorsum of the foot—it is absent in 14 per cent of normal subjects. The *posterior tibial artery* (Fig. 147) may be felt a finger's-breadth below and behind the medial malleolus. In about 5 per cent of healthy subjects this artery is replaced by the peroneal artery.

The absence of one or both pulses at the ankle is not, therefore, in itself diagnostic of vascular disease.

The *short saphenous vein* commences as a continuation of the veins on the lateral side of the dorsum of the foot, runs proximally behind the lateral malleolus, and terminates by draining into the popliteal vein

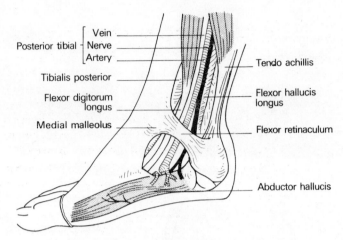

Posterior tibial { Vein / Nerve / Artery

Tibialis posterior

Flexor digitorum longus

Medial malleolus

Tendo achillis

Flexor hallucis longus

Flexor retinaculum

Abductor hallucis

**Fig. 147.** The structures passing behind the medial malleolus.

behind the knee. The *long saphenous vein* arises from the medial side of the dorsal network of veins, passes upwards in front of the medial malleolus, with the saphenous nerve anterior to it, to enter the femoral vein in the groin, one inch below the inguinal ligament and immediately medial to the femoral pulse.

These veins are readily studied in any patient with extensive varicose veins and are usually visible, in their lower part, in the thin normal subject on standing. (The word 'saphenous' is derived from the Arabic for 'visible'.)

From the practical point of view, the position of the long saphenous vein immediately in front of the medial malleolus is perhaps the most important single anatomical relationship; no matter how collapsed or obese, or how young and tiny the patient, the vein can be relied upon to be available at this site when urgently required for transfusion purposes (Fig. 149).

### Nerves

Only one nerve can be felt in the lower limb; this is the *common peroneal nerve* which can be rolled against the bone as it winds round the neck of the fibula (Fig. 150). Not unnaturally, it may be injured at this site in adduction injuries to the knee or compressed by a tight plaster cast or firm bandage, with a resultant foot drop.

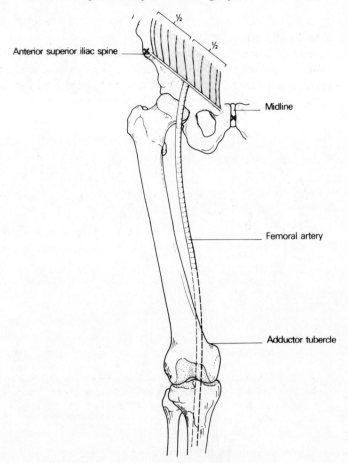

Anterior superior iliac spine

Midline

Femoral artery

Adductor tubercle

**Fig. 148.**   The surface markings of the femoral artery.

The *femoral nerve* emerges from under the inguinal ligament 12 mm lateral to the femoral pulse. After a course of only about 5 cm the nerve breaks up into its terminal branches.

The surface markings of the *sciatic nerve* (Fig. 151) can be represented by a line which commences at a point midway between the posterior superior iliac spine and the ischial tuberosity, curves outwards and downwards through a point midway between the greater trochanter and ischial tuberosity and then continues vertically downwards in the mid-line of the

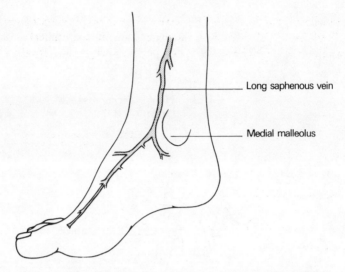

**Fig. 149.** The relationship of the saphenous vein to the medial malleolus.

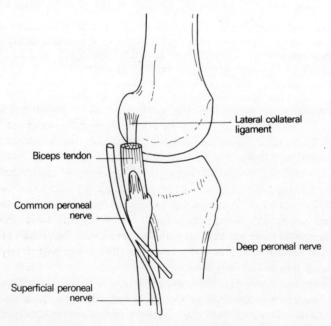

**Fig. 150.** The close relationship of the common peroneal nerve to the neck of the fibula; at this site it may be compressed by a tight bandage or plaster cast.

posterior aspect of the thigh. The nerve ends at a variable point above the popliteal fossa by dividing into the medial and lateral popliteal nerves, now re-named the tibial and common peroneal nerves respectively.

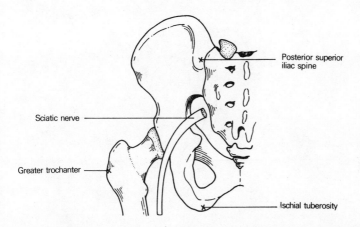

**Fig. 151.**   The surface markings of the sciatic nerve. Join the mid-point between the ischial tuberosity and posterior superior iliac spine to the mid-point between the ischial tuberosity and the greater trochanter by a curved line; continue this line vertically down the leg—it represents the course of the sciatic nerve.

It would seem inconceivable that a nerve with such constant and well-defined landmarks could be damaged by intramuscular injections, yet this has happened so frequently that it has seriously been proposed that this site should be prohibited. The explanation is, I believe, a psychological one. The standard advice is to employ the upper outer quadrant of the buttock for these injections, and when the full anatomical extent of the buttock—extending upwards to the iliac crest and outwards to the greater trochanter—is implied, perfectly sound and safe advice this is. Many nurses, however, have an entirely different mental picture of the buttock; a much smaller and more aesthetic affair comprising merely the hillock of the natus. An injection into the upper outer quadrant of this diminutive structure lies in the immediate area of the sciatic nerve!

A better surface marking for the 'safe area' of buttock injections can be defined as that area which lies under the outstretched hand when the thumb and thenar eminence are placed along the iliac crest with the tip of the thumb touching the anterior superior iliac spine (Fig. 152).

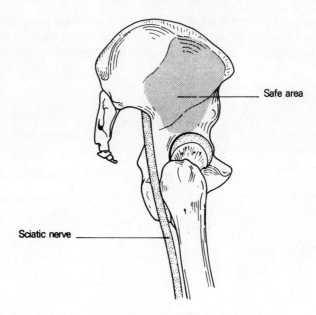

**Fig. 152.** The 'safe area' for injections in the buttock.

# The Bones and Joints of the Lower Limb

**The os innominatum** *(see* Pelvis)

**The femur** (Fig. 153 and 154)

The femur is the largest bone in the body. It is 18 in (45.5 cm) in length, a measurement it shares with the vas, the spinal cord and the thoracic duct and which is also the distance from the teeth to the cardia of the stomach.

The femoral *head* is two-thirds of a sphere and faces upwards, medially and forwards. It is covered with cartilage except for its central *fovea* where the ligamentum teres is attached.

The *neck* is 2 in (5 cm) long and is set at an angle of 125° to the shaft. In the female, with her wider pelvis, the angle is smaller.

The junction between the neck and the shaft is marked anteriorly by the *trochanteric line*, laterally by the *greater trochanter*, medially and somewhat posteriorly by the *lesser trochanter* and posteriorly by the prominent *trochanteric crest*, which unites the two trochanters.

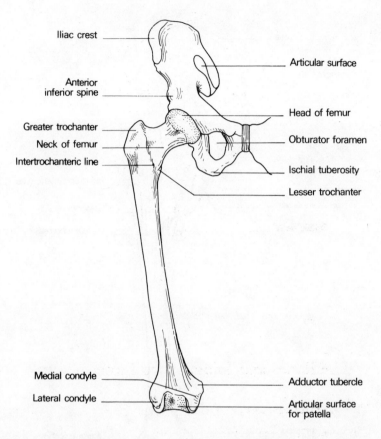

Iliac crest

Articular surface

Anterior
inferior spine

Head of femur

Greater trochanter

Obturator foramen

Neck of femur

Intertrochanteric line

Ischial tuberosity

Lesser trochanter

Medial condyle

Adductor tubercle

Lateral condyle

Articular surface
for patella

Fig. 153.    The anterior aspect of the right femur.

The blood supply to the femoral head is derived from vessels travelling up from the diaphysis along the cancellous bone, from vessels in the hip capsule, where this is reflected onto the neck in longitudinal bands or retinacula, and from the artery in the ligamentum teres; this third source is negligible (Fig. 155).

The femoral *shaft* is circular in section at its middle but is flattened posteriorly at each extremity. Posteriorly also it is marked by a strong crest, the *linea aspera*. Inferiorly this crest splits into the medial and lateral *supracondylar lines* leaving a flat *popliteal surface* between them. The medial supracondylar line ends distally in the *adductor tubercle*.

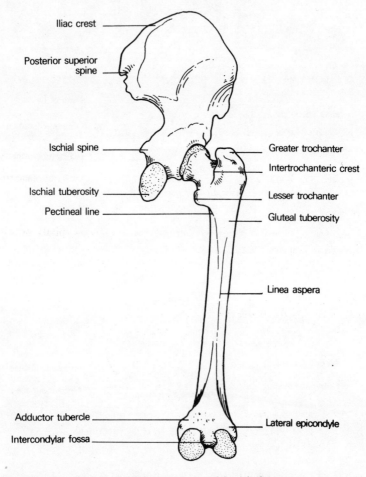

Fig. 154. The posterior aspect of the right femur.

The lower end of the femur bears the prominent *condyles* which are separated by a deep *intercondylar notch* posteriorly but which blend anteriorly to form an articular surface for the patella. The lateral condyle is the more prominent of the two and acts as a buttress to assist in preventing lateral displacement of the patella.

CLINICAL FEATURES

1. The upper end of the femur is a common site for fracture in the elderly.

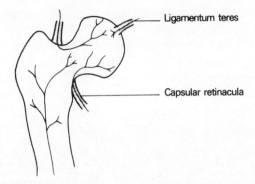

Ligamentum teres

Capsular retinacula

Fig. 155. The sources of blood supply of the femoral head—along the ligamentum teres, through the diaphysis and via the retinacula.

The neck may break immediately beneath the head (sub-capital), near its mid-point (cervical) or adjacent to the trochanters (basal), or the fracture line may pass between, along or just below the trochanters (Fig. 156).

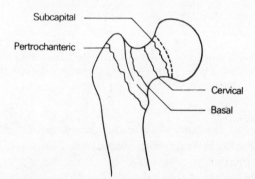

Subcapital

Pertrochanteric

Cervical

Basal

Fig. 156. The head and neck of the femur, showing the terminology of the common fracture sites.

Fractures of the femoral neck will interrupt completely the blood supply from the diaphysis and, should the retinacula also be torn, avascular necrosis of the head will be inevitable. The nearer the fracture to the femoral head, the more tenuous the retinacular blood supply and the more likely it is to be disrupted.

In contrast, the trochanteric fractures, being outside the joint capsule, leave the retinacula undisturbed; avascular necrosis therefore never follows such injuries (Fig. 157).

There is a curious age pattern of hip injuries; children may sustain greenstick fractures of the femoral neck, schoolboys may displace the epiphysis of the femoral head, in adult life the hip dislocates and, in old age, fracture of the neck of the femur again becomes the usual lesion.

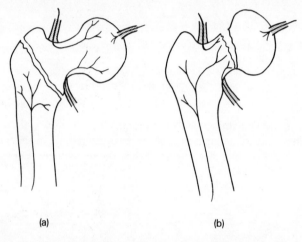

(a)                                      (b)

**Fig. 157.** (a) A petrochanteric fracture does not damage the retinacular blood supply—aseptic bone necrosis does not occur. (b) A subcapital fracture cuts off most of the retinacular supply to the head—aseptic bone necrosis is common.

2. Fractures of the femoral shaft are accompanied by considerable shortening due to the longitudinal contraction of the extremely strong surrounding muscles.

The proximal segment is flexed by ilio-psoas and abducted by gluteus medius and minimus, whereas the distal segment is pulled medially by the adductor muscles. Reduction requires powerful traction, to overcome the shortening, and then manipulation of the distal fragment into line with the proximal segment; the limb must therefore be abducted and also pushed forwards by using a large pad behind the knee.

Fractures of the lower end of the shaft, immediately above the condyles, are relatively rare; fortunately so because they may be extremely difficult to treat since the small distal fragment is tilted backwards by gastrocnemius, the only muscle which is attached to it. The sharp proximal edge of this distal fragment may also tear the popliteal artery which lies directly behind it (Fig. 158).

3. The angle subtended by the femoral neck to the shaft may be decreased, producing a *coxa vara* deformity. This may result from adduction fractures, slipping of the femoral epiphysis or bone-softening diseases. *Coxa vulga*, where the angle is increased, is much rarer but occurs in impacted abduction fractures. Note, however, that in children the normal angle between the neck and shaft is about 160°.

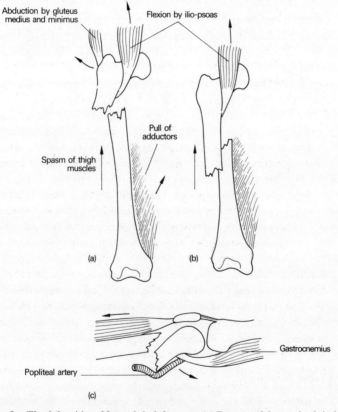

**Fig. 158.** The deformities of femoral shaft fractures. (a) Fracture of the proximal shaft—the proximal fragment is flexed by iliopsoas and abducted by gluteus medius and minimus. (b) Fracture of the mid-shaft—flexion of the proximal fragment by ilio-psoas. (c) Fracture of the distal shaft—the distal fragment is angulated backward by gastrocnemius—the popliteal artery may be torn in this injury. (In all these fractures over-riding of the bone ends is produced by muscle spasm.)

## The patella

The patella is a sesamoid bone in the expansion of the quadriceps tendon, which continues from the apex of the bone as the ligamentum patellae.

The posterior surface of the patella is covered with cartilage and articulates with the two femoral condyles by means of a larger lateral and smaller medial facet.

### CLINICAL FEATURES

1.   Lateral dislocation of the patella is resisted by the prominent articular surface of the lateral femoral condyle and by the medial pull of the lower-most fibres of vastus medialis which insert almost horizontally along the medial margin of the patella. If the lateral condyle of the femur is under-developed, or if there is a considerable genu valgum (knock-knee deform-ity), recurrent dislocations of the patella may occur (Fig. 159).

2.   A direct blow on the patella may split or shatter it but the fragments are not avulsed apart because the quadriceps expansion remains intact.

The patella may also be fractured transversely by violent contrac-tion of the quadriceps, for example in trying to stop a backward fall. In this case, the tear extends outwards into the quadriceps expansion, allow-ing the upper bone fragment to be pulled proximally; there may be a gap of over 2 in (5 cm) between the bone ends. Reduction is impossible by closed manipulation and operative repair of the extensor expansion is imperative.

Occasionally this same mechanism of sudden forcible quadriceps contraction tears the quadriceps expansion above the patella, ruptures ligamentum patellae or avulses the tibial tubercle.

It is interesting that following complete excision of the patella for a comminuted fracture, knee function and movement may return to 100 per cent efficiency; it is difficult, then, to ascribe any particular function to this bone.

### The tibia (Fig. 160)

The upper end of the tibia is expanded into the *medial and lateral condyles*, the former having the greater surface area of the two. Between the condyles is the *intercondylar area* which bears, at its waist, the *intercondylar eminence*, projecting upwards slightly on either side as the *medial* and *lateral intercondylar tubercles*.

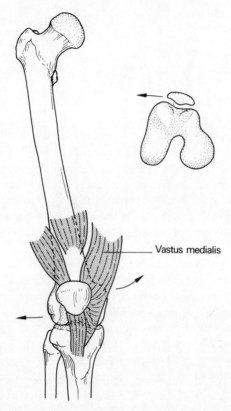

Vastus medialis

Fig. 159. Factors in the stability of the patella. (i) The medial pull of vastus medialis. (ii) The high patellar articular surface of the lateral femoral condyle. These resist the tendency for lateral displacement of the patella which results from the valgus angulation between the femur and the tibia.

The *tubercle* of the tibia is at the upper end of the anterior border of the shaft and gives attachment to the ligamentum patellae.

The anterior aspect of this tubercle is subcutaneous, only excepting the infrapatellar bursa immediately in front of it.

The shaft of the tibia is triangular in cross-section, its anterior border and antero-medial surface being subcutaneous throughout their whole extent.

The posterior surface of the shaft bears a prominent oblique line at its upper end termed the *soleal line*, which not only marks the tibial origin

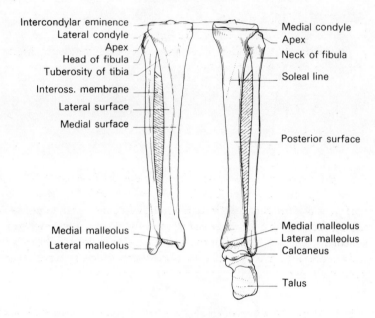

Intercondylar eminence
Lateral condyle
Apex
Head of fibula
Tuberosity of tibia
Inteross. membrane
Lateral surface
Medial surface

Medial condyle
Apex
Neck of fibula
Soleal line

Posterior surface

Medial malleolus
Lateral malleolus

Medial malleolus
Lateral malleolus
Calcaneus

Talus

Fig. 160. The right tibia and fibula.

of the soleus but also delimits an area above into which is inserted the popliteus.

The lower end of the tibia is expanded and quadrilateral in section, bearing an additional surface, the *fibular notch*, for the lower tibio-fibular joint.

The *medial malleolus* projects from the medial extremity of the bone and is grooved posteriorly by the tendon of tibialis posterior.

The inferior surface of the lower end of the tibia is smooth, cartilage-covered and forms, with the malleoli, the upper articular surface of the ankle joint.

CLINICAL FEATURES

1. The upper end of the tibial shaft is one of the commonest sites for acute osteomyelitis. Fortunately, the capsule of the knee joint is attached closely around the articular surfaces so that the upper extremity of the tibial diaphysis is extracapsular; involvement of the knee joint therefore only occurs in the late and neglected case.

2. The shaft of the tibia is subcutaneous and unprotected antero-medially throughout its course and is particularly slender in its lower

third. It is not surprising that the tibia is the commonest long bone to be fractured and to suffer compound injury.

3. The extensive subcutaneous surface of the tibia makes it a delightfully accessible donor site for bone-grafts.

## The fibula (Fig. 160)

The fibula serves three functions. It is:

1. an origin for muscles;
2. a part of the ankle joint;
3. a pulley for the tendons of peroneus longus and brevis.

It comprises the *head* with a *styloid process* (into which is inserted the tendon of biceps), the *neck* (around which passes the common peroneal nerve), the *shaft* and the *lower end* or *lateral malleolus*. The latter bears a medial roughened surface for the lower tibio-fibular joint, below which is the articular facet for the talus. A groove on the posterior aspect of the malleolus lodges the tendons of peroneus longus and brevis.

## A note on growing ends and nutrient foramina in the long bones

The shaft of every long bone bears one or more *nutrient foramina* which are obliquely placed; this obliquity is due to unequal growth at the upper and lower epiphyses. The artery is obviously dragged in the direction of more rapid growth and the direction of slope of entry of the nutrient foramen therefore points *away* from the more rapid growing end of the bone.

The direction of growth of the long bones can be remembered by a little jingle which runs:

'From the knee, I flee
To the elbow, I grow.'

With one exception, the epiphysis of the growing end of a long bone is the first to appear and last to fuse with its diaphysis; this exception is the epiphysis of the upper end of the fibula which, although at the growing end, appears *after* the distal epiphysis and fuses after the latter has blended with the shaft.

The site of the growing end is of considerable practical significance. For example, if a child has to undergo an above-elbow amputation, the humeral upper epiphyseal line continues to grow and the elongating bone may well push its way through the stump end, requiring reamputation.

## The bones of the foot

These are best considered as a functional unit and are therefore dealt with together under 'the arches of the foot' (*see* p. 251).

## The hip (Figs. 161, 162)

The hip is the largest joint in the body. To the surgeon, the examiner and, therefore, the student it is also the most important.

It is a perfect example of a ball-and-socket joint. Its articular surfaces are the femoral head and the horse-shoe shaped articular surface of the

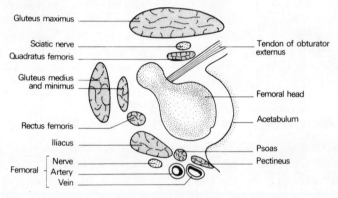

**Fig. 161.** The immediate relations of the hip joint (in diagrammatic horizontal section).

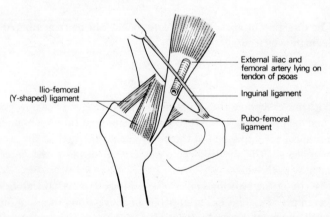

**Fig. 162.** The anterior aspect of the hip. Note that the psoas tendon and the femoral artery are intimate anterior relations of the joint.

acetabulum, which is deepened by the fibrocartilaginous *labrum aceta-bulare*. The non-articular lower part of the acetabulum, the *acetabular notch*, is closed off below by the *transverse acetabular ligament*. From this notch is given off the *ligamentum teres*, passing to the fovea on the femoral head.

The *capsule* of the hip is attached proximally to the margins of the acetabulum and to the transverse acetabular ligament. Distally, it is attached along the trochanteric line, the bases of the greater and lesser trochanters and, posteriorly, to the femoral neck about $\frac{1}{2}$ in (12 mm) from the trochanteric crest. From this distal attachment, capsular fibres are reflected onto the femoral neck as *retinacula* and provide one pathway for the blood supply to the femoral head (*see* femur) (Fig. 155).

Note that acute osteomyelitis of the upper femoral metaphysis will involve the neck which is intracapsular and which will therefore rapidly produce a secondary pyogenic arthritis of the hip joint.

Three ligaments reinforce the capsule.

1.   *The ilio-femoral* (Y-shaped ligament of Bigelow)—which arises from the anterior inferior iliac spine, bifurcates, and is inserted at each end of the trochanteric line (Fig. 162).

2.   *The pubo-femoral*—arising from the ilio-pubic junction to blend with the medial aspect of the capsule.

3.   *The ischio-femoral*—arising from the ischium to be inserted into the base of the greater trochanter.

Of these, the ilio-femoral is by far the strongest and resists hyperexten-sion strains on the hip. In posterior dislocation it usually remains intact.

The *synovium* of the hip covers the non-articular surfaces of the joint and occasionally bulges out anteriorly to form a bursa beneath the psoas tendon where this crosses the front of the joint.

*Movements.* The hip is capable of a wide range of movements—flexion, extension, abduction, adduction, rotation and circumduction.

The principal muscles acting on the joint are:

•flexors—ilio-psoas, assisted by rectus femoris
                          sartorius
                          pectineus
•extensors—gluteus maximus
                      the hamstrings
•adductors—adductor longus, brevis and magnus assisted by gracilis
                                                                      and pectineus
•abductors—gluteus medius and minimus
                        tensor fasciae latae

•lateral rotator—principally gluteus maximus
•medial rotators—principally ilio-psoas assisted by anterior fibres of
       gluteus medius and minimus

*Relations* (Fig. 161)
The hip joint is surrounded by muscles.
•Anteriorly—ilio-psoas and pectineus, together with the femoral artery
and vein.
•Laterally—tensor fasciae latae, gluteus medius and minimus.
•Posteriorly—the tendon of obturator internus with the gemelli, quadratus femoris, the sciatic nerve and, more superficially, gluteus maximus.
•Superiorly—the reflected head of rectus femoris lying in contact with
the joint capsule.
•inferiorly—the obturator externus, passing back to be inserted into the
trochanteric fossa.
   *Surgical exposure* of the hip joint therefore inevitably involves considerable and deep dissection.

*The lateral approach* comprises splitting down through the fibres of
tensor fasciae latae, gluteus medius and minimus onto the femoral neck.
Further access may be obtained by detaching the greater trochanter
with the gluteal insertions.
   *The anterior approach* passes between gluteus medius and minimus
laterally and sartorius medially, dividing the reflected head of rectus
femoris to expose the anterior aspect of the hip joint. More room may
be obtained by detaching these glutei from the external aspect of the
ilium.
*The posterior approach* is through an angled incision commencing at the
posterior superior iliac spine, passing to the greater trochanter and then
dropping vertically downwards from this point.
   Gluteus maximus is split in the line of its fibres and then incised along
its tendinous insertion. Gluteus medius and minimus are detached from
their insertions into the greater trochanter and an excellent view of the
hip joint thus obtained.

*Nerve supply*
*Hilton's law* states that the nerves crossing a joint supply the muscles
acting on it and the joint itself. The hip is no exception and receives
fibres from the femoral, sciatic and obturator nerves. It is important to
note that these nerves also supply the knee joint and, for this reason, it is
not uncommon for a patient, particularly a child, to complain bitterly of

pain in the knee and for the cause of the mischief, the diseased hip, to be overlooked.

## CLINICAL FEATURES

### Trendelenburg's test

The stability of the hip in the standing position depends on two factors, the strength of the surrounding muscles and the integrity of the lever system of the femoral neck and head within the intact hip joint. When standing on one leg, the abductors of the hip on this side (gluteus medius and minimus and tensor fasciae latae) come into powerful action to maintain fixation at the hip joint; so much so that the pelvis actually rises

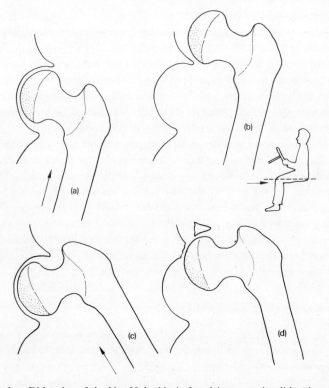

**Fig. 163.** Dislocation of the hip. If the hip is forced into posterior dislocation while adducted (a) there is no associated fracture of the posterior acetabular lip (b) Dislocation in the abducted position (c) can only occur with a concomitant acetabular fracture (d). (The manikin indicates the plane of these diagrams.)

slightly on the opposite side. If, however, there is any defect in these muscles or lever mechanism of the hip joint, the weight of the body in these circumstances forces the pelvis to tilt downwards on the opposite side.

This positive Trendelenburg test is seen if the hip abductors are paralysed (e.g. poliomyelitis), if there is an old unreduced or congenital dislocation of the hip, if the head of the femur has been destroyed by disease or removed operatively (pseudarthrosis), if there is an un-united fracture of the femoral neck or if there is a very severe degree of coxa vara.

The test may be said to indicate 'a defect in the osseo-muscular stability of the hip joint'.

A patient with any of the conditions enumerated above walks with a characteristic '*dipping gait*'.

### Dislocation of the hip (Fig. 163)

The hip is usually dislocated backwards and is produced by a force applied along the femoral shaft with the hip in the flexed position (e.g. the knee striking against the opposite seat when a train runs into the buffers). If the hip is also in the adducted position, the head of the femur is unsupported posteriorly by the acetabulum and dislocation can occur without an associated acetabular fracture. If the hip is abducted, dislocation must be accompanied by a fracture of the posterior acetabular lip.

The sciatic nerve, a close posterior relation of the hip, is in danger of damage in these injuries, as will be appreciated by a glance at Fig. 151.

Reduction of a dislocated hip is quite simple providing that a deep anaesthetic is used to relax the surrounding muscles; the hip is flexed, rotated into the neutral position and lifted back into the acetabulum. Occasionally forcible abduction of the hip will dislocate the hip forwards. Violent force along the shaft (e.g. a fall from a height) may thrust the femoral head through the floor of the acetabulum, producing a *central dislocation of the hip*.

### The knee joint (Figs. 164, 165)

The knee is a hinge joint made up of the articulations between the femoral and tibial condyles and between the patella and the patellar surface of the femur.

The *capsule* is attached to the margins of these articular surfaces but communicates above with the *suprapatellar bursa* (between the lower femoral shaft and the quadriceps), posteriorly with the bursa under the medial head of gastroenemius and often, through it, with the bursa under

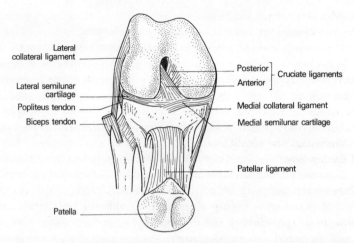

Fig. 164. The knee—anterior view; the knee is flexed and the patella has been turned downwards.

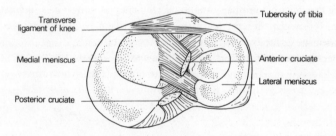

Fig. 165. The knee in transverse section.

semimembranosus. It may also communicate with the bursa under the lateral head of gastrocnemius. The capsule is also perforated posteriorly by popliteus, which emerges from it much in the same way that the long head of biceps bursts out of the shoulder joint.

The capsule of the knee joint is reinforced on each side by the *medial* and *lateral collateral ligaments*, the latter passing to the head of the fibula and lying free from the capsule.

Anteriorly the capsule is considerably strengthened by the *ligamentum patellae*, and, on each side of the patella, by the *medial* and *lateral patellar retinacula* which are expansions from vastus medialis and lateralis.

Posteriorly the tough *oblique ligament* arises as an expansion from the insertion of semimembranosus and blends with the joint capsule.

*Internal structures*

Within the joint are a number of important structures.

*The cruciate ligaments* are extremely strong connections between the tibia and femur. They arise from the anterior and posterior intercondylar areas of the superior aspect of the tibia, taking their names from their tibial origins, and pass upwards to attach to the intercondylar notch of the femur.

The anterior ligament resists hyperextension of the knee, the posterior becomes taut in hyperflexion.

*The semilunar cartilages(menisci)* are crescent-shaped and are triangular in cross-section, the medial being larger and less curved than the lateral. They are attached by their extremities to the tibial intercondylar area and by their periphery to the capsule of the joint, although the lateral cartilage is only laxly adherent and the popliteus tendon intervenes between it and the lateral collateral ligament.

They deepen, although to only a negligible extent, the articulations between the tibial and femoral condyles. If both menisci are removed the knee can regain complete functional efficiency although it is interesting that, following surgery, a rim of fibro-cartilage regenerates from the connective tissue margin of the excised menisci.

An *infrapatellar pad of fat* fills the space between the ligamentum patellae and the femoral intercondylar notch. The synovium covering this pad projects into the joint as two folds termed the *alar folds*.

*Movements of the knee*

The principal knee movements are flexion and extension, but note on yourself that rotation of the knee is possible when this joint is in the flexed position. In full extension, i.e. in the standing position, the knee is quite rigid because the medial condyle of the tibia, being rather larger than the lateral condyle, rides forward on the medial femoral condyle, thus 'screwing' the joint firmly together. The first step in flexion of the fully extended knee is 'unscrewing' or internal rotation. This is brought about by *popliteus* which arises from the lateral side of the lateral condyle of the femur, emerges from the joint capsule posteriorly and is inserted into the back of the upper end of the tibia.

The principal muscles acting on the knee are:
- extensor—quadriceps femoris
- flexors—hamstrings assisted by gracilis, gastrocnemius and sartorius
- medial rotator—popliteus.

CLINICAL FEATURES

1.   The stability of the knee depends upon the strength of its surrounding muscles and of its ligaments. Of the two, the muscles are by far the more important. Providing quadriceps femoris is powerfully developed, the knee will function satisfactorily even in the face of considerable ligamentous damage. Conversely, the most skilful surgical repair of torn ligaments is doomed to failure unless the muscles are functioning strongly; without their support, reconstructed ligaments will merely stretch once more.

2.   When considering soft tissue injuries of the knee joint, think of three Cs that may be damaged—the collateral ligaments, the cruciates and the cartilages.

*The collateral ligaments* are taut in full extension of the knee and are therefore only liable to injury in this position. The medial ligament may be partly or completely torn when a violent abduction strain is applied, whereas an adduction force may damage the lateral ligament. If one or other collateral ligament is completely torn the extended knee can be rocked away from the affected side.

*The cruciate ligaments* may both be torn (along with the collateral ligaments) in severe abduction or adduction injuries. The anterior cruciate, which is taut in extension, may be torn by violent hyperextension of the knee or in anterior dislocation of the tibia on the femur. The posterior cruciate tears in a posterior dislocation.

If both the cruciate ligaments are torn, unnatural antero-posterior mobility of the knee can be demonstrated.

If there is only increased forward mobility, the anterior cruciate ligament has been divided or is lax. Increased backward mobility implies a lesion of the posterior cruciate.

*The semilunar cartilages* can only tear when the knee is flexed and is thus able to rotate. If you place a finger on either side of the ligamentum patellae on the joint line and then rotate your flexed knee internally and then externally you will note how the lateral and medial cartilages are respectively sucked into the knee joint. If the flexed knee is forcibly abducted and externally rotated the medial cartilage will be drawn between, and then split by, the granding surfaces of the medial condyles of the femur and tibia. This occurs when a footballer twists his flexed knee while running or when a miner topples over in the crouched position while hewing coal in a narrow seam. A severe adduction and internal rotation strain may similarly tear the lateral cartilage but this injury is less common.

The knee 'locks' in this type of injury because the torn and displaced segment of cartilage lodges between the condyles and prevents full extension of the knee.

## The tibio-fibular joints

The tibia and fibula are connected by:

1. *the superior tibio-fibular joint*, a synovial joint between the head of the fibula and the lateral condyle of the tibia;

2. *the interosseous membrane*, which is crossed by the anterior tibial vessels above and pierced by the perforating branch of the peroneal artery below;

3. *the inferior tibio-fibular joint*, a fibrous joint between the triangular areas of each bone immediately above the ankle joint.

## The ankle (Fig. 166)

The ankle is a hinge joint between a mortice formed by the malleoli and lower end of the tibia and the body of the talus.

The *capsule* of the joint fits closely around its articular surfaces, and, as in every hinge joint, it is weak anteriorly and posteriorly but reinforced laterally and medially by *collateral ligaments*.

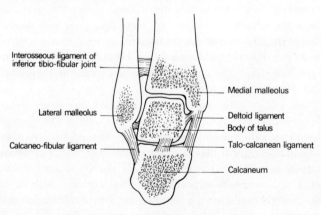

Fig. 166. The ankle in coronal section.

*Movements of the ankle*
The ankle joint is capable of being flexed and extended (plantar- and dorsi-flexion).

The body of the talus is slightly wider anteriorly and, in full extension, becomes firmly wedged between the malleoli. Conversely, in flexion, there is slight laxity at the joint and some degree of side to side tilting is possible: test this fact on yourself.

The principal muscles acting on the ankle are:

•dorsi-flexors—tibialis anterior assisted by extensor digitorum longus and extensor hallucis longus
•plantar flexors—gastrocnemius and soleus assisted by tibialis posterior, flexor hallucis longus and flexor digitorum longus.

CLINICAL FEATURES

1. The collateral ligaments of the ankle can be sprained or completely torn by forcible abduction or adduction, the lateral ligament being far the more frequently affected. If the ligament is completely disrupted the talus can be tilted in its mortice; this is difficult to demonstrate clinically and is best confirmed by taking an A.P. radiograph of the ankle while forcibly inverting the foot.

2. The most usual ankle fracture is that produced by an abduction-external rotation injury; the patient catches his foot in a rabbit hole, his body and his tibia internally rotate while the foot is rigidly held. First there is a torsional spiral fracture of the lateral malleolus, then avulsion of the medial collateral ligament, with or without avulsion of a flake of the medial malleolus and, finally, as the tibia is carried forwards, the posterior margin of the lower end of the tibia shears off against the talus. These stages are termed 1st, 2nd and 3rd degree Pott's fractures. Notice that, with widening of the joint, there is forward dislocation of the tibia on the talus, producing characteristic prominence of the heel in this injury.

## The joints of the foot

Inversion and eversion of the foot take place at the *talo-calcaneal* articulations and at the *mid-tarsal joints* between the calcaneum and the cuboid and between the talus and the navicular.

Loss of these rotatory movements of the foot, e.g. after injury or because of arthritis, results in quite severe disability because the foot cannot adapt itself to walking on rough or sloping ground.

Inversion is brought about by tibialis anterior and posterior assisted by the long extensor and flexor tendons of the hallux; eversion is the duty of the peronei.

The other tarsal joints allow slight gliding movements only and, individually, are not of clinical importance. The arrangement of the metacarpo-phalangeal and interphalangeal joints is on the same basic plan as in the upper limb.

## The arches of the foot

On standing the heel and the metatarsal heads are the principal weight-bearing points, but a moment's study of footprints on the wet bathroom floor will show that the lateral margin of the foot and the tips of the phalanges also touch the ground.

The bones of the foot are arranged in the form of two longitudinal arches. The *medial arch* comprises calcaneum, talus, navicular, the three cuneiforms and the three medial metatarsals; the apex of this arch is the talus. The *lateral arch*, which is lower, comprises the calcaneum, cuboid and the lateral two metatarsals.

The foot plays a double role; it functions as a rigid support for the weight of the body in the standing position, and as a mobile springboard during walking and running.

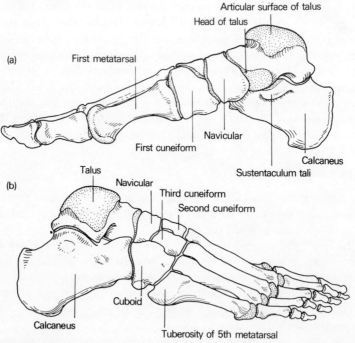

**Fig. 167.** The longitudinal arches of the right foot. (a) Medial view. (b) Lateral view.

When one stands, the arches sink somewhat under the body's weight, the individual bones lock together, the ligaments linking them are at maximum tension and the foot becomes an immobile pedestal. When one walks, the weight is released from the arches, which unlock and become a mobile lever-system in the spring-like actions of locomotion.

The arches are maintained:

1. By the shape of the interlocking bones;
2. By the ligaments of the foot;
3. By muscle action.

The ligaments concerned are (Fig. 168):

1. the dorsal, plantar and interosseous ligaments between the small bones of the forefoot;
2. the *spring ligament,* which passes from the *sustentaculum tali* of the calcaneum forward to the *tuberosity of the navicular* and which supports the inferior aspect of the head of the talus;
3. the *short plantar ligament* which stretches from the plantar surface of the calcaneum to the cuboid;
4. the *long plantar ligament* which arises from the plantar surface of the calcaneum, covers the short plantar ligament, forms a tunnel for peroneus longus tendon with the cuboid, and is inserted into the bases of the 2nd, 3rd and 4th metatarsals.

These ligaments are reinforced in their action by the *plantar fascia,*

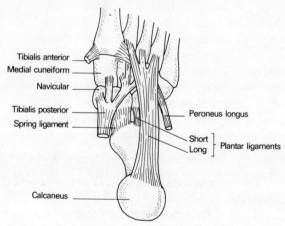

**Fig. 168.** Plantar aspect of the left foot to show the attachments of the important ligaments and long tendons.

which is the condensed deep fascia of the sole of the foot. This arises from the plantar aspect of the calcaneum and is attached to the deep transverse ligaments linking the heads of the metatarsals; it also continues forward into each toe to form the fibrous flexor sheaths, in a similar arrangement to that of the palmar fascia of the hand.

The principal muscles concerned in the mechanism of the arches of the foot are peroneus longus, tibialis anterior and posterior, flexor hallucis longus and the intrinsic muscles of the foot.

*Peroneus longus* tendon passes obliquely across the sole in a groove on the cuboid bone and is inserted into the lateral side of the base of the 1st metatarsal and the medial cuneiform. Into the medial aspect of these two bones is inserted the tendon of *tibialis anterior* so that these muscles form, in effect, a stirrup between them which supports the arches of the foot.

The medial arch is further reinforced by flexor hallucis longus, whose tendon passes under the sustentaculum tali of the calcaneum, and by *tibialis posterior*, two-thirds of whose fibres are inserted into the tuberosity of the navicular and support the spring ligament.

The longitudinally-running intrinsic muscles of the foot also act as ties to the longitudinal arches.

In the process of walking, the heel· is raised from the ground, the metataraso-phalangeal joints flex to give a 'push off' movement; the foot then leaves the ground completely and is dorsi-flexed to clear the toes.

Just before the toes of one foot leave the ground, the heel of the other makes contact.

Forward progression is produced partly by the 'push off' of the toes, partly by powerful plantar-flexion of the ankle and partly by the forward swing of the hips accentuated by swinging movements of the pelvis. Paraplegics can be taught to walk purely by this pelvic swing action, even though paralysed from the waist downwards.

When one foot is off the ground, dropping of the pelvis to the unsupported side is prevented by the hip abductors (gluteus medius and minimus and tensor fasciae latae). Their paralysis is one cause of a 'dipping gait' and of a positive Trendelenburg sign (*see* p. 244).

# Three Important Zones of the Lower Limb— The Femoral Triangle, the Adductor Canal and the Popliteal Fossa

**The femoral triangle** (Fig. 169)

This triangle is bounded:

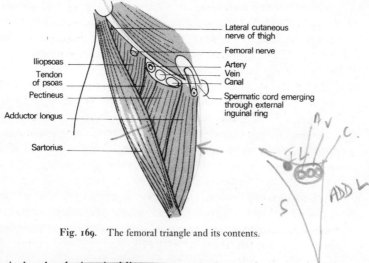

Fig. 169. The femoral triangle and its contents.

•superiorly—by the inguinal ligament,
•medially—by the medial border of adductor longus,
•laterally—by the medial border of sartorius.

Its *floor* consists of iliacus, the tendon of psoas, pectineus and adductor longus.

The *roof* is formed by the superficial fascia, containing the superficial inguinal lymph nodes and the saphenous vein with its tributaries, and the deep fascia (fascia lata) which is pierced by the saphenous vein at the saphenous opening.

The *contents* of the triangle are the femoral vein, artery and nerve together with the deep inguinal glands.

Some of these structures must now be considered in greater detail.

### The fascia lata

The deep fascia of the thigh, or fascia lata, extends downwards to ensheath the whole lower limb except over the subcutaneous surface of the tibia (to whose margins it adheres), and at the saphenous opening. Above it is attached all around to the root of the lower limb; that is to say, to the inguinal ligament, pubis, ischium, sacro-tuberous ligament, sacrum and coccyx and the iliac crest. The fascia of the thigh is particularly dense laterally (the *ilio-tibial tract*), where it receives tensor fasciae latae, and posteriorly where the greater part of gluteus maximus is inserted into it.

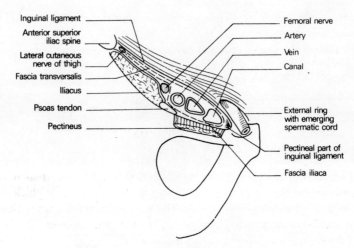

**Fig. 170.** The femoral canal and its surrounds.

The tough lateral fascia of the thigh is an excellent source of this material for hernia and dural repairs.

### The femoral sheath and femoral canal (Fig. 170)

The femoral artery and vein enter the femoral triangle from beneath the inguinal ligament within a fascial tube termed the *femoral sheath*. This is derived from the extraperitoneal intra-abdominal fascia, its anterior wall arising from the transversalis fascia and its posterior wall from the fascia covering the iliacus.

The medial part of the femoral sheath contains a small, almost vertically-placed gap, the *femoral canal*, which is about $\frac{1}{2}$ in (12 mm) in length and which just admits the tip of the little finger. Because of the greater width of the female pelvis, the canal is somewhat larger in the female and femoral herniae are, in consequence, commoner in this sex.

The boundaries of the femoral canal are:
- anteriorly—the inguinal ligament,
- medially—the sharp edge of the pectineal part of the inguinal ligament (Gimbernat's ligament),
- laterally—the femoral vein,
- posteriorly—the pectineal ligament (of Astley Cooper) which is the thickened periosteum along the pectineal border of the superior pubic

ramus and which continues medially with the pectineal part of the inguinal ligament.

The canal contains a plug of fat and a constant lymph node—the *node of the femoral canal* or Cloquet's gland.

The canal has two functions; first as a dead space for expansion of the distended femoral vein and second as a lymphatic pathway from the lower limb to the external iliac nodes.

### Femoral hernia

The great importance of the femoral canal is, of course, that it is a potential point of weakness in the abdominal wall through which may develop a femoral hernia. Unlike the indirect inguinal hernia, this is never due to a congenital sac and, although cases do occur rarely in children, it is never found in the newborn.

As the hernia sac enlarges, it emerges through the saphenous opening then turns upwards along the pathway presented by the superficial epigastric and superficial circumflex iliac vessels so that it may come to project above the inguinal ligament. There should not, however, be any difficulty in differentiating between an irreducible femoral and inguinal hernia; the neck of the former must always lie below and lateral to the pubic tubercle whereas the sac of the latter extends above and medial to this landmark (Fig. 171).

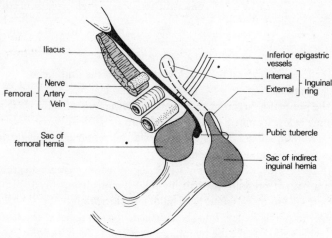

**Fig. 171.** The relationship of an indirect inguinal and a femoral hernia to the pubic tubercle; the inguinal hernia emerges above and medial to the tubercle, the femoral hernia lies below and lateral to it.

The neck of the femoral canal is narrow and bears a particularly sharp medial border; for this reason, irreducibility and strangulation occur more commonly at this site than at any other. In order to enlarge the opening of the canal at operation on a strangulated case, this sharp edge of Gimbernat's ligament may require incision; there is a slight risk of damage to the abnormal obturator artery in this manoeuvre and it is safer to enlarge the opening by making several small nicks into the ligament.

*Note*
Normally there is an anastomosis between the pubic branch of the inferior epigastric artery and the pubic branch of the obturator artery. Occasionally the obturator artery is entirely replaced by this branch from the inferior epigastric— the *abnormal abturator artery*. This aberrent vessel usually passes laterally to the femoral canal and is out of harm's way; more rarely it passes behind Gimbernat's ligament and it is then in surgical danger.

## The lymph nodes of the groin and the lymphatic drainage of the lower limb

The lymph nodes of the groin are arranges in a superficial and a deep group. The *superficial nodes* lie in two chains, a *longitudinal chain* along the saphenous vein, receiving the bulk of the superficial lymph drainage of the lower limb, and a *horizontal chain*, just distal to the inguinal ligament. These horizontal nodes receive lymphatics from the skin and superficial tissues of:

1. the lower trunk and back, below the level of the umbilicus,
2. the buttock,
3. the perineum, scrotum and penis (or lower vagina and vulva) and the anus below its muco-cutaneous junction.

In addition, some lymphatics drain via the round ligament to these nodes from the fundus of the uterus.

(All these sites, as well as the whole leg, must be examined carefully when a patient presents with an inguinal lymphadenopathy.)

The two groups of superficial nodes drain through the saphenous opening in the fascia lata into the *deep nodes* lying medial to the femoral vein, which also receive the lymph drainage from the tissues of the lower limb beneath the deep fascia. In addition, a small area of skin over the heel and lateral side of the foot drains by lymphatics along the short saphenous vein to nodes in the popliteal fossa and then, along the femoral vessels, directly to the deep nodes at the groin.

The deep groin nodes drain to the external iliac nodes by lymphatics which travel partly in front of the femoral artery and vein and partly through the femoral canal.

CLINICAL FEATURES

1.   Minor sepsis and abrasions of the leg are so common that it is usual to find that the inguinal nodes are palpable in perfectly healthy people.

2.   Secondary involvement of the inguinal nodes by malignant deposits may be dealth with by *block dissection of the groin*. This involves removal of the superficial and deep fascial roof of the femoral triangle, the saphenous vein and its tributaries and the fatty and lymphatic contents of the triangle, leaving only the femoral artery, vein and nerve. The inguinal ligament is detached so that, in addition, an extra-peritoneal removal of the external iliac nodes can be carried out.

3.   In making a differential diagnosis of a lump in the femoral triangle, think of each anatomical structure and of the pathological conditions to which it may give rise, thus:

•skin and soft tissues—lipoma, sebaceous cyst, sarcoma;
•artery—aneurysm of the femoral artery;
•vein—varix of the saphenous vein;
•nerve—neuroma of the femoral nerve or its branches;
•femoral canal—femoral hernia;
•psoas sheath—psoas abscess;
•lymph nodes—any of the causes of lymphadenopathy.

## The adductor canal (of Hunter)-or subsartorial canal (Fig. 172)

This canal leads on from the apex of the femoral triangle. Its boundaries are:

•posteriorly—adductor longus and magnus.
•antero-laterally—vastus medialis.
•antero-medially—the sartorius, which lies on a fascial sheet forming the roof of the canal.

The contents of the canal are the femoral artery, the femoral vein (which lies behind the artery) and the saphenous nerve.

John Hunter described the exposure and ligation of the femoral artery in this canal for aneurysm of the popliteal artery; this method has the advantage that the artery at this site is healthy and will not tear when tied, as may happen if ligation is attempted immediately above the aneurysm.

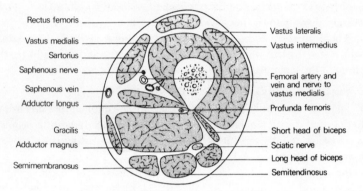

Rectus femoris

Vastus medialis

Sartorius

Saphenous nerve

Saphenous vein

Adductor longus

Gracilis

Adductor magnus

Semimembranosus

Vastus lateralis

Vastus intermedius

Femoral artery and vein and nerve to vastus medialis

Profunda femoris

Short head of biceps

Sciatic nerve

Long head of biceps

Semitendinosus

Fig. 172. Cross-section through the thigh in the region of the adductor, or subsartorial, canal of Hunter.

## The popliteal fossa (Fig. 173)

The popliteal fossa is the distal continuation of the adductor canal. This 'fossa' is, in fact, a closely-packed compartment which only becomes the rhomboid-shaped space of anatomical diagrams when opened up at operation or by dissection.

Its *boundaries* are:

• supero-laterally—biceps tendon.
• supero-medially—semimembranosus reinforced by semitendinosus.
• infero-medially and infero-laterally—the medial and lateral heads of gastrocnemius.

The *roof* of the fossa is deep fascia which is pierced by the short saphenous vein as this enters the popliteal vein.

Its *floor*, from above down, is formed by:

• the popliteal surface of the femur.
• the posterior aspect of the knee joint, and
• the popliteus muscle covering the upper posterior surface of the tibia.

From without in, the popliteal fossa contains the popliteal nerves, vein and artery.

The *common peroneal (lateral popliteal) nerve* passes out of the fossa along the medial border of the biceps tendon; the *tibial (medial popliteal) nerve* is first lateral to the popliteal vessels and then crosses superficially to these vessels to lie on their medial side.

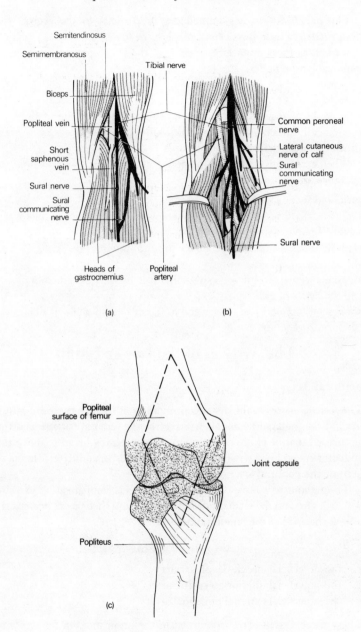

**Fig. 173.** The popliteal fossa. (a) Superficial dissection. (b) Deep dissection. (c) Floor.

The *popliteal vein* lies immediately superficial to the artery; the *popliteal artery* itself lies deepest of all in the fossa.

As well as these important structures, the fossa contains fat and the popliteal lymph nodes.

CLINICAL FEATURES

The popliteal fossa is another good example of the value of thinking anatomically when considering the differential diagnosis of a mass situated in a particular anatomical area.

When examining a lump in the popliteal region, let these possibilities pass through your mind:

- skin and soft tissues—lipoma, sarcoma;
- vein—varicosities of the short saphenous vein in the roof of the fossa;
- artery—popliteal aneurysm;
- lymph nodes—infection secondary to suppuration in the foot;
- knee joint—joint effusion;
- tendons—enlarged bursae, especially those beneath semimembranosus and the heads of gastrocnemius;
- bones—a tumour of the lower end of femur or upper end of tibia.

# The Arteries of the Lower Limb

## Femoral Artery

The *femoral artery* is the distal continuation of the external iliac artery beyond the inguinal ligament. It traverses the femoral triangle and the adductor canal of Hunter then terminates a hand's-breadth above the adductor tubercle by passing through the hiatus in adductor magnus to become the popliteal artery (Fig. 148).

Throughout its course the femoral artery is accompanied by its vein, which lies first on the medial side of the artery and then passes posteriorly to it at the apex of the femoral triangle.

*Branches.* In the groin, the femoral artery gives off:

1. the superficial circumflex iliac artery;
2. the superficial inferior epigastric artery;
3. the superficial external pudendal artery.

These three vessels are encountered in the groin incision for repair of an inguinal hernia.

The *profunda femoris* arises from the femoral artery 2 in (5 cm) distal to the inguinal ligament. It is conventional to call the femoral artery above this branch the *common femoral*, and below it, the *superficial femoral artery*.

The profunda passes deep to adductor longus and gives off *medial* and *lateral circumflex branches* and 4 *perforating branches*. These are important both as the source of blood supply to the great muscles of the thigh and as collateral channels which link the rich arterial anastomoses around the hip and the knee.

CLINICAL FEATURES

1. Recapitulate the surface markings of the femoral artery—the upper two-thirds of a line connecting the mid-inguinal point with the adductor tubercle, the hip being held somewhat flexed and externally rotated (Fig. 148).

The femoral artery in the upper 4 in (10 cm) of its course lies in the femoral triangle where it is quite superficial and, in consequence, easily injured. A laceration of the femoral artery at this site is an occupational hazard of butchers and bullfighters.

2. The femoral artery at the groin is readily punctured by a hypodermic needle and is the most convenient site from which to obtain arterial blood samples. Arteriography of the peripheral leg vessels is also easily performed at this point. A Seldinger catheter can be passed proximally through a femoral artery puncture in order to carry out aortography or selective renal, coeliac and mesenteric angiography.

3. Arteriosclerotic changes, with consequent thrombotic arterial occlusion, frequently commence at the lower end of the femoral artery, perhaps as a result of compression of the diseased vessel by the margins of the hiatus in adductor magnus. Collateral circulation is maintained via anastomoses between the branches of profunda femoris and the popliteal artery. If degenerative changes are slight above and below the femoral block, it is possible to by-pass the occluded segment by means of a graft between the common femoral and popliteal arteries.

## Popliteal artery

*The popliteal artery* continues on from the femoral artery at the adductor hiatus and terminates at the lower border of the popliteus muscle. It lies deep within the popliteal fossa (see above), being covered superficially by the popliteal vein and, more superficially still, crossed by the tibial (medial popliteal) nerve.

The popliteal artery gives off *muscular* branches, *geniculate* branches (to the knee joint) and terminal branches, the *anterior* and *posterior tibial arteries*.

CLINICAL FEATURES

1. Aneurysm of the popliteal artery, once common, is now rare. Its frequency in former days was associated with the repeated traumata of horse-riding and the wearing of high riding-boots.

Pressure of the aneurysm on the adjacent vein may cause venous thrombosis and peripheral oedema; pressure on the tibial nerve may cause severe pain in the leg.

2. The popliteal artery is exposed by deep dissection in the mid-line within the popliteal fossa, care being taken not to injure the more super-ficial vein and nerve.

## Posterior tibial artery

*The posterior tibial artery* is the larger of the terminal branches of the popliteal artery. It descends deep to soleus, where it can be exposed by splitting gastocnemius and soleus in the mid-line, then becomes super-ficial in the lower third of the leg and passes behind the medial malleolus between the tendons of flexor digitorum longus and flexor hallucis longus. It is accompanied by its corresponding vein and by the posterior tibial nerve (Fig. 174).

Below the ankle, the posterior tibial artery divides into the *medial* and *lateral plantar arteries* which constitute the principal blood supply to the foot.

As well as branches to muscles and skin and a large nutrient branch to the tibia, the posterior tibial artery gives off the *peroneal artery* about $\frac{1}{2}$ in (4 cm) from its origin. The peroneal artery runs down the posterior aspect of the fibula, close to the medial margin of the bone, supplying adjacent muscles and giving a nutrient branch to the fibula. Above the ankle it gives off its *perforating branch* which pierces the interosseous membrane, descends over the lateral malleolus and anastomoses with the arteries of the dorsum of the foot.

## Anterior tibial artery

*The anterior tibial artery* arises at the bifurcation of the popliteal artery. It passes forwards between the tibia and fibula over the upper margin of

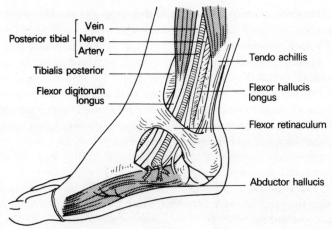

Posterior tibial — Vein
Nerve
Artery

Tibialis posterior

Flexor digitorum
longus

Tendo achillis

Flexor hallucis
longus

Flexor retinaculum

Abductor hallucis

Fig. 174. The relations of the posterior tibial nerve as it passes behind the medial malleolus.

the interosseous membrane and descends on this structure in the anterior compartment of the leg.

At first deeply buried, it becomes superficial just above the ankle between the tendons of extensor hallucis longus and tibialis anterior, being crossed superficially by the former immediately proximal to the line of the ankle joint.

The artery continues over the dorsum of the foot as the *dorsalis pedis*; this gives off the *arcuate artery* which, in turn, supplies cutaneous branches to the backs of the toes. Dorsalis pedis itself plunges between the 1st and 2nd metatarsals to join the lateral plantar artery in the formation of the *plantar arch*, from which branches run forwards to supply the plantar aspects of the toes.

# The Veins of the Lower Limb

The veins of the lower limb are divided into the deep and superficial groups according to their relationship to the investing deep fascia of the leg. The *deep veins* accompany the corresponding major arteries. The *superficial veins* are the *long* and *short saphenous veins* and their tributaries (Fig. 175).

*The short saphenous vein* commences at the ankle behind the lateral malleolus where it drains the lateral side of the dorsal venous plexus of the foot. It courses over the back of the calf, perforates the deep fascia over the popliteal fossa and terminates in the popliteal vein. One or more branches run upwards and medially from it to join the long saphenous vein.

*The long saphenous vein* drains the medial part of the venous plexus on the dorsum of the foot and passes upwards immediately in front of the medial malleolus (Fig. 149); here branches of the saphenous nerve lie in front of and behind the vein. The vein then ascends over the posterior parts of the medial condyles of the tibia and femur to the groin where it pierces the deep fascia at the saphenous opening 1 in (2.5 cm) below the inguinal ligament, to enter the femoral vein.

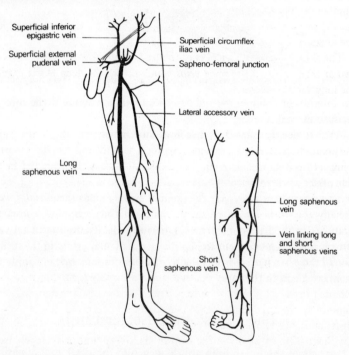

Superficial inferior epigastric vein

Superficial external pudenal vein

Superficial circumflex iliac vein

Sapheno-femoral junction

Lateral accessory vein

Long saphenous vein

Long saphenous vein

Vein linking long and short saphenous veins

Short saphenous vein

Fig. 175. The superficial veins of the lower limb.

The long saphenous vein is joined by one or more branches from the short saphenous, and by the *lateral accessory vein* which usually enters the main vein at the mid-thigh, although it may not do so until the saphenous opening is reached.

At the groin a number of tributaries from the lower abdominal wall, thigh and scrotum enter the saphenous vein; these tributaries are variable in number and arrangement but usually comprise (Fig. 175):

1. the superficial epigastric vein;
2. the superficial circumflex iliac vein;
3. the superficial external pudendal vein.

The superficial epigastric vein communicates with the lateral thoracic tributary of the axillary vein via the *thoraco-epigastric vein*. This dilates (and may become readily visible coursing over the trunk), following obstruction of the inferior vena cava. The long saphenous vein communicates with the deep venous system not only at the groin but also at a number of points along its course through *perforating veins*; one is usually present a hand's-breadth above, another a hand's-breadth below the knee.

The skin of the medial aspect of the leg is drained to the deep veins by two or three *direct perforating veins* which pierce the deep fascia behind the long saphenous vein.

CLINICAL FEATURES

1. We have already noted (under surface anatomy of the lower limb) the great importance of the constant position of the long saphenous vein lying immediately in front of the medial malleolus. Knowledge that a vein *must* be present at this site, even if not visible in an obese or collapsed patient, may be life-saving when urgent transfusion is required. Occasionally the immediately adjacent saphenous nerve is caught up by a ligature during this procedure—the patient, if conscious, will complain bitterly of pain if this is done.

2. The saphenous veins frequently become dilated, incompetent and varicose. Usually this is idiopathic but may result from the increased venous pressure caused by more proximal venous obstruction (a pelvic tumour or the pregnant uterus, for example) or may be secondary to obstruction of the deep venous pathway of the leg by thrombosis.

3. Stagnation of blood in the skin of the lower limb may result from venous thrombosis or valve incompetence; the skin, in consequence, is poorly nourished and easily breaks down into a *varicose ulcer* if subjected to even minor trauma. This is especially liable to occur over the subcutaneous antero-medial surface of the tibia where the cutaneous blood supply is least generous.

4. In operating upon varicose veins it is important that all tributaries at the groin are ligated as well as the main saphenous trunk; if one tributary escapes, it in turn becomes dilated and produces recurrence of the varices.

# The Course and Distribution of the Principal Nerves of the Lower Limb

The nerves of the lower limb are derived from the lumbar and sacral plexuses.

### The lumbar plexus (Fig. 176)

The lumbar plexus originates from the anterior primary rami of L1-4. The trunks of the plexus traverse psoas major and emerge from its lateral border. There are two exceptions: the obturator nerve appears at the medial border of psoas tendon, and the genito-femoral nerve emerges on the anterior aspect of the muscle.

The principal branches of the plexus are the *femoral nerve* and the *obturator nerve*.

*The femoral nerve* (L2-4) passes through the substance of psoas then under the inguinal ligament a finger's-breadth lateral to the femoral artery, to break up into its terminal branches after a course in lower limb of only some 2 in (5 cm).

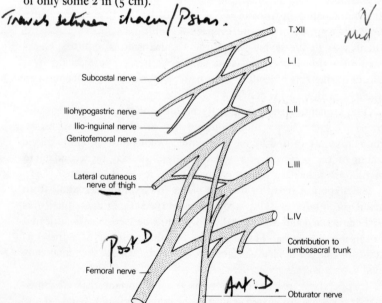

Fig. 176. Plan of the lumbar plexus (muscular branches have been omitted for clarity.)

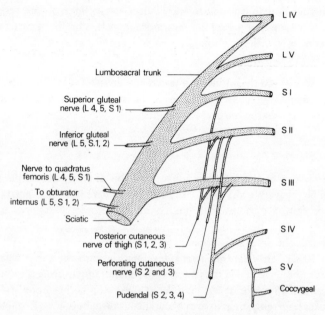

Fig. 177. Plan of the sacral plexus.

Its *branches* are:

• muscular—to the anterior compartment of the thigh (quadriceps, sartorius and pectineus);

• cutaneous—the medial and intermediate cutaneous nerves of the thigh and the *saphenous nerve*, which traverses the adductor canal to supply the skin of the medial side of the leg, ankle and foot to the great toe;

• articular—to the hip and knee joints.

The femoral nerve supplies the skin of the medial and anterior aspects of the thigh via its medial and intermediate cutaneous branches, but the lateral aspect is supplied by the *lateral cutaneous nerve of the thigh*. This arises directly from the lumbar plexus and enters the thigh usually by passing deep to the inguinal ligament. Occasionally the nerve pierces the ligament and may then be pressed upon by it with resultant pain and anaesthesia over the upper outer thigh (*meralgia paraesthetica*). This is relieved by dividing the deeper fasciculus of the inguinal ligament where the nerve passes over it.

*The obturator nerve* (L2-4) emerges from the medial aspect of the psoas

*Lateral to { I.I vessels. Sup. to Artery ovary.*

270 *The Lower Limb*

and runs downwards and forwards, deep to the internal iliac vessels, to reach the superior part of the obturator foramen. This the nerve traverses, in company with the obturator vessels, to enter the thigh.

Its *branches* are:

- •muscular—to the adductor muscles and gracilis;
- •cutaneous—to an area of skin over the medial aspect of the thigh;
- •articular—to the hip and knee joints.


### CLINICAL FEATURES

(1) Spasm of the adductor muscles of the thigh in spastic paraplegia can be relieved by division of the obturator nerve (*obturator neurectomy*). This can be performed through a mid-line lower abdominal incision exposing the nerve trunk extraperitoneally on each side as it passes towards the obturator foramen.

(2) Rarely an *obturator hernia* develops through the canal where the obturator nerve and vessels traverse the membrane covering the obturator foramen. Pressure of a strangulated obturator hernia upon the nerve causes referred pain in its area of cutaneous distribution, so that intestinal obstruction associated with pain along the medial side of the thigh should suggest this diagnosis.

(3) The femoral and obturator nerves, as well as the sciatic nerve and its branches, supply sensory fibres to both the hip and the knee; it is not uncommon for hip disease to present disguised as pain in the knee.

**The sacral plexus** (Fig. 177)
This plexus originates from the anterior primary rami of L4-5, S1-4. Note that L4 is shared by both plexuses, a branch from it joining L5 to form the *lumbo-sacral trunk* which carries its contribution to the sacral plexus.

The sacral nerves emerge from the anterior sacral foramina and unite in front of piriformis where they are joined by the lumbo-sacral trunk.

Branches from the plexus supply:

- •the pelvic muscles;
- •the muscles of the hip;
- •the skin of the buttock and back of the thigh.

The plexus itself terminates as the *pudendal nerve* and the *sciatic nerve*.

*The pudendal nerve* (S2, 3, 4) provides the principal innervation of the perineum. It has a complex course, passing from the pelvis, briefly through the gluteal region, along the side-wall of the ischiorectal fossa and through the deep perineal pouch to end by supplying the skin of the external genitalia (Fig. 178).

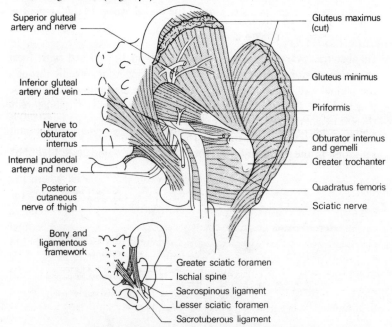

Superior gluteal artery and nerve

Inferior gluteal artery and vein

Nerve to obturator internus

Internal pudendal artery and nerve

Posterior cutaneous nerve of thigh

Bony and ligamentous framework

Gluteus maximus (cut)

Gluteus minimus

Piriformis

Obturator internus and gemelli

Greater trochanter

Quadratus femoris

Sciatic nerve

Greater sciatic foramen
Ischial spine
Sacrospinous ligament
Lesser sciatic foramen
Sacrotuberous ligament

Fig. 178. The boundaries and contents of the sciatic foramina.

It arises as the lower main division of the sacral plexus although it is dwarfed by the giant sciatic nerve. It leaves the pelvis through the greater foramen below the piriformis muscle. It crosses the dorsum of the ischial spine and immediately disappears through the lesser sciatic foramen into the perineum. The nerve now traverses the lateral wall of the ischiorectal fossa in company with the internal pudendal vessels, and lies within a distinct fascial compartment on the medial aspect of obturator internus termed the pudendal canal (Alcock's canal, see Fig. 97). Within the canal it first gives off the *inferior rectal nerve*, which crosses the fossa to innervate the external anal sphincter and the perianal skin, and then divides into the *perineal nerve* and the *dorsal nerve of the penis (or clitoris)*.

*The perineal nerve* is the larger of the two. It bifurcates almost at once; its deeper branch supplies the sphincter urethrae and the other muscles of the anterior perineum (the ischio-cavernosus, bulbospongiosus and the superficial and deep transverse perinei). Its more superficial branch innervates the skin of the posterior aspect of the scrotum.

*The dorsal nerve of the penis (or clitoris)* traverses the deep perineal pouch, pierces the perineal membrane and then penetrates the suspensory ligament of the penis to supply the dorsal aspect of this structure.

### CLINICAL FEATURES

In obstetric practice the pudendal nerve can be blocked with local anaesthetic prior to forceps delivery by inserting a long needle through the vaginal wall and guided by a finger to the ischial spine, which can be palpated per vaginam. When the procedure is carried out bilaterally there is loss of the anal reflex (which is a useful test that a successful block has been achieved), relaxation of the pelvic floor muscles and loss of sensation to the vulva and lower one third of the vagina (see Fig. 96b).

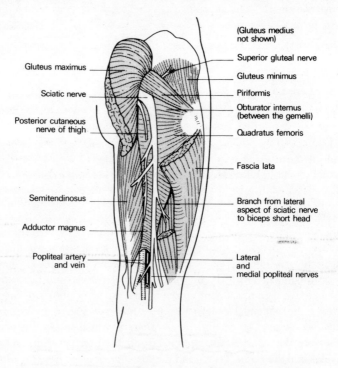

Gluteus maximus

Sciatic nerve

Posterior cutaneous nerve of thigh

Semitendinosus

Adductor magnus

Popliteal artery and vein

(Gluteus medius not shown)

Superior gluteal nerve

Gluteus minimus

Piriformis

Obturator internus (between the gemelli)

Quadratus femoris

Fascia lata

Branch from lateral aspect of sciatic nerve to biceps short head

Lateral and medial popliteal nerves

**Fig. 179.** Dissection of the sciatic nerve in the thigh and popliteal fossa.

*The sciatic nerve*

The sciatic nerve (L4, 5, S1–3) is the largest nerve in the body (Fig. 179) It is broad and flat at its origin, although peripherally it becomes rounded.

The nerve emerges from the greater sciatic foramen distal to piriformis and under cover of gluteus maximus, crosses the posterior surface of the ischium and descends on adductor magnus. Here it lies deep to the hamstrings and is crossed by the long head of biceps.

The sciatic nerve terminates by dividing into the *medial* and *lateral popliteal nerves* now renamed the *tibial* and *common peroneal* nerves respectively (*see* Fig. 173). The level of this division is variable—usually it is at the mid-thigh, but the popliteal nerves may be separate even at their origins from the sacral plexus.

*Branches.* The trunk of the sciatic nerve supplies the hamstring muscles (biceps, semimembranosus, semitendinosus) and also the adductor magnus—the latter being innervated also by the obturator nerve.

All the muscle branches apart from the one to the short head of biceps arise on the medial side of the nerve; its lateral border is therefore the side of relative safety in its operative exposure.

CLINICAL FEATURES

1. The sciatic nerve may be wounded in penetrating injuries or in posterior dislocation of the hip associated with fracture of the posterior lip of the acetabulum, to which the nerve is closely related.

Damage to the sciatic nerve is followed by paralysis of the hamstrings and all the muscles of the leg and foot (supplied by its distributing branches); there is loss of all movements in the lower limb below the knee joint with foot drop deformity. Sensory loss is complete below the knee, except for an area along the medial side of the leg, over the medial malleolus and down to the hallux, which is innervated by the saphenous branch of the femoral nerve.

2. The sciatic nerve is accompanied by a companion artery (derived from the inferior gluteal artery) which bleeds quite sharply when the nerve is divided during an above-knee amputation. The artery must be neatly isolated and tied without any nerve fibres being incorporated in the ligature, since this would be followed by severe pain in the stump.

*The tibial nerve*

The tibial (*medial popliteal*) nerve (L4, 5, S1–3) is the larger of the two terminal branches of the sciatic nerve; it traverses the popliteal fossa superficial to the popliteal vein and artery, which it crosses from the lateral to the medial side.

Its *branches* in the popliteal fossa are:

•muscular—to gastrocnemius, soleus and popliteus.

•cutaneous—the *sural nerve* which descends over the back of the calf, behind the lateral malleolus to the 5th toe. It receives a communicating branch from the lateral popliteal nerve and supplies the lateral side of the leg, foot and 5th toe.

•articular—to the knee joint.

It then descends deep to soleus, in company with the posterior tibial vessels, passes on their lateral side behind the medial malleolus to end by dividing into the *medial* and *lateral plantar nerves*.

*Branches in the leg.* The tibial nerve supplies flexor hallucis longus, flexor digitorum longus and tibialis posterior. Its terminal plantar branches supply the intrinsic muscles and skin of the sole of the foot, the medial plantar nerve having an equivalent distribution to that of the median nerve in the hand, the lateral plantar nerve being comparable to the ulnar nerve.

### The common peroneal nerve

*The common peroneal (lateral popliteal) nerve* (L4, 5, S1, 2) is the smaller of the terminal branches of the sciatic nerve. It enters the upper part of the popliteal fossa, passes along the medial border of the biceps tendon then curves around the neck of the fibula where it lies in the substance of peroneus longus and divides into its terminal branches, the *deep peroneal (anterior tibial)* and *superficial peroneal (musculo-cutaneous) nerves* (Fig. 150).

*Branches.* While still in the popliteal fossa, the common peroneal nerve gives off the *lateral cutaneous nerve of the calf*, a *sural communicating branch* and twigs to the knee joint but has no muscular branches.

*The deep peroneal nerve* pierces extensor digitorum longus then descends, in company with the anterior tibial vessels, over the interosseous membrane and then over the ankle joint.

Its *branches* are:

•muscular—to the muscles of the anterior compartment of the leg— extensor digitorum longus, extensor hallucis longus, tibialis anterior, peroneus tertius—and extensor digitorum brevis.

•cutaneous—to a small area of skin in the web between the 1st and 2nd toes.

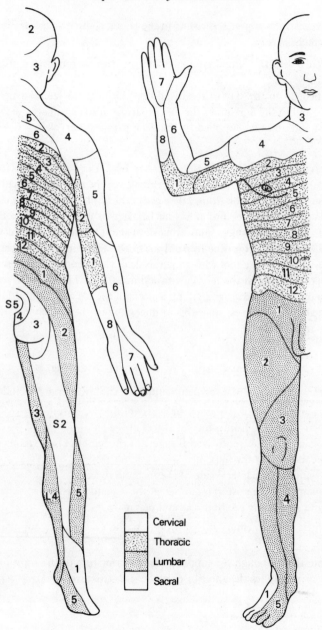

**Fig. 180.** The segmental cutaneous nerve supply of the skin.

*The superficial peroneal nerve* runs in the lateral compartment of the leg. Its branches are:

•muscular—to the lateral compartment muscles (peroneus longus and brevis).
•cutaneous—to the skin of the distal two-thirds of the lateral aspect of the leg and to the dorsum of the foot (apart from the small area between the 1st and 2nd toes supplied by the deep peroneal nerve).

### CLINICAL FEATURES

The common peroneal (lateral popliteal) nerve is in a particularly vulnerable position as it winds around the neck of the fibula. It may be damaged at this site by the pressure of a tight bandage or plaster cast or may be torn in severe adduction injuries to the knee. Damage to this nerve is followed by foot drop (due to paralysis of the ankle and foot extensors) and inversion of the foot (due to paralysis of the peroneal muscles with unopposed action of the foot flexors and invertors). There is also anaesthesia over the anterior and lateral aspects of the leg and foot, although the medial side escapes, since this is innervated by the saphenous branch of the femoral nerve.

### Segmental cutaneous supply of the lower limb (Fig. 180)

The arrangement of root segments supplying the lower limb is as follows:

•L1, 2 and 3—supply the anterior aspect of the thigh from above down.
•L4—supplies the fronto-medial aspect of the leg.
•L5—supplies the fronto-lateral aspect of the leg but also extends onto the medial side of the foot.
•S1—supplies the lateral side of the foot and the sole.
•S2—supplies the posterior surface of the leg and thigh.
•S3 and 4—supply the buttocks and perianal region.

A little aid to memory is that 5 supplies the 1st toe and 1 supplies the 5th.

Note that although S3 supplies the posterior part of the scrotum (or vulva) L1 supplies the anterior part of these structures.

# PART V
## THE HEAD AND NECK

§5 anus.
L, ant. vulva/scrotum
S3 post.

# The Surface Anatomy of the Neck

In the mid-line, from above down, can be felt (Fig. 181):

1. the *hyoid bone*—at the level of C3;
2. the notch of the *thyroid cartilage*—at the level of C4;
3. the *cricoid cartilage*—terminating in the trachea at C6;
4. the rings of the *trachea*, over the third and fourth of which can be rolled the *isthmus of the thyroid* gland;
5. the *suprasternal notch*.

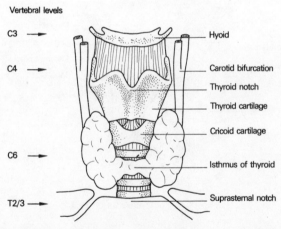

Fig. 181. Structures palpable on the anterior aspect of the neck, together with their corresponding vertebral levels.

Note that the lower border of the cricoid is an important level in the neck; it corresponds not only to the level of the 6th cervical vertebra but also to:

1. The junction of the larynx with the trachea.
2. The junction of the pharynx with the oesophagus.
3. The level at which the inferior thyroid artery and the middle thyroid vein enter the thyroid gland.

4. The level at which the vertebral artery enters the transverse foramen in the 6th cervical vertebra.

5. The level at which the superior belly of the omohyoid crosses the carotid sheath.

6. the level of the middle cervical sympathetic ganglion.

7. The site at which the carotid artery can be compressed against the transverse process of C6. (The carotid tubercle.)

By pressing the jaw laterally against the resistance of one's hand the opposite *sternomastoid* is tensed. This muscle helps define the *posterior triangle* of the neck, bounded by sternomastoid, trapezius and the clavicle, and the *anterior triangle*, defined by sternomastoid, the mandible and the mid-line (Fig. 182).

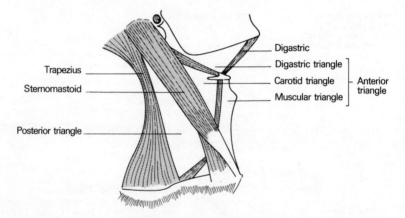

**Fig. 182.** The triangles of the neck.

Violently clench the jaws; the *platysma* then comes into view as a sheet of muscle, passing from the mandible down over the clavicles, lying in the superficial fascia of the neck. The *external jugular vein* lies immediately deep to platysma, crosses the sternomastoid into the posterior triangle, perforates the deep fascia just above the clavicle and enters the subclavian vein. It is readily visible in a thin subject on straining and is seen from the audience when a singer hits a sustained high note or when an orthopaedic surgeon reduces a fracture.

The *common carotid* artery pulse can be felt by pressing backwards against the long anterior tubercle of the transverse process of C6. The line of the carotid sheath can be marked out by a line joining a point mid-

way between the tip of the mastoid process and the angle of the jaw to the
sterno-clavicular joint. Along this line, the carotid bifurcates into the
*external and internal carotid* arteries at the level of the upper border of the
thyroid cartilage; at this level the vessels lie just below the deep fascia
where their pulsation is palpable and often visible.

## The fascial compartments of the neck (Fig. 183)

The fascial planes of the neck are of considerable importance to the surgeon; they form convenient lines of cleavage through which he may
separate the tissues in operative dissections and they delimit the spread
of pus in neck infections.

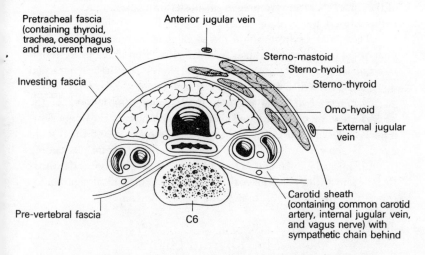

**Fig. 183.** Transverse section of the neck through C6—showing the fascial planes and also
the contents of the pretracheal fascia (or 'visceral compartment of the neck').

The *superficial fascia* is a thin fatty membrane enclosing the platysma.
The *deep fascia* can be divided into three layers.
1. *The enveloping fascia* which invests the muscles of the neck. It is
attached to all the bony landmarks at the upper and lower margin of the
neck; above to the mandible, zygomatic arch, mastoid process and
superior nuchal line; below to the manubrium, clavicle, acromion and
scapular spine. Posteriorly the ligamentum nuchae provides a longitudinal line of attachment for it.

This enveloping fascia splits to enclose the trapezius, the sterno-mastoid, the strap muscles and the parotid and submandibular glands.

The external jugular vein pierces the deep fascia above the clavicle. If the vein is divided here it is held open by the deep fascia which is attached to its margins, air is sucked into the vein lumen during inspiration and a fatal air embolism may ensue.

2. *The prevertebral fascia* passes across the vertebrae and prevertebral muscles behind the oesophagus, the pharynx and the great vessels. Above it is attached to the base of the skull. Laterally, the fascia covers the scalene muscles, and the emerging brachial plexus and subclavian artery. These structures carry with them a sheath formed from the prevertebral fascia, which becomes the axillary sheath.

Inferiorly, the fascia blends with the anterior longitudinal ligament of the upper thoracic vertebrae in the posterior mediastinum.

Pus from a tuberculous cervical vertebra bulges behind this dense fascial layer and may form a mid-line swelling in the posterior wall of the pharynx. The abscess may then track laterally, deep to the prevertebral fascia, to a point behind the sternomastoid. Rarely, pus has even tracked down along the axillary sheath into the arm.

3. *The pretracheal fascia* encloses the 'visceral compartment of the neck'. Extending from the hyoid above to the fibrous pericardium below, it encloses larynx and trachea, pharynx and oesophagus and the thyroid gland. A separate tube of fascia forms the *carotid sheath*, containing carotid, internal jugular and vagus nerve and bearing the cervical sympathetic chain in its posterior wall. (Some points of clinical significance concerning this fascia are to be found under 'Thyroid'.)

# The Thyroid Gland

The thyroid is made up of (Fig. 184):
1.   the *isthmus*—overlying the 2nd to 4th rings of the trachea;
2.   the *lateral lobes*—each extending from the side of the thyroid cartilage downwards to the 6th tracheal ring;
3.   an inconstant *pyramidal lobe* projecting upwards from the isthmus, usually on the left side.

*Relations* (Fig. 183)
The gland is enclosed in the pretracheal fascia, covered by the strap muscles and overlapped by the sternomastoids. The anterior jugular

veins course over the isthmus. When the thyroid enlarges, the strap muscles stretch and adhere to the gland so that, at operation, they often appear to be thin layers of fascia.

On the deep aspect of the thyroid lie the larynx and trachea, with the pharynx and oesophagus behind and the carotid sheath on either side. Two nerves lie in close relationship to the gland; in the groove between the trachea and oesophagus lies the *recurrent laryngeal nerve* and deep to the upper pole lies the *external branch* of the *superior laryngeal nerve* passing to the crico-thyroid muscle.

*Blood supply*

Three arteries supply and three veins drain the thyroid gland (Fig. 184);
• *the superior thyroid artery* — arises from the external carotid and passes to the upper pole.

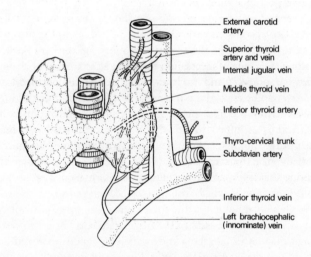

Fig. 184. The thyroid and its blood vessels.

• *the inferior thyroid artery* — arises from the thyro-cervical trunk of the 1st part of the subclavian artery and passes behind the carotid sheath to the back of the gland.
• *the thyroidea ima artery* — is inconstant; when present it arises from the aortic arch or the innominate artery.
• *the superior thyroid vein* — drains the upper pole to the internal jugular vein.

•*the middle thyroid vein*—drains from the lateral side of the gland to the internal jugular.

•*the inferior thyroid veins*—often several—drain the lower pole to the brachiocephalic (innominate) veins.

As well as these named branches, numerous small vessels pass to the thyroid from the pharynx and trachea so that even when all the main vessels are tied, the gland still bleeds when cut across during a partial thyroidectomy.

## Development

The thyroid develops from a bud which pushes out from the floor of the pharynx; this outgrowth then descends to its definitive position in the neck. It normally loses all connection with its origin which is commemorated, however, by the foramen caecum at the junction of the middle and posterior thirds of the tongue.

### CLINICAL FEATURES

1.  The development of the thyroid accounts for the rare occurrence of the whole or a part of the gland remaining as a swelling at the tongue base (*lingual thyroid*) and for the much commoner occurrence of a *thyroglossal cyst* or *sinus* along the pathway of descent. Such a sinus can be dissected from the mid-line of the neck along the front of the hyoid (in such intimate contact with it that the centre of the hyoid must be excised during the dissection) then backwards through the muscles of the tongue to the foramen caecum (Fig. 185).

Descent of the thyroid may go beyond the normal position in the neck down into the superior mediastinum (*retrosternal goitre*).

2.  A benign enlargement of the thyroid may compress or displace any of its close relations; the trachea and oesophagus may be narrowed, with resulting difficulty in breathing and swallowing, and the carotid may be displaced posteriorly. A carcinoma of the thyroid invades its neighbours rather than displacing them—eroding into trachea or oesophagus, surrounding the carotid sheath and occasionally causing severe haemorrhage therefrom. The recurrent laryngeal nerve and the cervical sympathetic chain may be involved, producing changes in the voice and Horner's syndrome respectively.

3.  We have already noted, in dealing with the fasciae of the neck, that the thyroid gland is enclosed in the pretracheal fascia. This thyroid capsule is much denser in front than behind and the enlarging gland therefore tends to push backwards, burying itself round the sides and even the

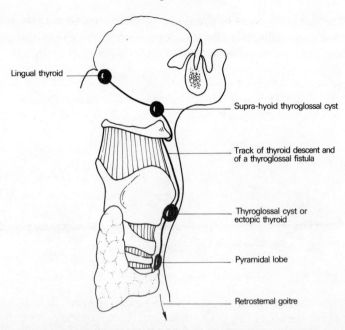

Lingual thyroid

Supra-hyoid thyroglossal cyst

Track of thyroid descent and
of a thyroglossal fistula

Thyroglossal cyst or
ectopic thyroid

Pyramidal lobe

Retrosternal goitre

Fig. 185. The descent of the thyroid, showing possible sites of ectopic thyroid tissue or thyroglossal cysts, and also the course of a thyroglossal fistula. (The arrow shows the further descent of the thyroid which may take place retrosternally into the superior mediastinum.)

back of the trachea and oesophagus. Because of the attachments of its fascial compartment, a large goitre will also extend downwards into the superior mediastinum ('plunging goitre').

Above, the pretracheal fascia blends with the larynx, accounting for the upward movement of the thyroid gland with each act of swallowing.

4. *Thyroidectomy* is carried out through a transverse 'collar' incision, two finger-breadths above the suprasternal notch. This lies in the line of the natural skin folds of the neck. Skin flaps are reflected, together with platysma, and the investing fascia opened longitudinally between the strap muscles and between the anterior jugular veins.

If more room is required in the case of a large goitre, the strap muscles are divided; this is carried out at their upper extremity because their nerve supply (the ansa hypoglossi) enters the lower part of the muscles and is hence preserved.

The pretracheal fascia is then divided, exposing the thyroid gland; unless this tissue plane deep to the fascia is found, dissection is a difficult and bloody procedure.

The thyroid is then mobilized and its vessels ligated *seriatim*. Both the recurrent and superior laryngeal nerves are at risk during this procedure and must be carefully avoided (Fig. 186).

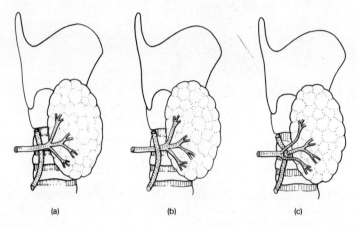

(a)                    (b)                    (c)

**Fig. 186.** The relationship of the recurrent laryngeal nerve to the thyroid gland and the inferior thyroid artery. (a) The nerve is usually deep to the artery but (b) may be superficial to it or (c) pass through its branches. In these diagrams the lateral lobe of the thyroid is pulled forwards, as it would be in a thyroidectomy.

### The parathyroid glands (Fig. 187)

These are usually four in number, a superior and inferior on either side, however, the numbers vary from two to six. Ninety per cent are in close relationship to the thyroid, 10 per cent are aberrant, the latter invariably being the inferior glands.

Each gland is about the size of a split pea and is of a yellowish-brown colour. The superior parathyroid is more constant in position than the inferior gland. It usually lies at the middle of the posterior border of the lobe of the thyroid above the level at which the inferior thyroid artery crosses the recurrent laryngeal nerve. The inferior parathyroid is most usually situated below the inferior artery near the lower pole of the thyroid gland. The next commonest site is within 1 cm of the lower pole of the thyroid gland. Aberrant inferior parathyroids may descend along the inferior thyroid veins in front of the trachea and may even track into the superior mediastinum in company with thymic tissue, for which there is an embryological explanation (see below). Less commonly, the inferior gland may lie behind and outside the fascial sheath of the thyroid and be

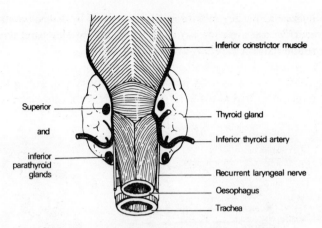

Fig. 187. The normal sites of the parathyroid glands.

found behind the oesophagus or even in the posterior mediastinum. Only on extremely rare occasions are the glands actually completely buried within thyroid tissue (Fig. 188).

### Development
The superior parathyroids differentiate from the 4th branchial pouch.

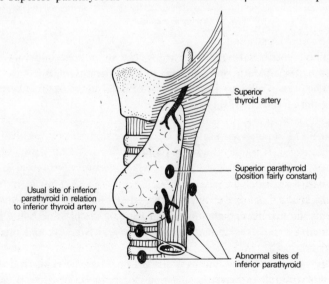

Fig. 188

The *inferior* gland develops from the 3rd pouch in company with the thymus (Fig. 189). As the latter descends, the inferior parathyroid is dragged down with it.

It is thus easily understood that the inferior parathyroid may be dragged beyond the thyroid into the mediastinum and why, although very rarely, parathyroid tissue is found actually within the thymus.

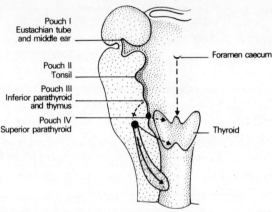

Pouch I
Eustachian tube
and middle ear

Foramen caecum

Pouch II
Tonsil

Pouch III
Inferior parathyroid
and thymus

Pouch IV
Superior parathyroid

Thyroid

Fig. 189. The derivatives of the branchial pouches. Note that the inferior parathyroid migrates downwards from the third pouch whereas the superior parathyroid (fourth pouch) remains stationary.

CLINICAL FEATURES
1.  These possible aberrant sites are, of course, of great importance in searching for a parathyroid adenoma in hyperparathyroidism.
2.  The parathyroids are usually safe in subtotal thyroidectomy because the posterior rim of the thyroid is preserved.

# The Palate

The palate separates the nasal and buccal cavities and comprises:
1.  *the hard palate*—which is vault-shaped and made up of the palatine plate of the maxilla and the horizontal plate of the palatine bone; it is bounded by the alveolar margin anteriorly and laterally, and merges posteriorly with:
2.  *the soft palate*—hanging as a curtain between the naso- and oro-pharynx; centrally it bears the *uvula* on its free posterior edge; laterally it blends into the anterior and posterior pillars of the fauces.

The hard palate is made up of bone, periosteum, and a squamous mucosa in which are embedded tiny accessory salivary glands.

The framework of the soft palate is formed by the aponeurosis of the tensor palati muscle which adheres to the posterior border of the hard palate. To this fibrous sheet are attached the palatine muscles covered by a mucous membrane which is squamous on its buccal aspect and ciliated columnar on its nasopharyngeal surface.

The sensory supply of the palate is largely from the maxillary division of V but fibres of IX supply its most posterior part.

Motor innervation to the palatine muscles is from XI transmitted by the vagus fibres in the pharyngeal plexus. The tensor palati is the exception to this rule and is supplied by the mandibular division of V.

In speaking, swallowing and blowing, the soft palate closes off the nasopharynx from the buccal cavity. If the palate is paralysed, as may occur after diphtheria, the voice is impaired and fluids regurgitate through the nose on swallowing.

## The development of the face, lips and palate with special reference to their congenital deformities (Fig. 190)

Around the primitive mouth, or stomodaeum, there develop:

1. *The fronto-nasal process* which projects down from the cranium. Two olfactory pits develop in it and rupture into the pharynx to form the nostrils. Definitively, this process forms the nose, the nasal septum, nostril, the philtrum of the upper lip (the small mid-line depression) and the premaxilla—the V-shaped anterior portion of the upper jaw which usually bears the four incisor teeth.

2. *The maxillary processes* on each side, which fuse with the fronto-

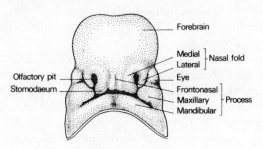

**Fig. 190.** The ventral aspect of a fetal head showing the three processes, fronto-nasal, maxillary and mandibular, from which the face, nose and jaws are derived.

nasal process and become the cheeks, upper lip (exclusive of the philtrum), upper jaw and palate (apart from the premaxilla).

3. *The mandibular processes* which meet in the mid-line to form the
lower jaw.

Abnormalities of this complex fusion process are numerous and
constitute one of the commonest groups of congenital deformities. It
is estimated that one child in 600 in England is born with some degree
of either cleft lip or palate (Fig. 191).

Frequently these anomalies are associated with other congenital conditions such as spina bifida, syndactyly (fusion of fingers or toes), etc.
Indeed, it is good clinical practice to search a patient with any congenital
defect for others.

The anomalies associated with defects of fusion of the face are:

1. *Macrostoma* and *microstoma* where either too little or too great a
closure of the stomodaeum occurs.

2. *Cleft upper lip* (or 'hare lip'). This is only very rarely like the upper
lip of a hare, i.e. a median cleft, although this may occur as a failure of
development of the philtrum from the fronto-nasal process. Much more
commonly, the cleft is on one or both sides of the philtrum, occurring as
failure of fusion of the maxillary and fronto-nasal processes. The cleft
may be a small defect in the lip or may extend into the nostril, split the
alveolus or even extend along the side of the nose as far as the orbit. There
may be an associated cleft palate.

3. *Cleft lower lip*—occurs very rarely but may be associated with a
cleft tongue and cleft mandible.

4. *Cleft palate* is a failure of fusion of the segments of the palate.

The following stages may occur (Fig. 191):

*a* A bifid uvula, of no clinical importance.

*b* A partial cleft, which may involve the soft palate only or the posterior
part of the hard palate also.

*c* A complete cleft. This may be unilateral, running the full length of
the maxilla and then alongside one face of the premaxilla, or bilateral in
which the palate is cleft with an anterior V separating the premaxilla
completely.

5. *Inclusion dermoids* may form along the lines of fusion of the face.
The most common of these is the *external angular dermoid* at the lateral
extremity of the upper eyebrow. Occasionally this dermoid extends
through the skull to attach to the underlying dura.

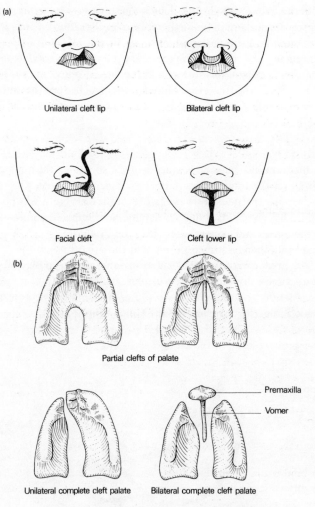

**Fig. 191.** Types of (a) Cleft lip and (b) Cleft palate.

# The Tongue and Floor of the Mouth

## The tongue

*The tongue* consists of a buccal and a pharyngeal portion separated by a V-shaped groove on its dorsal surface, the *sulcus terminalis*. At the

apex of this groove is a shallow depression, the *foramen caecum*, marking
the embryological origin of the thyroid (*see* p. 284). Immediately in front
of the sulcus lie a row of large *vallate papillae*.

The under aspect of the tongue bears the median *frenulum linguae*;
the mucosa is thin on this surface and the lingual veins can thus be seen
on either side of the frenulum. More laterally can be seen the fimbriated
fold on each side, overlying the deep artery of the tongue and the lingual
nerve.

### Structure

The thick stratified squamous mucosa of the dorsum of the tongue bears
papillae over the anterior two-thirds back as far as the sulcus terminalis.
These papillae (particularly the vallate) bear the taste buds. The posterior
one-third has no papillae but carries numerous lymphoid nodules which,
with the tonsils and adenoids, make up the *'lymphoid ring of Waldeyer'*.

Small glands are scattered throughout the submucosa of the dorsum;
these are predominantly serous anteriorly and mucous posteriorly.

The tongue is divided by a median vertical fibrous septum, as indicated
on the dorsum by a shallow groove. On each side of this septum are the
intrinsic and extrinsic muscles of the tongue (Fig. 192).

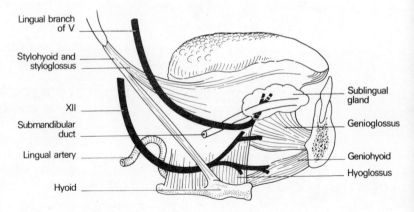

**Fig. 192.** Lateral view of the tongue, its extrinsic muscles and its nerves.

The *intrinsic muscles* are disposed in vertical, longitudinal and transverse bundles; they alter the shape of the tongue.

The *extrinsic muscles* move the tongue as a whole. They pass to the
tongue from the symphysis of the mandible, the hyoid, styloid process

and the soft palate; respectively the *genioglossus, hyoglossus, styloglossus* and *palatoglossus.*

### Blood supply

Blood is supplied from the lingual branch of the external carotid artery. There is little cross circulation across the median raphe which is therefore a relatively avascular plane.

### Lymph drainage (Fig. 193)

The drainage zones of the mucosa of the tongue can be grouped into three:

1.   the tip drains to the submental nodes.
2.   the anterior two-thirds drains to the submental and submandibular nodes and thence to the *lower* nodes of the deep cervical chain along the carotid sheath.
3.   the posterior one-third drains to the *upper* nodes of the deep cervical chain.

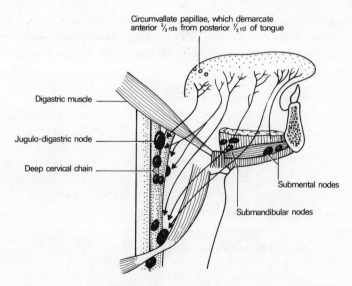

Circumvallate papillae, which demarcate
anterior ⅔rds from posterior ⅓rd of tongue

Digastric muscle

Jugulo-digastric node

Deep cervical chain

Submental nodes

Submandibular nodes

**Fig. 193.** Diagram of the lymph drainage of the tongue. Note two points: (i) The anterior part of the tongue tends to drain to the nodes farthest down the deep cervical chain, whereas the posterior part drains to the upper chain. (ii) The anterior two-thirds of the tongue drain unilaterally, the posterior one-third bilaterally.

There is a rich anastomosis across the mid-line between the lymphatics of the posterior one-third of the tongue so that a tumour on one side readily metastasizes to contralateral nodes. In contrast, there is little cross communication in the anterior two-thirds, where growths more than $\frac{1}{2}$ in (12 mm) from the mid-line do not metastasize to the opposite side of the neck till late in the disease.

## Nerve supply
The anterior two-thirds of the tongue receives its sensory supply from the lingual branch of V which also transmits the gustatory fibres of the chorda tympani (VII).

Common sensation and taste to the posterior one-third, including the vallate papillae, are derived from IX. A few fibres of the superior laryngeal nerve (X) carry sensory fibres from the posterior part of the tongue.

All the muscles of the tongue except palatoglossus are supplied by XII; palatoglossus, a muscle of the soft palate, is innervated by XI via the pharyngeal plexus.

## Development (Fig. 194)
A small nodule, the *tuberculum impar*, is the first evidence of the developing tongue in the floor of the pharynx. This is soon covered over by two swellings derived from the first branchial arch. These fuse in the mid-line

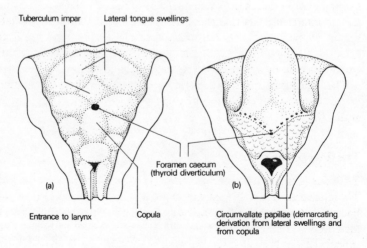

Fig. 194. Development of the tongue.

to form the definitive anterior two-thirds of the tongue supplied by V and reinforced by chorda tympani.

Posteriorly, this mass meets the *copula*, a central swelling in the pharyngeal floor which represents the second, third and fourth arches and which forms the posterior one-third of the tongue (nerve supply IX and X).

The tongue muscles derive from the occipital myotomes which migrate forward dragging with them their nerve supply (XII, the hypoglossal nerve).

CLINICAL FEATURES

1. Damage to the hypoglossal nerve is readily detected clinically by hemiatrophy of the tongue and deviation of the projected organ towards the paralysed side.

2. If the unconscious or deeply anaesthetized patient is laid on his back, the posterior aspect of the tongue drops back to produce a laryngeal obstruction. This can be prevented either by lying the patient on his side with the head down ('the tonsil position'), when the tongue flops forward with the weight of gravity, or by pushing the mandible forwards by pressure on the angle of the jaw on each side; this is effective because genio-glossus, attached to the symphysis menti, drags the tongue forward along with the lower jaw.

3. Although lymphatics pierce the floor of the mouth (i.e. the mylohyoid muscle) to reach the submental and submandibular lymph nodes, it is an interesting fact that these tissues are not affected by lymphatic spread of malignant cells (although they may be invaded by direct extension of growth). It seems that the nodes are involved by lymphatic emboli and not by a permeation of the lymphatic channels.

The bilateral lymphatic spread of growths of the posterior one-third of the tongue is one factor contributing to the poor prognosis of tumours at this site.

## The floor of the mouth

The floor of the mouth is formed principally by the mylohyoid muscles. These stretch as a diaphragm from their origin along the mylohyoid line on the medial aspect of the body of the mandible on each side, to their insertion along a median raphe and into the hyoid bone. They support the tongue as a muscular sling (Fig. 195).

On the lower aspect of this diaphragm, on each side, are the anterior belly of the digastric muscle, the superficial part of the submandibular gland and the submandibular lymph nodes, all covered by deep fascia and platysma.

Lying above mylohyoid are the tongue muscles, as a central mass, with the sublingual salivary gland and the deep part of the submandibular gland and its duct lying beneath the mucosa of the mouth floor on either side.

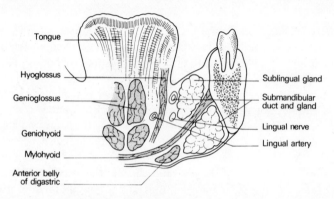

Fig. 195.   Coronal section of the floor of the mouth.

CLINICAL FEATURES

*Ludwig's angina* is a cellulitis of the floor of the mouth, usually originating from a carious molar tooth. The infection spreads above the mylohyoid; oedema forces the tongue upwards and the mylohyoid itself is pushed downwards so that there is swelling both below the chin and within the mouth. There is considerable danger of spread of infection backwards with oedema of the glottis and asphyxia.

Drainage is carried out by a deep incision below the mandible which must divide the mylohyoid muscle.

# The Pharynx

*The pharynx* is a musculo-fascial tube, incomplete anteriorly, which extends from the base of the skull to the oesophagus and which acts as a common entrance to the respiratory and alimentary tracts.

From above downwards, it is made up of three portions (Fig. 196):

1. *the nasopharynx*—lying behind the nasal fossae and above the soft palate;
2. *the oropharynx*—lying behind the anterior pillars of the fauces;
3. *The laryngopharynx*—lying behind the larynx;

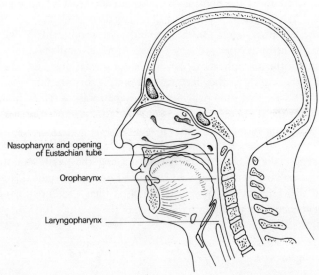

Nasopharynx and opening of Eustachian tube

Oropharynx

Laryngopharynx

Fig. 196. A saggital section through the head and neck to show the subdivisions of the pharynx.

## The nasopharynx

The nasopharynx lies above the soft palate which cuts it off from the rest of the pharynx during deglutition and therefore prevents regurgitation of food through the nose.

Two important structures lie in this compartment.

*The nasopharyngeal tonsil* ('*the adenoids*') which consists of a collection of lymphoid tissue beneath the epithelium of the roof and posterior wall of this region. It helps to form a continuous lymphoid ring with the tonsils and the lymphoid nodules on the dorsum of the tongue (Waldeyer's ring).

*The orifice of the pharyngo-tympanic or auditory tube* (Eustachian canal) which lies on the side-wall of the nasopharynx level with the floor of the nose. The posterior lip of this opening is prominent, due to the underlying cartilage of the Eustachian tube, and is termed the *Eustachian cushion*, behind which lies the slit-like *pharyngeal recess*.

CLINICAL FEATURES

1.   The nasopharynx may be inspected indirectly by a mirror passed through the mouth (posterior rhinoscopy) or studied through a rhinoscope passed along the floor of the nose. Under anaesthesia, it can be palpated by a finger passed behind the soft palate.

2.   The nasopharyngeal tonsils (adenoids) are prominent in children but usually atrophy after puberty. When chronically inflamed they may all but fill the nasopharynx, causing mouth-breathing and also, by blocking the auditory tube, deafness and middle ear infection.

3.   The Eustachian tube provides a ready pathway of sepsis from the pharynx to the middle ear and accounts for the frequency with which otitis media complicates infections of the throat.

4.   The middle ear can be intubated through a catheter passed into the Eustachian tube. The catheter is passed along the nasal floor to the posterior wall of the nasopharynx. Its curved tip is then rotated laterally so that it lies in the pharyngeal recess; it is then withdrawn over the Eustachian cushion to slip into the orifice of the auditory tube.

## The oropharynx   *Soft palate → Epiglottis*

This part of the pharynx lies behind the mouth and tongue. Its anterior boundaries are the anterior pillars of the fauces and it extends from the palate above to the tip of the epiglottis below. Its most important contents are the tonsils.

### The tonsils

The tonsil lies in the *tonsillar fossa* between the anterior and posterior pillars of the fauces. The anterior pillar, or *palato-glossal fold*, forms the boundary between the buccal cavity and the oropharynx; it fuses with the lateral wall of the tongue and contains the palato-glossus muscle. The posterior pillar, or *palato-pharyngeal fold*, blends with the wall of the pharynx and contains the palato-pharyngeus (Fig. 197).

The floor of the tonsillar fossa is formed by the superior constrictor of the pharynx separated from the tonsil by the *tonsillar capsule*, which is a thick condensation of the pharyngeal submucosa (the pharyngobasilar fascia). This capsule is itself separated from the superior constrictor by a film of loose areolar tissue.

The tonsil consists of a collection of lymphoid tissue covered by a squamous epithelium; a unique histological combination which makes it easy to "spot" in examinations. This epithelium is pitted by *crypts*, up to twenty in number, and often bears a deep *intratonsillar cleft* in its upper part.

The lymphoid material may extend up to the soft palate, down to the tongue or into the anterior faucial pillar. From late puberty onwards this lymphoid tissue undergoes progressive atrophy.

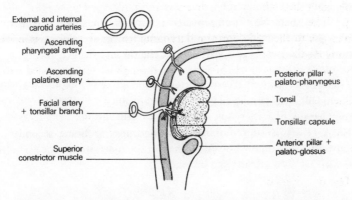

**Fig. 197.** Diagram of the tonsil and its relations—in horizontal section.

*Blood supply* is principally from the tonsillar branch of the facial artery entering at the lower pole of the tonsil, although twigs are also derived from the lingual, ascending palatine and ascending pharyngeal arteries.

The venous drainage passes to the pharyngeal plexus. An important constant vein, the *paratonsillar vein*, descends from the soft palate across the lateral aspect of the tonsillar capsule. It is nearly always divided in tonsillectomy and may give rise to troublesome haemorrhage.

*Lymph drainage* is via lymphatics which pierce the superior constrictor muscle and pass to the nodes along the internal jugular vein, especially the *tonsillar or jugulo-digastric node* at the angle of the jaw. Since this node is affected in tonsillitis it is the most common lymph node in the body to undergo pathological enlargement.

*Embryologically* the tonsil derives from the second internal branchial cleft (*see* Fig. 189).

CLINICAL FEATURES

1.   *Tonsillectomy* may be carried out by dissection or by the guillotine; both depend on removing the lymphoid tissue and underlying fascial capsule from the loose areolar tissue clothing the superior constrictor in the floor of the tonsillar fossa. In dissection, an incision is made in the mucosa of the anterior pillar immediately in front of the tonsil; the gland is then freed by blunt dissection until it remains attached only by its pedicle of vessels near its lower pole. This pedicle is then crushed and divided by means of a wire snare.

In the second method, the guillotine is applied so that the tonsil bulges through the ring in the instrument. The tonsil is then removed by closing the blade of the guillotine.

Unless there have been repeated infections, the superior constrictor lies separated from the tonsil and its capsule by loose areolar tissue which prevents the pharyngeal wall being dragged into danger during tonsillectomy.

Similarly the internal carotid artery, although only 1 in (2.5 cm) behind the tonsil, is never injured in this operation since it lies safely freed from the pharynx by fatty tissue around the carotid sheath.

2.   A *quinsy* is suppuration in the peritonsillar tissue secondary to tonsillitis. It is drained by an incision in the most prominent part of the abscess where softening can be felt.

## The laryngopharynx *epiglottis - L6.*

The laryngopharynx extends from the level of the tip of the epiglottis to the termination of the pharynx in the oesophagus at the level of C6.

The inlet of the larynx, defined by the epiglottis, ary-epiglottic folds and the arytenoids, lies anteriorly. The larynx itself bulges into this part of the pharynx leaving a deep recess anteriorly on either side, the *piriform fossa*, in which sharp ingested foreign bodies (for example, fish-bones), may lodge.

## The structure of the pharynx

The pharynx is made up of mucosa, submucosa, muscle and a loose areolar sheath. The mucosa is a ciliated columnar epithelium in the nasopharynx but elsewhere it is stratified and squamous. Beneath this, the submucosa is thick and fibrous (*the pharyngo-basilar fascia*) and it is this layer which forms the capsule of the tonsil.

The three *pharyngeal constrictor muscles* (superior, middle and inferior) are arranged like flower pots placed one inside the other, but are open in front at the entries of the nasal, buccal and laryngeal cavities.

Each constrictor muscle is attached anteriorly to the side-wall of these cavities and fans out to insert into a median raphe along the posterior aspect of the pharynx, extending from the base of the skull to the oesophagus (Fig. 198).

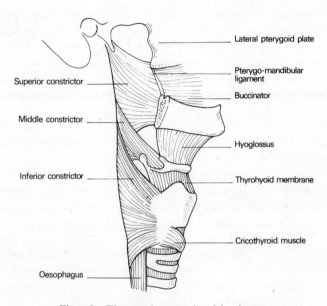

**Fig. 198.** The constrictor muscles of the pharynx.

Covering these muscles is an areolar sheath continuous with that covering the buccinator and hence termed the *bucco-pharyngeal fascia.*

*Blood supply.* The pharynx receives its arterial supply mainly from the superior thyroid and ascending pharyngeal branches of the external carotid.

A pharyngeal venous plexus lies in the areolar sheath of the pharynx and drains into the internal jugular vein.

*Nerve supply.* The pharyngeal branches of IX and X constitute the principal sensory and motor supply of the pharynx respectively. The maxillary division of V supplies the sensory innervation of the nasopharynx.

## The mechanism of deglutition

Cine-radiographic studies have enabled accurate analysis of deglutition
to be made. The bolus of food is pushed backwards by the tongue squeezing against the hard palate. The soft palate closes off the nasopharynx
and also assists in the formation of the food bolus against the tongue. The
arched dorsal surface of the tongue then lowers to form a steep slope
down which the food descends.

During this phase the hyoid and larynx are elevated, the laryngeal
aditus contracts and the vocal cords close. The entry into the larynx is
thus protected, but this is not always complete and some food may pass
into the laryngeal vestibule. The epiglottis stands erect while the bolus
streams by, guiding it along both piriform fossae and away from the
laryngeal orifice. The epiglottis then flaps back over the laryngeal inlet
but only after the main bolus has passed by, perhaps to prevent deposition
of crumbs of food over the inlet during re-establishment of the airway.

The tongue then passes forward again, the epiglottis rights itself
(probably by elastic recoil) and the larynx descends.

The food bolus meantime traverses the oesophagus partly by gravity
and partly by peristalsis (swallowing a glass of beer is thus possible at
a party when standing on one's head).

CLINICAL FEATURES
*Pharyngeal pouch.* The inferior constrictor muscle is made up of an upper
oblique and a lower transverse part, the former arising from the side of the

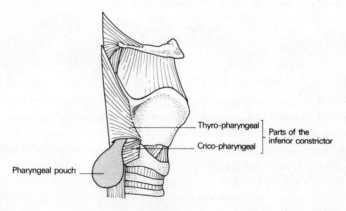

Fig. 199.   A pharyngeal pouch emerging between the two components of the inferior
constrictor muscle.

thyroid cartilage and the latter from the cricoid (the cricopharyngeus).

Posteriorly there is a potential gap between these two components termed the pharnygeal dimple or *Killian's dehiscence*. The mucosa and submucosa of the pharynx may bulge through this weak area to form a pharyngeal pouch (Fig. 199), possibly as a result of muscle inco-ordination or of spasm of the crico-pharyngeus. This diverticulum first protrudes posteriorly; as it enlarges, backward extension is prevented by the prevertebral fascia and it therefore has to project to one side of the pharynx—usually to the more exposed left.

With further enlargement, the pouch pushes the oesophagus aside and lies directly in line with the pharynx; most food then passes into the pouch with resulting severe dysphagia and cachexia. Spill of the pouch contents into the larynx is very liable to cause inhalation of food material into the bronchi with respiratory infection and lung abscess as possible consequences.

# The Larynx

The larynx has a triple function, that of an open valve in respiration, that of a partially closed valve whose orifice can be modulated in phonation and that of a closed valve, protecting the trachea and bronchial tree during deglutition. Coughing is only possible when the larynx can be closed effectively.

The structures which form its framework are the hyoid, epiglottis, thyroid cartilage, cricoid and the arytenoids (Fig. 200).

*The hyoid bone* is U-shaped; from it is slung the rest of the larynx via the *thyro-hyoid membrane* and *muscle*. The hyoid bone itself is attached to the mandible and tongue by the hyoglossus, the mylohyoid, geniohyoid and digastric muscles, to the styloid process by the stylohyoid ligament and muscle and to the pharynx by the middle constrictor. Three of the four strap muscles of the neck, the omohyoid, sternohyoid and thyrohyoid, find attachment to it, only the sternothyroid failing to gain it.

*The epiglottis* is a leaf-shaped elastic cartilage lying behind the root of the tongue. It is attached anteriorly to the body of the hyoid and below to the back of the thyroid cartilage immediately above the vocal cords. The sides of the epiglottis are connected to the arytenoids by the *aryepiglottic folds* which run backwards to form the margins of the entrance, or *aditus*, of the larynx.

The upper anterior surface of the epiglottis projects above the hyoid bone; the epiglottic mucosa is reflected forward to the base of the tongue and is raised up into a median *glosso-epiglottic* fold and lateral *pharyngo-epiglottic folds*. The depression on either side between these folds is termed the *vallecula*.

*The thyroid cartilage* is shield-like, being made up of two lateral plates meeting in the mid-line in the prominent 'V' of the 'Adam's apple'.

*The cricoid* is signet-ring shaped, deepest behind. It is attached to the thyroid cartilage above via the *crico-thyroid membrane*; this membrane is free at its upper border where it forms the *vocal ligament* within the vocal cord on each side. Inferiorly, the cricoid is attached to the trachea by the *crico-tracheal membrane*.

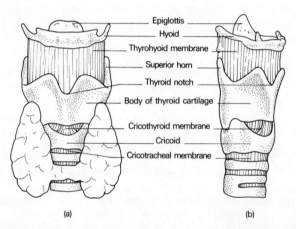

Epiglottis
Hyoid
Thyrohyoid membrane
Superior horn
Thyroid notch
Body of thyroid cartilage
Cricothyroid membrane
Cricoid
Cricotracheal membrane

(a)                                    (b)

**Fig. 200.**    External view of the larynx: (a) Anterior aspect; (b) antero-lateral aspect.

*The arytenoids* sit one on each side of the posterior 'signet' of the cricoid cartilage.

Passing forward from the arytenoid to the back of the thyroid cartilage, just below the epiglottic attachment, are two folds of mucosa. The upper is the *vestibular fold*, containing a small amount of fibrous tissue and forming on each side the *false vocal cord*. The lower fold (the *vocal fold or cord*) contains the *vocal ligament* (Fig. 201).

The mucosa is firmly adherent to the vocal ligament without there being any intervening submucosa. This accounts for the pearly-white, avascular appearance of the vocal cords as seen on laryngoscopy. Oedema of the larynx cannot involve the true cords since there is no submucous tissue in which fluid can collect.

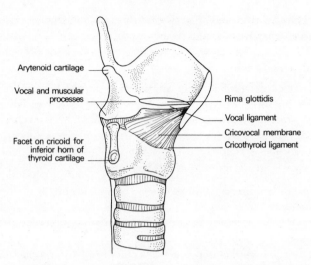

Fig. 201. The internal structure of the larynx—the lamina of the thyroid cartilage has been cut away.

These folds demarcate the larynx into three zones:

1.   The supraglottic compartment (*vestibule*) above the false cords.
2.   The glottic compartment between the false and true cords.
3.   The infraglottic compartment between the true cords and the first ring of the trachea.   SUBGLOTIC

On either side of the vestibule the pharynx forms a recess, the *piriform fossa*, in which swallowed foreign bodies tend to lodge.

*The muscles of the larynx* function to open the glottis in inspiration, close the vestibule and glottis in deglutition and alter the tone of the true vocal cords in phonation.

The crico-thyroid is the only external muscle of the larynx and tenses the vocal cord by a slight tilting action on the cricoid. It is supplied by the superior laryngeal nerve.

The remaining muscles constitute a single encircling sheet whose various attachments are denoted by the names of its separate parts: the thyro-arytenoid, posterior and lateral crico-arytenoid, the ary-epiglottic, thyro-epiglottic and inter-arytenoid muscles. These are all supplied by the recurrent laryngeal nerve.

All muscles except one have a sphincter action, the exception is the posterior crico-arytenoid on each side which, by rotating the arytenoids outwards, separates the vocal cords.

*Blood supply*
The larynx receives a superior and inferior laryngeal artery from the superior and inferior thyroid artery respectively. These vessels accompany the superior and recurrent laryngeal nerves.

*Lymph drainage*
Above the vocal cords the larynx drains to the upper deep cervical lymph nodes, some lymphatics passing via small nodes lying on the thyro-hyoid membrane.

Below the cords, drainage is to the lower deep cervical nodes, partially via nodes on the front of the larynx and trachea.

The vocal cords themselves act as a complete barrier separating the two lymphatic areas, but posteriorly there is free communication between them; a laryngeal carincoma may thus seed throughout the lymphatic drainage area of the larynx.

*Nerve supply*
The nerve supply of the larynx is of great practical importance and comprises the superior and recurrent laryngeal branches of the vagus nerve.

The *superior laryngeal nerve* passes deep to the internal and external carotid arteries where it divides; its internal branch pierces the thyro-hyoid membrane together with the superior laryngeal vessels to supply the mucosa of the larynx down to the vocal cords. The external branch passes deep to the superior thyroid artery to supply the crico-thyroid muscle.

The *recurrent laryngeal nerve* has a different course on each side. The right arises from the vagus as this crosses the front of the subclavian artery, passes deep to and behind this vessel, then ascends behind the common carotid to lie in the tracheo-oesophageal groove accompanied by the inferior laryngeal vessels (Fig. 183). The nerve then passes deep to the inferior constrictor muscle of the pharynx to enter the larynx behind the crico-thyroid articulation.

The left nerve arises on the arch of the aorta, winds below it, deep to the ligamentum arteriosum, and ascends to the trachea. It then lies in the tracheo-oesophageal groove and is distributed as on the right side.

The recurrent nerves supply all the intrinsic laryngeal muscles, apart from the crico-thyroid, and the mucosa below the vocal cords.

## CLINICAL FEATURES

1. The laryngeal nerves bear relationships to the thyroid arteries which are of considerable practical importance in thyroidectomy. The external branch of the superior laryngeal nerve lies immediately deep to the superior thyroid artery and may be injured in ligating this vessel.

The recurrent laryngeal nerve, lying in the tracheo-oesophageal groove, is usually behind the terminal branches of the inferior thyroid artery. Occasionally, however, the nerve lies in front of these vessels or passes through them (Fig. 186). Moreover, when a large thyroid is pulled forward during thyroidectomy, the nerve becomes dragged forward with it and is therefore placed in further jeopardy. To avoid nerve damage during ligation of the inferior thyroid artery, this procedure should be carried out well laterally, just as the artery emerges from behind the carotid sheath and before it takes up its intimate and inconstant relationship to the nerve.

2. Damage to the superior nerve causes some weakness of phonation due to the loss of the tightening effect of the crico-thyroid muscle on the cord.

3. Complete division of a recurrent laryngeal nerve causes the cord on the affected side to take up the neutral (or paramedian) position between abduction and adduction. Usually the other cord is able to compensate in a remarkable way and speech is not greatly affected; if both nerves are divided, however, the voice is completely lost and breathing becomes difficult through the only partially opened glottis.

4. If the recurrent nerve is only bruised or partially damaged, the abductors (posterior crico-arytenoids) are affected more than the adductors; this is known as *Semon's law*. The affected cord adopts the mid-line, adducted, position. In bilateral incomplete paralysis, therefore, the cords come together, stridor is intense and tracheotomy may become essential.

5. The left recurrent laryngeal nerve, in its thoracic course, may become involved in a bronchial or oesophageal carcinoma, or in a mass of enlarged mediastinal glands, or may become stretched over an aneurysm of the aortic arch. The enlarged left atrium in advanced mitral stenosis may produce a recurrent laryngeal palsy by pushing up the left pulmonary artery which compresses the nerve against the aortic arch.

Either nerve, in the neck, may be damaged by an extending thyroid carcinoma or malignant lymph nodes. For these reasons, loss of voice must always be regarded as an ominous symptom requiring careful investigation.

**6.**   The larynx can be inspected either directly, by means of the laryngo-scope, or indirectly through a laryngeal mirror. The base of the tongue, valleculae, epiglottis, ary-epiglottic folds and piriform fossae are viewed, then the false cords, which are red and widely apart, then, between these, the pearly white true cords (Fig. 202).

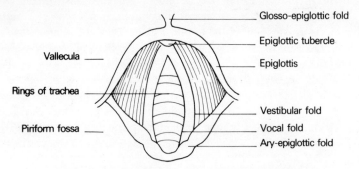

Fig. 202.   Diagram of the larynx as seen at laryngoscopy.

For the passage of the laryngoscope, endotracheal tube or broncho-scope it is essential to know the position which brings the axes of the mouth, oropharynx and larynx into line; this is achieved by bringing the neck forward and at the same time extending the head fully at the atlanto-occipital joint—it is the position in which one sniffs at the fresh air after a long day in the operating theatre.

# The Salivary Glands

### The parotid gland

This is the largest of the salivary glands, lying wedged between the mandible and sterno-mastoid and overflowing both these bounding structures (Fig. 203).

*Relations*

- Above—lie the external auditory meatus and temporo-mandibular joint.
- Below—it overflows the posterior belly of digastric.
- Anteriorly—it overflows the mandible with the overlying masseter.
- Posteriorly—it overflows the sterno-mastoid.

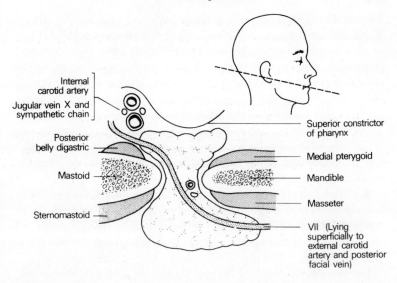

**Fig. 203.** The parotid and its surrounds in a schematic horizontal section—the facial nerve is the most superficial of the structures traversing the gland. (The line of section is shown in the inset head.)

Medially—lies the styloid process and its muscles separating the parotid from the internal jugular vein, internal carotid artery, last four cranial nerves and the lateral wall of the pharynx.

The gland itself is enclosed in a split in the investing fascia, lying both on and below which are the parotid lymph nodes. Antero-inferiorly this parotid fascia is thickened and is the only structure separating the parotid from the submandibular gland (the stylo-mandibular ligament).

Traversing the gland (from without in) are:

1.   The facial nerve (see below).
2.   The posterior facial vein formed by the junction of the superficial temporal and maxillary veins.
3.   The external carotid artery, dividing at the neck of the mandible into its superficial temporal and maxillary terminal branches.

The *parotid duct* (of Stensen) is 2 in (5 cm) long. It arises from the anterior part of the gland, runs over the masseter a finger's breadth below the zygomatic arch to pierce the buccinator and open opposite the second upper molar tooth. The duct can easily be felt by a finger rolled over the masseter if this muscle is tensed by clending the teeth.

*The relations of the facial nerve to the parotid*

The facial nerve is unique in traversing the substance of a gland, a fact of considerable importance to the surgeon. This co-existence is explained embryologically; the parotid gland develops in the crotch formed by the two major branches of the facial nerve. As the gland enlarges it overlaps these nerve trunks, the superficial and deep parts fuse and the nerve comes to lie buried within the gland. The fanciful comparison between the nerve and the two parotid lobes and sandwich-filling between two slices of bread is not valid because the two lobes of the parotid come to fuse intimately with each other both around and between the branches of the nerve.

The facial nerve emerges from the stylomastoid foramen, winds laterally to the styloid process and can then be exposed surgically in the inverted V between the bony part of the external auditory meatus and the mastoid process. This has a useful surface marking, the intertragic notch of the ear, which is situated directly over the facial nerve.

Just beyond this point the nerve dives into the posterior aspect of the parotid gland and bifurcates almost immediately into its two main divisions (occasionally it divides before entering the gland). The upper division divides into *temporal* and *zygomatic* branches, the lower division gives the *buccal mandibular* and *cervical* branches (Fig. 204).

These two divisions may remain completely separate within the parotid, may form a plexus of intermingling connections, or, most usually, display a number of cross-communications which can be safely divided during dissection without jeopardy.

The branches of the nerve then emerge on the anterior aspect of the parotid to lie on the masseter, thence to pass to the muscles of the face. No branches emerge from the superficial aspect of the gland, which can therefore be completely exposed with impunity.

CLINICAL FEATURES

1.   A malignant tumour of the parotid gland, unlike benign lesions, may involve VII and produce a facial palsy.

2.   In removing a benign mixed salivary tumour of the parotid, the facial nerve is exposed posteriorly in the wedge-shaped space between the bony canal of the external auditory meatus and the mastoid process. It is then traced into the gland, its main divisions defined and the tumour excised with a wide margin of normal gland, carefully preserving the exposed nerves.

It is interesting that giant mixed tumours 'extrude' clear away from

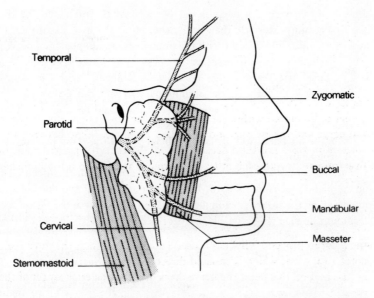

Temporal

Zygomatic

Parotid

Buccal

Mandibular

Cervical

Masseter

Sternomastoid

Fig. 204. The named branches of the facial nerve which traverse the parotid gland.

the facial nerve and can be excised with an adequate margin without even seeing the nerve.

3. The parotid duct and its ramifications can be demonstrated radiologically by injecting radio-opaque dye through a cannula placed in the mouth of the duct (a parotid sialogram).

## The submandibular gland

The submandibular gland is made up of a large superficial and small deep lobe which connect with each other around the posterior border of the mylohyoid.

The superficial lobe of the gland lies at the angle of the jaw, wedged between the mandible and the mylohyoid and overlapping the digastric muscle (Fig. 195). Posteriorly it comes into contact with the parotid gland, separated only by a condensation of its fascial sheath (the stylomandibular ligament).

Superficially the gland is covered by platysma and by its capsule of deep fascia, but it is crossed by the cervical branch of VII and by the anterior facial vein. Its deep aspect lies against the mylohoid for the most part, but posteriorly the gland rests against the hyoglossus muscle and here comes into contact with the lingual and the hypoglossal nerve; both of which lie on hyoglossus as they pass forward to the tongue.

The facial artery also comes into close relationship with the gland, approaching it posteriorly, the arching over its superior aspect (which it grooves), to attain the inferior border of the mandible and thence to ascend onto the face in front of the masseter.

From the medial aspect of the superficial part of the gland projects its deep prolongation along the hyoglossus.

*The submandibular duct* (Wharton's duct) arises from this deep part of the gland and runs forward, beneath the mucosa of the floor of the mouth along the side of the tongue, to open immediately at the side of the frenulum linguae (Fig. 192). Here its orifice is readily visible and saliva can be seen trickling from it.

The sublingual gland (*vide infra*) lies immediately lateral to the submandibular duct.

The lingual nerve reaches the tongue by passing from the lateral side of the duct below and then medial to it—thus 'double-crossing' it.

*The submandibular lymph nodes* lie partly embedded within the gland and partly between it and the mandible.

CLINICAL FEATURES

1.  The rather complex relations of this gland have been given at some longth because excision of the gland for calculus or tumour is not uncommon. This operation is carried out through a skin crease incision below the angle of the jaw.

The mandibular branch of VII passes behind the angle of the jaw rather less than 1 inch from it before arching upwards over the body of the mandible to supply the depressor of the lip. The incision must therefore be placed rather more than 2.5 cm below the angle of the jaw in order to preserve this nerve.

2.  The presence of small lymph nodes actually within the substance of the gland makes removal of the gland an imperative part of block dissection of the neck.

3.  In differentiating between an enlarged submandibular gland and a mass of submandibular lymph nodes, one remembers that the gland lies not only below the mandible but also extends into the floor of the mouth; it can therefore be palpated bimanually between a finger in the mouth

and a finger below the angle of the jaw. Enlarged lymph nodes are felt only at the latter site.

4. A stone in Wharton's duct can be felt bimanually in the floor of the mouth and can be seen if sufficiently large.

### The sublingual gland

This is an almond-shaped salivary gland lying immediately below the mucosa of the floor of the mouth and immediately in front of the deep part of the submandibular gland. Laterally it rests against the sublingual groove of the mandible while medially it is separated from the base of the tongue by the submandibular duct and its close companion, the lingual nerve (Fig. 192).

The gland opens by a series of ducts into the floor of the mouth.

The sublingual gland produces a mucous secretion, the parotid a serous secretion and the submandibular gland a mixture of the two.

As well as these main salivary glands, small accessory glands are found scattered over the palate, lips, cheek, tonsil and tongue. These glands are occasional sites for development of a mixed salivary tumour.

# The Major Arteries of the Head and Neck

### The common carotid arteries

The *left common carotid artery* arises from the aortic arch in front and to the right of the origin of the left subclavian artery. It passes behind the left sterno-clavicular joint, lying in its thoracic course at first in front and then to the left side of the trachea, with the left lung and pleura, the vagus and the phrenic nerve as its lateral relations.

The *right common carotid artery* begins behind the right sterno-clavicular joint at the bifurcation of the innominate artery.

In the neck, both common carotids have essentially similar courses and relationships; they ascend in the carotid fascial sheath which contains also the internal jugular vein laterally, and the vagus nerve between and rather behind the artery and vein. The cervical sympathetic chain ascends immediately posterior to the carotid sheath. These structures form a quartet which should always be considered in this inseparable manner; the relations of any one are those of the other three (Figs. 183, 205).

In the neck, each common carotid artery lies on the cervical transverse processes separated from them by the prevertebral muscles. Medially are larynx and trachea, pharynx and oesophagus, together with the thyroid gland, which overlaps onto the anterior aspect of the carotid. Superficially, the artery is covered by the sterno-mastoid and, in its lower part, by the strap muscles.

The common carotid artery gives off no side branches but terminates at the level of the upper border of the thyroid cartilage (at the vertebral level C4) into the external and internal carotids which are more or less equal in size.

## The external carotid artery

This artery lies first deep to the anterior border of sterno-mastoid and then quite superficially in the anterior triangle of the neck, where its pulsations are usually visible as well as palpable. At first it is slightly deep to the internal carotid, then passes anterior and lateral to it. The jugular vein is first lateral to the external carotid then posterior to it, coming into lateral relationship to the internal carotid. The pharynx lies medially.

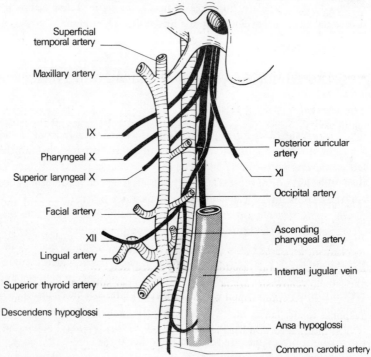

Superficial temporal artery

Maxillary artery

IX

Pharyngeal X

Superior laryngeal X

Facial artery

XII

Lingual artery

Superior thyroid artery

Descendens hypoglossi

Posterior auricular artery

XI

Occipital artery

Ascending pharyngeal artery

Internal jugular vein

Ansa hypoglossi

Common carotid artery

**Fig. 205.** The carotid arteries, their branches and their related nerves.

The external carotid artery ascends beneath the XII nerve and the posterior belly of the digastric to enter the parotid gland, within which it lies deep to the facial nerve and posterior facial vein (Fig. 203).

The artery ends within the parotid gland at the level of the neck of the mandible by dividing into the superficial temporal and internal maxillary arteries.

*Branches* (Fig. 205)

1. *The superior thyroid artery* (giving off the superior laryngeal artery).
2. *The lingual artery*, passing deep to the hyoglossus to supply the tongue.
3. *The facial artery*, which gives off its important branch to the tonsil, loops over the submandibular gland, hooks round the mandible (against which it can be felt pulsating), and ascends onto the face.
4. *The occipital artery*, running along the inferior border of the digastric muscle's posterior belly to the back of the scalp, where its pulse is often palpable.
5. *The posterior auricular artery*.
6. *The ascending pharyngeal artery*.

Its terminal branches are:

• the *superficial temporal* artery, which is palpable on the zygomatic process;
• the *internal maxillary* artery, which supplies the upper and lower jaws, nasal cavity and the muscles of mastication and also gives off the *middle meningeal artery*.

## The internal carotid artery

This artery commences at the bifurcation of the common carotid and, at its origin, is dilated into the *carotid sinus*. This area receives a rich nerve supply from IX and acts as a pressor-receptor; through this mechanism a rise of blood pressure brings about reflex slowing of the heart and peripheral vasodilatation. Tucked deep to the bifurcation is the small, yellowish *carotid body* which is also supplied by IX. This is a chemo-receptor which produces a reflex increase in respiration in response to any fall in the oxygen tension of the blood.

The internal carotid lies first lateral to the external carotid but rapidly passes medial and posterior to it, to ascend along the side wall of the pharynx. It does so with the internal jugular vein, vagus and cervical sympathetic chain in the same relationship to it that they bear to the common carotid artery. At first the artery is covered superficially only by the sterno-mastoid, XII, and the common facial vein; it then passes under the posterior belly of the digastric muscle and the parotid gland to the base of the skull. It is separated from the external carotid artery not only by the parotid but also by the styloid process and the muscles arising from it, by IX and by the pharyngeal branches of X.

At the base of the skull, the internal carotid artery enters the *carotid canal* in the petrous temporal bone. Only at the skull base does the jugular vein lose its close lateral relation to the internal carotid, passing posterior to the artery into the jugular foramen. At this point the two vessels are separated by the emerging last four cranial nerves.

The artery gives off no branches in the neck.

The internal carotid, on entering the skull, commences an extraordinary twisted course. It passes forwards through the temporal bone, upwards into the cavernous sinus, forward in this, upwards through the roof of the sinus to lie medial to the anterior clinoid process, turn, back on itself above the cavernous sinus, then passes up once more, lateral to the optic chiasma, to end by dividing into the *anterior and middle cerebral arteries*. There are thus six bends in the intracranial course of this artery (readily appreciated by studying a lateral carotid arteriogram) which are believed to lessen the pulsating force of the arterial systolic blood pressure on the delicate cerebral tissues.

The *ophthalmic artery* originates from the internal carotid immediately after its emergence from the cavernous sinus, enters the orbit through the optic foramen below and lateral to the optic nerve and supplies the orbital contents and the skin above the eyebrow (via the supratrochlear and supra-orbital branches). Its most important branch, however, is the *central artery of the retina* which is the sole blood supply to this structure.

The two terminal branches of the internal carotid are distributed as follows (Fig. 206):

The *anterior cerebral artery* winds round the genu of the corpus callosum to supply the medial and supero-lateral aspect of the cerebral hemisphere.

The *middle cerebral artery* enters the lateral cerebral sulcus, gives off central branches to supply the internal capsule ('the *artery of cerebral haemorrhage*') and feeds most of the lateral aspect of the cerebral cortex.

The *arterial circle of Willis* (Fig. 207) is completed in front by the *anterior communicating artery*, which links the two anterior cerebral

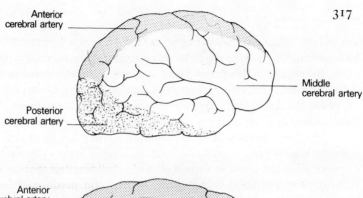

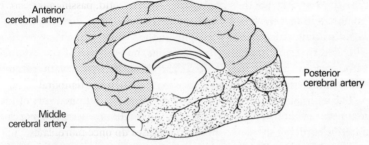

**Fig. 206.** The arterial supply of the cerebral cortex. *A.* Lateral aspect. *B.* Medial aspect.

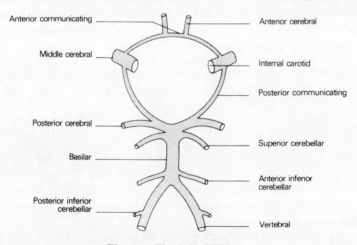

**Fig. 207.** The circle of Willis.

arteries, and behind by a *posterior communicating artery* on each side, passing backwards from the internal carotid to anastomose with the posterior cerebral, a branch of the *basilar artery*; the latter being formed by the junction of the two *vertebral arteries*.

CLINICAL FEATURES

The common carotid artery can be exposed through a transverse incision over the origin of the sterno-mastoid immediately above the sterno-clavicular joint. The carotid sheath lies immediately deep to the junction between the sternal and clavicular heads of the sternomastoid and is revealed either by retracting this muscle laterally or by splitting between its heads. Opening the sheath then reveals the artery lying medial to the internal jugular vein.

Ligation of the common carotid artery is performed for intra-cranial aneurysm arising on the internal carotid. This operation is effective because it lowers the blood flow through the aneurysm, allowing thrombosis to occur. Adequate blood supply to the brain on the affected side is provided by free communication between the branches of the external carotid arteries on each side. Within the cranium, cross-circulation occurs through the circle of Willis.

The internal and external carotids, as well as the terminal part of the common carotid artery, can be exposed through an incision along the anterior border of the sterno-mastoid passing downwards from the angle of the jaw. The sterno-mastoid is retracted, the common facial vein divided, but the hypoglossal nerve, crossing the external and internal carotids just below the posterior belly of the digastric, is carefully preserved.

It may be surprisingly difficult to differentiate between the external and internal carotids at operation; the former is the anterior and rather deeper-placed vessel at origin and, moreover, is the only carotid in the neck which gives off branches.

## The subclavian arteries (Fig. 208)

The *left subclavian artery* arises from the arch of the aorta, immediately behind the commencement of the left common carotid artery. It ascends against the mediastinal surface of the left lung and pleura laterally and the trachea and oesophagus medially to lie behind the sterno-clavicular joint.

The *right subclavian artery* is formed behind the right sterno-clavicular joint by the bifurcation of the brachiocephalic (innominate) artery; beyond this point, the course of the two arteries is much the same.

The cervical course of the subclavian arteries is conveniently divided by the scalenus anterior muscle into three parts:

The *first* part arches over the dome of the pleura and lies deeply placed beneath the sterno-mastoid and the strap muscles. It is crossed at its origin by the carotid sheath and, more laterally, by the phrenic and vagus nerves. At this site, on the right side, the vagus gives off its recurrent laryngeal branch which hooks behind the artery.

On the left side, the thoracic duct crosses the first part of the artery to open into the commencement of the left brachiocephalic (innominate) vein.

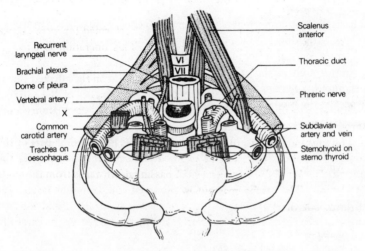

Fig. 208. The root of the neck.

The *second* part of the artery lies behind scalenus anterior which separates it from the subclavian vein. Behind lie scalenus medius and also the middle and upper trunks of the brachial plexus.

The *third* part extends to the lateral border of the first rib against which it can be compressed and its pulse easily felt since here it is just below the deep fascia. Immediately behind the artery is the lower trunk of the brachial plexus which is, in fact, responsible for the 'subclavian groove' on the first rib.

*Branches of the subclavian artery*

*1st part:*

1.  The vertebral artery
2.  The thyro-cervical trunk—
3.  The internal mammary artery

{ *a* inferior thyroid artery
  *b* transverse cervical artery
  *c* suprascapular artery

*2nd part:*

The costo-cervical trunk (supplying deep structures of the neck and giving off the 1st and 2nd posterior intercostal arteries).

*3rd part:*

Gives no constant branch.

### The vertebral artery

This is the most important of the lateral branches of the subclavian artery. It crosses the dome of the pleura, traverses the transverse foramina of the upper six cervical vertebrae, then turns posteriorly and medially over the posterior arch of the atlas to enter the cranial cavity at the foramen magnum by piercing the dura mater. It then runs on the antero-lateral aspect of the medulla to join its fellow in front of the pons to form the *basilar artery* (Fig. 207).

The following are the important branches of the vertebral artery:

1.  Anterior and posterior spinal arteries.
2.  Posterior inferior cerebellar artery.

(From the basilar):

3.  Anterior inferior cerebellar artery.
4.  Superior cerebellar artery.
5.  Posterior cerebral artery (supplying the occipital lobe and medial aspect of the temporal lobe) (Fig. 206).

### CLINICAL FEATURES

1.  The right subclavian artery is grafted end-to-side into the right pulmonary artery to short-circuit the pulmonary stenosis of the tetralogy of Fallot (Blalock's operation). It is important to note, therefore, that variations occur in the origins of the right subclavian artery, which may arise directly from the aortic arch either as its first or as its last branch. In the latter case, the right subclavian artery passes behind the trachea and oesophagus in its course to the neck; this vessel may then compress the oesophagus and produce difficulty in swallowing (dysphagia lusoria). Occasionally the left subclavian artery has a common origin with the left carotid from the aortic arch.
2.  An aneurysm of the subclavian artery is not rare; it never involves the thoracic part of the subclavian and its site of election is the third part of the artery. The close relation of the subclavian artery to the brachial

plexus accounts for the pain, weakness and numbness in the arm which accompany this lesion. Oedema of the arm may result from compression of the subclavian vein.

3.   A cervical rib may elevate the subclavian artery and render it unduly palpable; under these circumstances it may closely simulate an aneurysm and, in fact, there may be aneurysmal dilation of the artery distal to the edge of the cervical rib. Vascular changes in the arm associated with a cervical rib are probably due to peripheral emboli thrown off from thrombi forming on the walls of the compressed subclavian artery.

# The Veins of the Head and Neck

## The cerebral venous system

The venous drainage of the brain follows two pathways:

1.   The superficial structures, e.g. the cerebral and cerebellar cortices, drain to the nearest available dural sinus (see below) by thin-walled veins.
2.   The deep structures drain through the *internal cerebral vein* on each side, which is formed at the interventricular foramen by the junction of the *choroid vein* (draining the choroid plexus of the lateral ventricle) with the *thalamo-striate vein* (draining the basal ganglia).

The two internal cerebral veins unite to form the *great cerebral vein* (the vein of Galen) which emerges from under the splenium of the corpus callosum to join the inferior sagittal sinus in the formation of the straight sinus.

## The venous sinuses of the dura (Fig. 209)

The venous sinuses lie between the layers of the dura. They receive the venous drainage of the brain and of the skull (*the diploic veins*) and disgorge ultimately into the internal jugular vein. They also communicate with the veins of the scalp, face and neck via *emissary veins* which pass through a number of the foramina in the skull.

*The superior sagittal sinus* lies along the attached edge of the falx cerebri and ends posteriorly (usually) in the right transverse sinus. Connecting with it are a number of venous lakes (*lacunae laterales*) into which project the *Pacchionian bodies* of arachnoid, filtering C.S.F. back into the blood.

*The inferior sagittal sinus* lies in the free margin of the falx cerebri and opens into the straight sinus.

*The straight sinus* lies in the tentorium cerebelli along the attachment of the falx cerebri. It is formed by the junction of the great cerebral vein with the inferior sagittal sinus and runs backwards to open (usually) into the left transverse sinus.

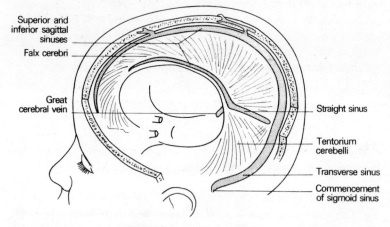

**Fig. 209.** The venous dural sinuses.

*The transverse sinuses* commence at the internal occipital protuberance and run in the tentorium cerebelli on either side along its attached margin. On reaching the mastoid part of the temporal bone each passes downwards, forwards and medially as the *sigmoid sinus* to emerge through the jugular foramen as the *internal jugular vein*.

*The cavernous sinuses* (Fig. 210) lie one on either side of the body of the sphenoid against the fibrous wall of the pituitary fossa and rest inferiorly on the greater sphenoid wing. They communicate freely with each other via the *intercavernous sinuses*.

Traversing the cavernous sinus are the carotid artery and the cranial nerves III, IV, V (ophthalmic and maxillary divisions) and VI. Lying above the cavernous sinus are three important structures—the optic tract, the uncus and the internal carotid artery, which first pierces the roof of the sinus then doubles back to lie against it.

The *ophthalmic veins* drain into the anterior aspect of the cavernous sinus which also links up, through these veins, with the pterygoid venous plexus and the anterior facial vein. The cavernous sinus also receives venous drainage from the brain (*the superficial middle cerebral vein*) and from the dura (*the spheno-parietal sinus*).

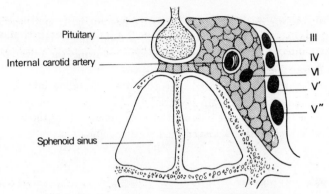

**Fig. 210.** The cavernous sinus—shown in coronal section.

Posteriorly the *petrosal sinuses* drain the cavernous sinus into the sigmoid sinus and into the commencement of the internal jugular vein.

CLINICAL FEATURES

1. The cavernous sinus is liable to sepsis and thrombosis as a result of spread of superficial infection from the lips and face via the anterior facial and ophthalmic veins, or from deep infections of the face via the pterygoid venous plexus around the pterygoid muscles, or from suppuration in the orbit or accessory nasal sinuses along the ophthalmic vein and its tributaries. A characteristic picture results—blockage of the venous drainage of the orbit causes oedema of the conjunctiva and eyelids and marked exophthalmos, which demonstrates transmitted pulsations from the internal carotid artery. Pressure on the contained cranial nerves results in ophthalmoplegia. Examination of the fundus shows papilloedema, venous engorgement and retinal haemorrhages, all resulting from the acutely obstructed venous drainage.

2. Fractures of the skull or penetrating injuries of the skull base may rupture the internal carotid artery within the cavernous sinus. A caroticocavernous arterio-venous fistula results with pulsating exophthalmos, a loud bruit easily heard over the eye and, again, ophthalmoplegia and marked orbital and conjunctival oedema due to the venous pressure within the sinus being raised to arterial level.

3. The sigmoid and transverse sinuses are often together termed '*the lateral sinus*' by clinicians. Close relationship to the mastoid and middle ear renders these sinuses liable to infective thrombosis secondary to otitis media.

Spread of infection or thrombosis from the lateral sinus to the sagittal sinus may cause impaired C.S.F. drainage into the latter and therefore the development of a hydrocephalus—this syndrome of raised C.S.F. pressure associated with sinus thrombosis following ear infection is termed *otitic hydrocephalus.*

It is also possible for sagittal sinus thrombosis to follow infections of the skull, nose, face or scalp because of its diploic and emissary vein connections; if there were no emissary veins, infections of the face and scalp would never have achieved their sinister reputation.

## The internal jugular vein

The internal jugular vein runs from its origin at the jugular foramen (where it continues the sigmoid sinus) to its termination behind the sternal extremity of the clavicle, where it joins the subclavian vein to form the brachiocephalic (innominate) vein.

It lies lateral first to the internal and then to the common carotid artery within the carotid sheath and its relations are therefore identical with these vessels (Fig. 205). The deep cervical chain of lymph nodes lies close against the vein and, if involved by malignant or inflammatory disease, may become densely adherent to the vein. Tearing of the jugular vein for this reason is far from rare in dissections of tuberculous cervical lymph nodes.

Tributaries:

1.  the pharyngeal venous plexus;
2.  the common facial vein;
3.  the lingual vein;
4.  the superior and middle thyroid veins.

The arrangement of the *superficial veins* of the head and neck are somewhat variable but the usual plan is as follows (Fig. 211).

The *superficial temporal* and *maxillary* veins join to form the *posterior facial vein.* This branches while traversing the parotid gland; its posterior division, together with the *posterior auricular vein*, forms the *external jugular vein*, whereas the anterior division joins the anterior facial vein to form the *common facial vein* which opens into the internal jugular.

*The external jugular vein* crosses the sterno-mastoid in the superficial fascia, traverses the roof of the posterior triangle then plunges through the deep fascia 1 in (2.5 cm) above the clavicle to enter the subclavian vein.

*The anterior jugular vein* runs down one on either side of the mid-line of the neck, crossing the thyroid isthmus. Just above the sternum it communicates with its fellow then passes outwards, deep to the sternomastoid, to enter the external jugular vein.

# The Lymph Nodes of the Neck

Although the lymph drainage of particular viscera is dealt with under appropriate headings (tongue, larynx, etc.) it is convenient to summarize here the arrangements of the lymph nodes of the head and neck as a whole (Fig. 212). These can be grouped into horizontally and vertically disposed aggregates.

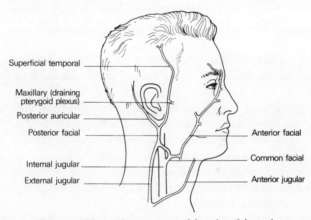

Superficial temporal

Maxillary (draining pterygoid plexus)

Posterior auricular

Posterior facial

Internal jugular

External jugular

Anterior facial

Common facial

Anterior jugular

Fig. 211. The usual arrangement of the veins of the neck.

*The horizontal nodes* form a number of groups which encircle the junction of the head with the neck and which are named, according to their position, the *submental, submandibular, superficial parotid* (or preauricular), *mastoid* and *suboccipital* nodes.

These nodes drain the superficial tissues of the head and efferents then pass to the deep cervical nodes (although some lymph vessels pass direct to the cervical nodes, by-passing the horizontal nodes).

*The vertical nodes* drain the deep structures of the head and neck. The most important is the *deep cervical group* which extends along the internal jugular vein from the base of the skull to the root of the neck (*see* Fig. 193). The lymph then passes via the jugular trunk to the thoracic duct or the right lymphatic duct.

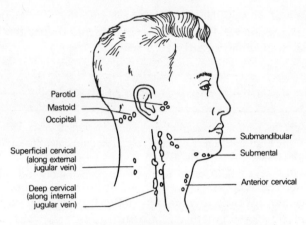

**Fig. 212.** Scheme of the lymph nodes of the head and neck.

*The superficial cervical nodes* lie along the external jugular vein, serve the parotid and lower part of the ear and drain into the deep cervical group.

Along the front of the neck lies another group of vertically disposed nodes, the *infrahyoid* (on the thyrohyoid membrane), the *prelaryngeal* and the *pre- and para-tracheal* nodes. These drain the thyroid, larynx, trachea and part of the pharynx and empty into the deep cervical group.

*The retropharyngeal nodes*, lying vertically behind the pharynx, drain the back of the nose, pharynx and Eustachian tube; their efferents pass cervical nodes.

Thus all structures in the head and neck drain through the deep cervical nodes either directly or ultimately.

**CLINICAL FEATURES.**

1. *Block dissection of the neck* for malignant disease is the removal of the lymph nodes of the anterior and posterior triangles of the neck and their associated lymph channels, together with those structures which must be excised in order to make this lymphatic ablation possible. It is sometimes combined with en-bloc removal of the primary tumour.

The usual incision is Y-shaped, its centre being at the level of the upper border of the thyroid cartilage, its lower limb running downwards to the mid-point of the clavicle, its anterior limb extending to the symphysis menti and its posterior limb to the mastoid process. The block of tissue removed extends from the mandible above to the clavicle below and from the mid-line anteriorly to the anterior border of the trapezius behind. It consists of all the structures between the platysma and pre-tracheal fascia enclosed by these boundaries, preserving only the carotid arteries, the vagus trunk, the cervical sympathetic chain and the lingual and hypoglossal nerves. The sterno-mastoid, omo-hyoid and digastric muscles are removed in the dissection. Excision also includes the external and internal jugular veins, around each of which lymph nodes are intim-ately related, and the submandibular gland and the lower pole of the parotid gland, since these both contain potentially involved lymph nodes.

The accessory nerve, passing across the posterior triangle, is usually sacrificed.

2.   Tuberculous disease of the neck usually involves the upper part of the deep cervical chain (from tonsillar infection). These infected nodes may adhere very firmly to the internal jugular, vein which may be wounded in the course of their excision.

# The cervical sympathetic trunk

The sympathetic chain continues upwards from the thorax by crossing the neck of the first rib, then ascends embedded in the posterior wall of the carotid sheath to the base of the skull (Fig. 213). It bears three ganglia:

*The superior cervical ganglion* (the largest) lies opposite C2 and 3 vertebrae and sends grey rami communicantes to C1–4 spinal nerves.

*The middle ganglion* lies level with C6 vertebra and sends grey rami to C5 and 6 nerves.

*The inferior ganglion* lies level with C7 and is tucked behind the verte-bral artery. Frequently it fuses with the first thoracic ganglion to form the *stellate ganglion* at the neck of the first rib. Grey rami pass from it to C7 and 8 nerves.

Note that these ganglia receive no white rami from the cervical nerves; their pre-ganglionic fibres originate from the upper thoracic white rami and then ascend in the sympathetic chain.

As well as *somatic branches* transmitted with the cervical nerves, the cervical chain gives off *cardiac branches* from each of its ganglia and also *vascular* plexuses along the carotid, subclavian and vertebral vessels. The sympathetic fibres to the dilator pupillae muscle travel in this plexus along the internal carotid artery. Grey rami pass from the superior ganglion to cranial nerves VII, IX, X and XI.

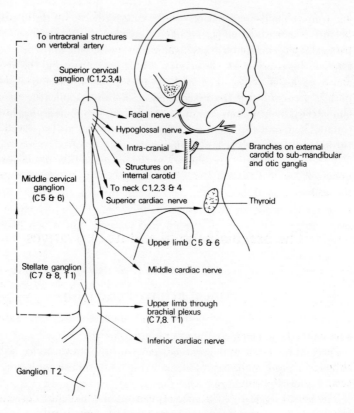

**Fig. 213.** The cervical sympathetic chain.

CLINICAL FEATURES

1. 'Cervical sympathectomy' is a misnomer; it is an upper thoracic sympathectomy carried out through a cervical incision. The sympathetic chain is divided below the 3rd thoracic ganglion and the grey and white rami to the 2nd and 3rd ganglia are also cut. In this way the sudomotor and vasoconstrictor pathways to the head and upper limb (from segments T2, 3 and 4) are divided, preserving the T1 connection and the stellate ganglion which are the sympathetic connections to the eyelid and pupil. The upper thoracic chain can also be removed via a transthoracic transpleural approach through the second intercostal space, the incision being situated on the medial wall of the axilla. The lung is allowed to collapse and the chain identified as it lies on the heads of the upper ribs. Resection of the T2–4 segment results in a warm, dry hand.

2. *Horner's syndrome* results from interruption of the sympathetic fibres to the eyelids and pupil. The pupil is constricted (unopposed para-sympathetic innervation via the oculo-motor nerve), there is ptosis (partial paralysis of levator palpebrae) and the face on the affected side is dry and flushed (sudomotor and vasoconstrictor denervation). Enophthalmos is said to occur, but this is not confirmed by exophthalmometry. The syndrome may follow spinal cord lesions at the T1 segment (tumour or syringomyelia), closed, penetrating or operative injuries to the stellate ganglion or the cervical sympathetic chain, or pressure on the chain or stellate ganglion produced by enlarged cervical lymph nodes, an upper mediastinal tumour, a carotid aneurysm or a malignant mass in the neck.

# The brachial system and its derivatives

Six visceral arches form on the lateral aspects of the foetal head separated, on the outside, by ectodermal branchial clefts and, on the inside, by 5 endodermal pharyngeal clefts (Fig. 189). In the human embryo the 5th and 6th arches do not appear externally and are represented only by a mesodermal core and by the 5th pharyngeal cleft.

Each arch has its own nerve supply, cartilage, muscle and artery, although considerable absorption and migration of these derivatives occur in development.

The embryological significance of many of the branchial derivatives has already been discussed under appropriate headings (the development of the face, tongue, thyroid, parathyroid and aortic arch) but Table 4 serves conveniently to bring these various facts together.

### Branchial cyst and fistula

The second branchial arch grows downwards to cover the remaining arches, leaving temporarily a space lined with squamous epithelium. This usually disappears but may persist and distend with cholesterol-containing fluid to form a *branchial cyst.*

If fusion fails to occur distally a sinus persists at the anterior border of the origin of the sterno-mastoid; this *branchial fistula* can be traced upwards between the internal and external carotids and may even open into the tonsillar fossa, demonstrating its association with the second branchial arch.

**Table 4.** Derivatives of the six arches and five clefts of the branchial system.

| Arch | Nerve Supply | Visceral Arch | External Cleft | Internal Cleft | Floor | Cartilage | Muscle | Artery |
|---|---|---|---|---|---|---|---|---|
| I | V | Lower face | External auditory meatus | Eustachian tube and middle ear | Anterior two-thirds of tongue | Meckel's cartilage. Incus and Malleus | Muscles of mastication and anterior belly of digastric | Disappears |
| II | VII | | | Tonsil | Contributes to anterior tongue. Thyroid develops as outgrowth between 1st and 2nd arch | Stapes, Styloid process and upper part of Hyoid | Muscles of facial expression and postr. belly of digastric | Disappears |
| III | IX | Grows down to cover remaining external clefts to form the skin of the neck | All covered by 2nd visceral arch | Thymus and inferior para-thyroid | Posterior third of tongue | Lower part of Hyoid | | Carotids |
| IV | X AND | | | Superior parathyroid | | | Muscles of Pharynx, Larynx and Palate | Rt. subclavian Left arch of aorta |
| V | XI | | | Disappears | Outgrowth of lung buds | Cartilages of the larynx | | Disappears |
| VI | | | | | | | | Pulmonary artery and ductus arteriosus |

# The surface anatomy and surface markings
# of the head

Many of the important landmarks of the skull are readily felt. Revise on your own skull the position of: the *external occipital protuberance* (the apex of this is termed the *inion*), the *nasion*, which is the depression between the two *supra-orbital margins*, and the *glabella*, which is the ridge above the nasion. Feel the sharp edge of the lateral margin of the orbit which is formed by the frontal process of the zygoma; behind the zygoma is the *zygomatic arch* with the *superficial temporal artery* crossing its posterior extremity and forming a convenient pulse which the anaesthetist can reach. Rather less easily felt is the *jugal point*, the junction between the zygoma and the zygomatic arch of the temporal bone; it is the mass of bone encountered by the finger running forward along the upper border of the zygomatic arch, and it is a surface marking for the middle meningeal artery (*vide infra*). The anterior edge of the *mastoid* is easily palpable but its posterior aspect and its tip are rather obscured by the insertion of the sternomastoid.

The whole of the superficial surface of the *mandible* is palpable apart from its coronoid process. The *condyloid process* can be felt by a finger placed immediately in front, or within, of the external auditory meatus while the mouth is opened and closed.

When the teeth are clenched, *masseter* and the *temporalis* can be felt contracting respectively over the ramus of the mandible and above the zygomatic arch. The *parotid duct* can be rolled over the tensed masseter and its orifice seen within the mouth at the level of the 2nd upper molar tooth.

A line drawn vertically between the first and second pre-molar teeth passes through the mental foramen, the infra-orbital foramen and the supra-orbital notch. Through these three orifices lying in plumb-line pass branches from each of the divisions of the trigeminal nerve, respectively the mental branch of the inferior dental nerve (V'''), the infra-orbital nerve (V'') and the supra-orbital nerve (V').

The *middle meningeal artery* can be represented by a line drawn upwards and somewhat forwards from a point along the zygomatic arch, two fingers' breadths behind the jugal point. The posterior branch of this artery passes backwards a thumb's breadth above, and roughly parallel to, the zygomatic arch.

The *central sulcus* of the cerebrum corresponds to a line drawn downwards and forwards from a point 1 cm behind the mid-point between the nasion and the inion.

# The scalp

The soft tissues of the scalp are arranged in five layers (Fig. 214), which may be remembered thus:

S — skin.
C— connective tissue.
A — aponeurosis.
L — loose connective tissue.
P — periosteum.

Each of these layers has features of practical importance.

*The skin* of the scalp is richly supplied with sebaceous glands and is the commonest site in the body for sebaceous cysts.

*The subcutaneous connective tissue* consists of lobules of fat bound in tough fibrous septa, very much like the connective tissue of the palm and the sole. Because of this dense encapsulation of fat, it is not surprising that lipomata are extremely rare at these three sites, and also that excess fat does not collect in any of these places even in the grossly obese.

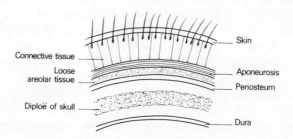

Fig. 214. The layers of the scalp.

The blood-vessels of the scalp lie in this layer. When the head is lacerated, the divided vessels retract between the fibrous septa and cannot be picked up individually by artery forceps in the usual way. Haemorrhage is arrested by pressing with the fingers firmly down onto the skull on either side of the wound (thus compressing the vessels), by placing series of artery forceps on the divided aponeurotic layer so that their weight again compresses these vessels and, finally, by suturing the laceration firmly in two layers (aponeurotic and cutaneous).

The haemorrhage from a scalp laceration or operation is profuse; this area has, in fact, the richest cutaneous blood supply of the body. For this reason, extensive avulsions of the scalp are usually viable providing even a narrow pedicle remains attached to the surrounding tissues.

The veins of the scalp connect with the intracranial venous sinuses via numerous *emissary veins* which pierce the skull and which also link these two venous systems with the *diploic veins* between the tables of the skull vault. A superficial infection of the scalp may spread via this system producing an osteitis of the skull, meningitis and venous sinus thrombosis.

*The aponeurotic layer* is the occipito-frontalis which is fibrous over the dome of the skull but muscular in the occipital and frontal regions. This muscle arises from the superior nuchal line of the occipital bone, gains a fascial insertion into the zygomatic arch, and inserts anteriorly into the subcutaneous tissues of the eyebrows and nose.

*The layer of loose connective tissue* beneath the aponeurosis accounts for the mobility of the scalp on the underlying bone; it is in this plane that the surgeon mobilizes scalp flaps, that machinery which has caught onto the hair avulses the scalp and that the Red Indians of bygone days scalped their victims.

Blood or pus collecting in this loose tissue tracks freely under the scalp but cannot pass into either the occipital or subtemporal regions because of the attachments of occipito-frontalis. Fluid can, however, track forward into the orbits and this accounts for the orbital haematoma that may form a few hours after a severe head injury or cranial operation.

The aponeurotic layer is under tension because of its muscular component and retracts on the underlying loose layer when divided; a gaping scalp wound must, therefore, have extended at least through the aponeurosis.

*The periosteum* adheres to the suture-lines of the skull; collections of pus or blood beneath this layer therefore outline the affected bone. This is particularly well seen in birth injuries involving the skull (*cephalo-haematoma*).

# The skull (Figs. 215, 216 and 217)

The important regional anatomy of the skull is dealt with under the appropriate headings (ear, nose, accessory sinuses, etc.).

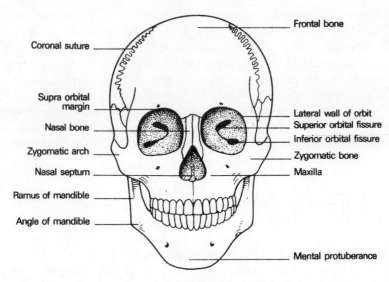

Coronal suture

Supra orbital margin

Nasal bone

Zygomatic arch

Nasal septum

Ramus of mandible

Angle of mandible

Frontal bone

Lateral wall of orbit

Superior orbital fissure

Inferior orbital fissure

Zygomatic bone

Maxilla

Mental protuberance

Fig. 215. The skull: anterior aspect.

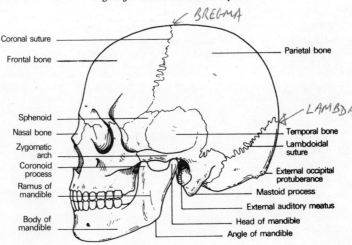

BREGMA

Coronal suture

Frontal bone

Sphenoid

Nasal bone

Zygomatic arch

Coronoid process

Ramus of mandible

Body of mandible

Parietal bone

LAMBDA

Temporal bone

Lambdoidal suture

External occipital protuberance

Mastoid process

External auditory meatus

Head of mandible

Angle of mandible

Fig. 216. Skull: lateral aspect.

Collected together in this section are some general facts of clinical relevance.

The bony vault of the skull is relatively elastic in consistency; thus a blow may injure the underlying brain without fracturing bone. Where

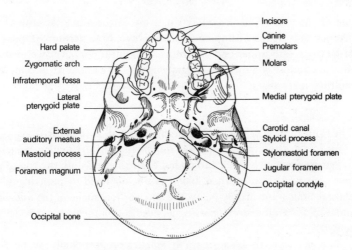

Fig. 217. Skull. Inferior aspect.

the cranium is protected by thick muscle (the lower part of the occipital bone and the squamous temporal) the skull is correspondingly thin; if held up to the light it can be seen to be translucent at these sites.

The palpable landmarks of the skull are enumerated in the section on the surface anatomy of the head. Radiologically, the *sutures* between the vault bones are important because they, as well as the vascular markings of the meningeal and diploic vessels, may be confused with fracture lines. The *coronal* suture divides the frontal from the parietal bones, the *sagittal* suture separates the parietal bones in the mid-line, the *lambdoid* suture marks off the occipital from the parietal and temporal bones and the *squamosal suture* separates squamous temporal bone from the parietal bone and greater wing of sphenoid.

In about 8 per cent of cases the *metopic suture* persists in the mid-line between the two frontal bones; in the rest, this suture fuses at about the 5th year.

Occasionally small separate areas of ossification develop between the parietal and occipital bones termed *Wormian bones* which, again, may cause radiological confusion.

The *lambda* is the point of junction of the lambdoid and sagittal sutures (the *posterior fontanelle* of infancy).

The *bregma* is the junction of the sagittal and coronal sutures (the infant's *anterior fontanelle*).

The *diploë*, between the inner and outer tables of the skull vault, is one of the sites of persistent red marrow in the adult skeleton. This distinction it shares with the pelvis, vertebrae, ribs, sternum, upper end of humerus and upper end of femur; a doubtful honour since to these sites are almost confined secondary deposits of carcinoma in bone and multiple myelomata.

## Development

The skull vault develops in membrane, the skull base in cartilage.

At birth, the square anterior fontanelle and triangular posterior fontanelle are widely open and do not normally fuse until the end of the first year of life. Up till then, blood can be obtained by puncturing the sagittal sinus immediately below the anterior fontanelle in the mid-line, and C.S.F. aspirated by passing a needle obliquely into the lateral ventricle.

The face at birth is considerably smaller proportionally to the skull than in the adult; this is due to the teeth being non-erupted and rudimentary and the nasal accessory sinuses being undeveloped; the sinuses are evident at about 8 years but only fully develop in the late 'teens.

The mastoid and its air cells develop at the end of the 2nd year; until then the facial nerve is relatively superficial near its origin from the skull and may be damaged by quite trivial injuries.

With advancing age, the relative vertical measurement of the face again diminishes as a result of loss of teeth and subsequent absorption of the alveolar margins.

### CLINICAL FEATURES

1. *Fractures of the skull*

Imagine the skull as a rather elastic sphere completely filled by semi-fluid material; a violent blow on such a structure will produce a splitting effect commencing at the site of the blow and tending to travel along lines of least resistance. The base of the skull is more fragile than the vault, and is thus commonly involved by such fractures. The petrous part of the temporal bone, however, forms a firm and rarely involved buttress of the skull base, the fracture line passing through less resistant areas, particularly the middle cranial fossa, the pituitary fossa and the various basal foramina.

A localized severe injury, in the adult, may produce a depressed comminuted fracture; the infant's skull is much more elastic and a similar injury here will result in a 'pond' depressed fracture, rather like the dimple produced by squeezing on a ping-pong ball.

**2.** *Localizing signs in cranial fractures*

Fractures of the anterior cranial fossa may involve the frontal, ethmoidal and sphenoidal sinuses and be accompanied by bleeding into the nose or mouth. C.S.F. leakage from the nose in such cases implies co-existing tearing of the meninges; the subarachnoid space is thus put in communication with the exterior via the nasal cavity with consequent risk of meningitis.

Fractures involving the roof of the orbit are frequently associated with blood tracking forward beneath the conjunctiva (sub-conjunctival haemorrhage); this must be differentiated from a small flame-shaped haemorrhage of the conjunctiva caused by direct injury to it.

A 'black eye' is not necessarily indicative of an anterior fossa fracture; it may be produced also by direct contusion of the soft tissues or by blood tracking down deep to the aponeurotic layer of the scalp (see scalp).

Anterior basal fractures may involve the cribriform plate (with anosmia—loss of smell—due to rupture of fibres of the olfactory bulb) or the optic foramen (with primary optic atrophy and blindness).

Fractures of the middle fossa may produce bleeding into the mouth (sphenoid involvement), bleeding or C.S.F. leakage from the ear, and facial and auditory nerve injury. Aural bleeding may, of course, be produced by direct injury to the ear, for example, rupture of the drum, without necessarily implying a skull fracture.

Posterior fossa fractures are occasionally accompanied by cranial nerve involvement. These fractures are suggested clinically by bruising over the mastoid region extending downwards over the sterno-mastoid.

# The accessory nasal sinuses

The nasal sinuses are air-containing sacs lined by ciliated epithelium and communicating with the nasal cavity through narrow, and therefore easily occluded, channels (Fig. 218). The maxillary antrum and sphenoid sinuses are present in a rudimentary state at birth, the rest become evident about the 8th year, but all become fully formed only in adolescence.

## The frontal sinuses

The frontal sinuses are contained in the frontal bone. In section each is roughly triangular, its anterior wall forming the prominence of the forehead, its postero-superior wall lying adjacent to the frontal lobe of

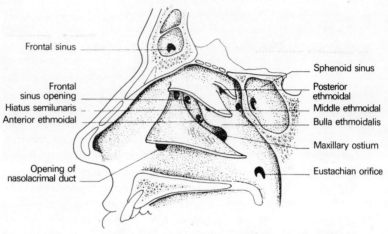

Frontal sinus

Frontal sinus opening
Hiatus semilunaris
Anterior ethmoidal

Opening of nasolacrimal duct

Sphenoid sinus

Posterior ethmoidal
Middle ethmoidal
Bulla ethmoidalis

Maxillary ostium

Eustachian orifice

**Fig. 218.** The lateral wall of the right nasal cavity; the conchae have been partially removed.

the brain, and its floor abutting against the ethmoid cells, the roof of the nasal fossa and the orbit.

The frontal sinuses are separated from each other by a median bony septum, and each in turn is further broken up by a number of incomplete septa. Each sinus drains into the anterior part of the middle nasal meatus.

CLINICAL FEATURES

1. The close relation of the frontal sinus to the frontal lobe of the brain explains how infection of this sinus may result in the development of a frontal lobe abscess.

2. A fracture involving the sinus, severe enough to tear the dura and pia-arachnoid, will place the subarachnoid space in communication with the nasal cavity. C.S.F. may then be detected trickling through the nostril, usually on the affected side (C.S.F. rhinorrhoea) although, as these sinuses may communicate, a contra-lateral leak sometimes occurs.

3. The neurosurgeon must take into account the considerable variations in size and extent of the frontal sinus when proposing to turn down a frontal skull-flap; obviously he will want to avoid opening the sinus because of the risk of infection. He therefore consults the radiographs of the patient's skull pre-operatively, which will clearly show the configuration of the sinuses.

**The maxillary sinus (antrum of Highmore)** (Fig. 219)

This is a pyramidal-shaped sinus occupying the cavity of the maxilla.

Its medial wall forms part of the lateral face of the nasal cavity and bears on it the inferior concha. Above this concha is the opening, or *ostium*, of the maxillary sinus into the middle meatus. This opening, unfortunately, is inefficiently placed as an adequate drainage point.

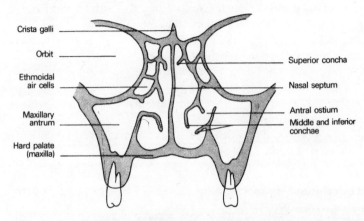

Fig. 219.    The maxillary antrum in coronal section. (Note the inefficient drainage of this antrum and its close inferior relationship to the teeth.)

The infra-orbital nerve lies in a groove which bulges down into the roof of the sinus, while its floor bears the impressions of the upper pre-molar and molar roots. These roots are separated only by a thin layer of bone which may, in fact, be deficient so that uncovered dental roots project into the sinus. Note that the floor of the sinus therefore corresponds to the level of the alveolus and not to the floor of the nasal cavity—it actually extends about $\frac{1}{2}$ in (12 mm) lower than the latter.

CLINICAL FEATURES

1.    The maxillary sinus, or antrum, may become infected either from the nasal cavity or from caries of the upper molar teeth.

Antral puncture can be carried out using a trocar and cannula passed through the nasal cavity in an outward and backward direction below the inferior concha.

More adequate drainage may require removing a portion of the medial wall of the sinus below the inferior concha or fenestrating the antrum in the gingivo-labial fold (Caldwell Luc operation). The old operation of draining the antrum via an extracted upper molar tooth is now seldom performed.

2.    The numerous symptoms and signs which may be produced by a carcinoma of the maxillary sinus are easily remembered anatomically.
*a*.    Medial invasion encroaches on the nasal cavity, producing obstruction of the nares and epistaxis. Blockage of the naso-lacrimal duct in this wall may cause epiphorea (leakage of tears down the face).
*b*.    Invasion of the orbit displaces the globe and causes diplopia. If the infra-orbital nerve becomes involved there will be facial pain and then anaesthesia of the skin over the maxilla.
*c*.    Invasion of the sinus floor may produce a visible bulge or even ulceration in the palatal roof.
*d*    Lateral spread may produce a swelling of the face or a palpable mass in the gingivo-labial fold.
*e*    Posterior spread may involve the palatine nerves and produce severe pain referred to the teeth of the upper jaw.

### The ethmoid sinuses

The ethmoid sinuses are made up of a group of 8–10 air cells within the lateral mass of the ethmoid and lie between the side walls of the upper nasal cavity and the orbits. Superiorly they lie on each side of the cribriform plate and are related above to the frontal lobes of the brain. These cells drain into the superior and middle meatus (Fig. 218).

CLINICAL FEATURES
As with the frontal sinus, infection (ethmoiditis) may result in a frontal cerebral abscess and an ethmoidal fracture may cause a C.S.F. leakage into the nasal cavity.

### The sphenoid sinuses

These lie one on either side of the mid-line, within the body of the sphenoid. They vary a good deal in size and may extend laterally into the greater wing of the sphenoid or backwards into the basal part of the occipital bone.

   Each sinus drains into the nasal cavity above the superior concha (the spheno-ethmoidal recess).

CLINICAL FEATURES
The pituitary gland may be excised through a trans-sphenoidal approach in patients with pituitary tumour or disseminated breast cancer.

# The mandible (Fig. 220)

The lower jaw comprises a horizontal *body* on each side which fuse at the *symphysis menti* (fusion occurring at the second year). From the posterior part of the body projects the vertical *ramus* which bears an

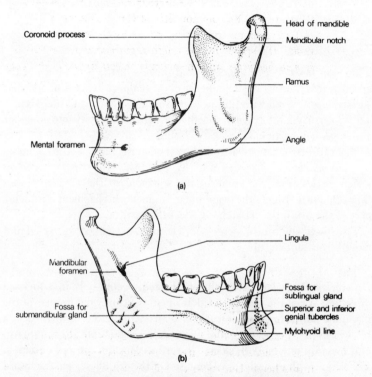

Fig. 220. The mandible. (a) Lateral aspect. (b) Medial aspect.

anterior *coronoid* and a posterior *condyloid* process or *head*. Between the two is the *mandibular notch*.

On the medial aspect of the ramus is the *mandibular foramen* for the inferior dental branch of the mandibular division of the trigeminal nerve, which traverses the body within the *mandibular canal*, then emerges as the mental nerve through the *mental foramen* on the lateral surface of the body below and between the two pre-molars. The nerve supply to the incisors and canine runs forward within the mandible beyond this point in the *incisive canal*.

The upper border of the body bears the *alveolar border* with 16 dental sockets or *alveoli*.

## Development

The mandible develops as membrane bone in the fibrous sheath of Meckel's cartilage (the cartilage of the first branchial arch). The cartilage itself is completely absorbed.

## The temporo-mandibular joint

This joint lies between the condyloid process of the mandible and the articular fossa and articular eminence of the temporal bone. The articular surfaces are covered with fibrous (*not* hyaline) cartilage and there is also a fibro-cartilaginous *articular disc* dividing the joint cavity into an upper and lower compartment.

The capsular ligament surrounding the joint is reinforced by a lateral *temporo-mandibular ligament* and by the *spheno-mandibular ligament* which passes from the spine of the sphenoid to the *lingular process* immediately in front of the mandibular foramen; this ligament represents part of the primitive 1st arch, or Meckel's cartilage.

The lower jaw can be depressed, elevated, protruded, retracted and moved from side to side.

### CLINICAL FEATURES

*Dislocation of the jaw*, when uncomplicated, occurs only in a forward direction. When the mouth is widely open, the condyloid process of the mandible slides forward onto the articular eminence; from thence, a blow or even a yawn, may cause forward dislocation into the infratemporal fossa on one or both sides. Upward dislocation can only occur in association with extensive comminution of the skull base, and backward dislocation with smashing of the bony external auditory canal and tympanic cavity which lie immediately behind the joint.

Reduction is effected by pressing down on the molar teeth with the thumbs placed in the mouth, at the same time pulling up the chin; the former stretches the masseter and temporalis muscles which are in spasm, the latter levers the mandibular head back into place.

## The teeth

There are 20 deciduous or 'milk' teeth replaced by 32 permanent teeth made up, in each half jaw, thus:

*Deciduous:* 2 incisors, 1  canine, 2 molars;
*Permanent:* 2 incisors, 1 canine, 2 premolars, 3 molars.

The times of eruption of the teeth are useful stepping-stones in a child's development as well as being of forensic interest.

As a rough guide, these times can be thought of in multiples of 6, thus:
- the 1st lower incisor deciduous tooth appears at 6 months,
- all the deciduous teeth have appeared by     24 months,
- the permanent 1st molar ⎫
- the permanent 1st incisor ⎭ appear at     6 years,
- the second permanent molar appears at     12 years,
- the third permanent molar appears at     (approx.) 24 years.

The lower teeth appear somewhat before their corresponding upper neighbours.

Each tooth is fixed in its socket by the *periodontal membrane* which is, in fact, periosteum. This layer is radio-translucent and is the dark line seen around the root of each tooth on radiography.

## Development
The enamel crown of the tooth develops from a downgrowth of the alveolar epithelium and represents the toughest tissue in the human body. The rest of the tooth (pulp, dentine and cement) differentiates from the underlying mesodermal connective tissue.

CLINICAL FEATURES
Osteomyelitis of the jaw following dental extractions is confined to the lower jaw and occurs only with the permanent dentition; the explanation of this is an anatomical one:

The lower jaw is supplied *only* by the inferior dental artery, which runs with the nerve in the mandibular canal; damage to this artery at extraction, or its thrombosis in subsequent infection, therefore produces bone necrosis. The upper jaw, on the other hand, receives segmental vertical branches from the superior dental vessels and ischaemia does not follow injury to an individual artery. The deciduous teeth of the lower jaw are placed well clear of the mandibular canal which is, in any case, protected by the un-erupted permanent teeth; damage to the artery cannot therefore occur during their removal.

# The vertebral column

The spinal, or vertebral, column is made up of 33 vertebrae of which 24 are discrete vertebrae and 9 are fused in the sacrum and coccyx.

In the embryo the spine is curved into a gentle C shape but, with the extension of the head and lower limbs that occurs when the child first holds up its head, then sits and then stands, secondary forward curvatures appear in the cervical and lumbar region which produce the sinusoidal curves of the fully developed spinal column.

The basic vertebral pattern (Fig. 221) is that of a *body* and of a *neural arch* surrounding the *vertebral canal*.

The neural arch is made up of a *pedicle* on either side, each supporting a *lamina* which meets its opposite posteriorly in the mid-line. The pedicle bears a notch above and below which, with its neighbour, forms the *intervertebral foramen*. The arch bears a posterior *spine*, lateral *transverse processes* and upper and lower *articular facets*.

The intervertebral foramina transmit the segmental spinal nerves as follows: C1 to C7 pass over the superior aspect of their corresponding cervical vertebrae, C8 passes through the foramen between C7 and T1, then all subsequent nerves pass between the vertebra of their own number and the one below.

Now to consider the individual vertebrae in turn.

### The cervical vertebrae (7)

These are readily identified by the *foramen transversarium* perforating the transverse processes. The spines are small and bifid (except C1 and C7 which are single) and the articular facets are relatively horizontal. (Fig. 222).

*The atlas* (C1) (Fig. 223) has no body. Its upper surface bears a superior articular facet on a thick lateral mass on each side which articulates with the occipital condyles of the skull.

Just posterior to this facet the upper aspect of the arch of the atlas is grooved by the vertebral artery as it passes medially and upwards to enter the foramen magnum.

*The axis* (C2) (Fig. 224) bears the *odontoid process* on the superior aspect of its body, representing the detached centrum of C1.

Nodding and lateral flexion movements occur at the atlanto-occipital joint, whereas rotation of the skull occurs at the atlanto-axial joint around the odontoid which acts as a pivot.

7/12/5/ 9(5,4)

33→ 72→
→9 fused

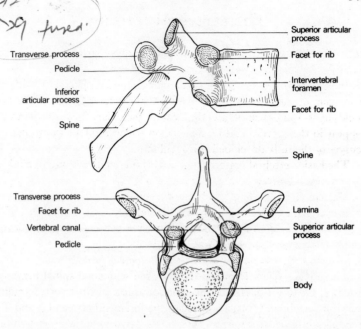

Fig. 221. A 'typical' thoracic vertebra.

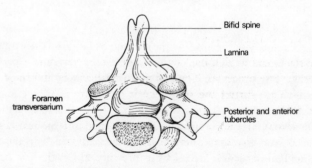

Fig. 222. Typical cervical vertebra.

$C7$ is the *vertebra prominens*, so called because of its relatively long and easily felt non-bifid spine; it is the first clearly palpable spine on running one's fingers downwards along the vertebral crests, although the spine of T1 immediately below it is, in fact, the most prominent one.

The vertebral artery enters its vertebral course nearly always at the foramen transversarium of C6; it is not surprising, therefore, that the

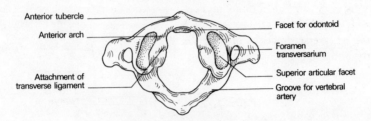

Anterior tubercle

Anterior arch

Attachment of
transverse ligament

Facet for odontoid

Foramen
transversarium

Superior articular facet

Groove for vertebral
artery

Fig. 223.   Atlas in superior view.

Odontoid process

Transverse process

Facet for anterior
arch of atlas

Foramen
transversarium

Body

Fig. 224.   The axis in oblique lateral view.

foramen of C7, which transmits only the vein, is small or even sometimes absent.

## The thoracic vertebrae (12)

These vertebrae are characterized by facets on the sides of their bodies for articulation with the heads of the ribs and by facets on their transverse processes (apart from those of the lower 2 or 3 vertebrae) for the rib tubercles. The spines are long and downward sloping and the articular facets are also relatively vertical.

The bodies of T5 to T8 are worth noting; they come into relationship with the descending aorta and are a little flattened by it on their left flank. If the descending aorta becomes aneurysmally dilated, these four vertebral bodies become eroded by its pressure, although their avascular intervertebral discs remain intact. You can make this diagnosis confidently when shown a specimen of four partly worn-away vertebrae with normal intervening discs.

## The lumbar vertebrae (5)

These are of great size with strong, square, horizontal spines and with articular facets which lie in the sagittal plane (Fig. 225).

L5 is distinguished by its massive transverse process which connects with the whole lateral aspect of its pedicle and encroaches on its body; the transverse processes of the other lumbar vertebrae attach solely to the junction of pedicle with lamina.

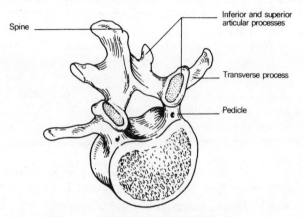

**Fig. 225.** A lumbar vertebra in anterosuperior view.

**The sacrum** (5 fused)
**The coccyx** (3, 4, or 5 fused)

These are considered with the bony pelvis (p. 133–4).

*Development*
Each vertebra ossifies from three primary centres, one for each side of the arch and one for the body. The body occasionally develops from two centres and failure of one of these to form results in formation of a hemivertebra with a consequent congenital scoliosis. Failure of the two arch centres to fuse posteriorly results in the condition of *spina bifida* which occurs particularly in the lumbar region. Usually this is not associated with any neurological abnormality (spina bifida occulta), although in such cases there is often an overlying dimple, lipoma or tuft of hair to warn the observant of a bony abnormality beneath. More rarely there is a gross defect of one or several arches with protrusion of the spinal cord or its coverings; this anomaly may be associated with hydrocephalus.

L5 may occasionally fuse wholly or in part with the sacrum (sacralization of the 5th lumbar vertebra) or, more rarely, the first segment of the sacrum may differentiate as a separate vertebra.

## The intervertebral joints

The spinal column is made up of individual vertebrae which articulate body to body and by their articular facets. Although movement between adjacent vertebrae is slight, the additive effect is considerable. Movement particularly occurs at the cervico-dorsal and dorso-lumbar junctions; these are the two common sites of vertebral injury.

The vertebral laminae are linked by the *ligamentum flavum* of elastic tissue, the spines by the tough *supraspinous* and relatively weak *interspinous ligaments*, and the articular facets by *articular ligaments* around their small synovial joints. All these ligaments serve to support the spinal column when it is in the fully flexed position.

Running the whole length of the vertebral bodies, along their anterior and posterior aspects respectively, are the tough *anterior* and *posterior longitudinal ligaments*.

The vertebral bodies are also joined by the extremely strong *intervertebral discs* (Fig. 226). These each consist of a peripheral *annulus fibrosus*, which adheres to the thin cartilage plate on the vertebral body above and below, and which surrounds the gelatinous semifluid *nucleus pulposus*. The intervertebral discs constitute approximately a quarter of the length of the spine as well as accounting for its secondary curvatures.

In old age, the discs atrophy, with resulting shrinkage in height and return of the curvature of the spine to the C shape of the newborn.

CLINICAL FEATURES

1.    Fractures of the spine most commonly involve T12, L1 and L2. The cause is usually a flexion-compression type of injury (for example, a fall from a height landing on the feet or buttocks, or a heavy weight falling on the shoulders), with resultant wedging of the involved vertebrae. If, in addition to compression, there is forceful forward movement, one vertebra may displace forward on its neighbour below with either dislocation or fracture of the articular facets between the two (fracture dislocation) and with rupture of the interspinous ligaments.

The cervical vertebrae may be fractured or, more commonly, dislocated by a fall on the head with acute flexion of the neck, as might happen on diving into shallow water. Dislocation may even result from the sudden forward jerk which may occur when a motor-car or aeroplane crashes. Note that the relatively horizontal intervertebral facets of the cervical vertebrae allow dislocation to take place without their being fractured,

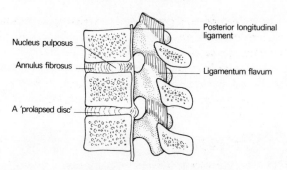

Nucleus pulposus

Annulus fibrosus

A 'prolapsed disc'

Posterior longitudinal ligament

Ligamentum flavum

Fig. 226. Longitudinal section through the lumbar vertebrae showing a normal and a prolapsed intervertebral disc.

whereas the relatively vertical thoracic and lumbar intervertebral facets nearly always fracture in forward dislocation of the dorso-lumbar region.
2. The comparatively thin posterior part of the annulus fibrosus may rupture, either due to trauma or to degenerative changes, allowing the nucleus pulposus to protrude posteriorly into the vertebral canal—the so-called *prolapsed intervertebral disc* (Fig. 226). This may sometimes occur at the lower cervical intervertebral discs (C5/6 and C6/7), very occasionally in the thoracic and upper lumbar region or, by far the most commonly, at the L4/5 or L5/S1 disc.

A prolapsed L4/5 disc produces pressure effects on the root of the 5th lumbar nerve, that of the L5/S1 disc on the 1st sacral nerve. Pain is referred to the back of the leg and foot along the distribution of the sciatic nerve. Hip flexion with the leg extended ('straight leg raising') is painful and limited due to the traction which this movement puts upon the already irritated and stretched nerve root. There may be weakness of ankle dorsiflexion and numbness over the lower and lateral part of the leg and medial side of the foot (L5) or the lateral side of the foot (S1). If S1 is affected, the ankle jerk may be diminished or absent.

Occasionally the disc prolapses directly backwards, and, if this is extensive, may compress the whole cauda equina, producing paraplegia.

# PART SIX
# THE CENTRAL NERVOUS SYSTEM

# The spinal cord

The spinal cord is 18 in (45 cm) long. It is continuous above with the medula oblongata at the level of the foramen magnum and ends below at the lower level of the 1st, or the upper level of the 2nd lumbar vertebra. Inferiorly it tapers into the *conus medullaris* from which a prolongation of pia mater, the *filum terminale*, descends to be attached to the back of the coccyx.

The cord bears a deep longitudinal *anterior fissure*, a narrower *posterior sulcus* and, on either side, a *postero-lateral sulcus* along which the *posterior (sensory) nerve roots* are serially arranged (Fig. 227).

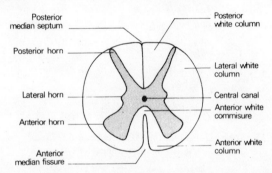

**Fig. 227.** The spinal cord—transverse section through a thoracic segment.

These posterior roots each bear a ganglion which constitutes the first cell-station of the sensory nerves.

The *anterior (motor) nerve roots* emerge serially along the anterolateral aspect of the cord on either side.

At each intervertebral foramen the anterior and posterior nerve roots unite to form a *spinal nerve* which immediately divides into its *anterior* and *posterior primary rami*, each transmitting both motor and sensory fibres.

The length of the roots increases progressively from above downwards due to the disparity between the length of the cord and the vertebral column; the lumbar and sacral roots below the termination of the cord at vertebral level L2 continue as a leash of nerve roots termed the *cauda equina*.

## Age differences

Up to the 3rd month of fetal life the spinal cord occupies the full extent

353

of the vertebral canal. The vertebrae then outpace the cord in the rapidity of their growth so that, at birth, the cord only reaches the level of the 3rd lumbar vertebra (Fig. 228).

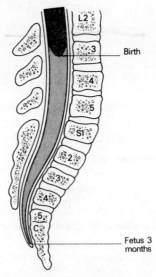

**Fig. 228.**   The relationship between the spinal cord and the vertebrae in the 3-month fetus and in the newborn child.

Further differential growth up to the time of adolescence brings the cord to its definitive position at the approximate level of the disc between the 1st and 2nd lumbar vertebrae (Fig. 229).

**Structure** (Fig. 228)

In transverse section of the cord is seen the *central canal* around which is the H-shaped *grey matter*, surrounded in turn by the *white matter* which contains the long ascending and descending tracts.

Within the *posterior horns* of the grey matter, capped by the *substantia gelatinosa*, terminate many of the sensory fibres entering from the posterior nerve roots. In the larger *anterior horns* lie the motor cells which give rise to the fibres of the anterior roots.

In the thoracic and upper lumbar cord are found the *lateral horns* on each side, containing the cells of origin of the sympathetic system.

The more important long tracts in the white matter are:

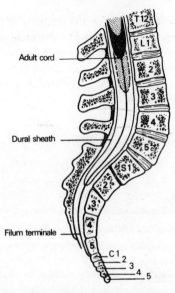

**Fig. 229.** The range of variation in the termination of the spinal cord in the adult.

## Descending tracts (Fig. 230)

1. *The pyramidal (lateral cerebrospinal or crossed motor) tract.* The motor pathway commences at the pyramidal cells of the motor cortex, decussates in the medulla, then descends in the pyramidal tract on the

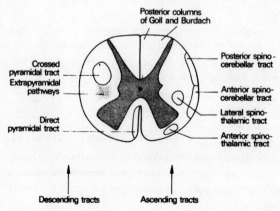

**Fig. 230.** The location of the important spinal tracts. (The descending tracts are shown on the left, the ascending tracts on the right.)

contralateral side of the cord. At each spinal segment, fibres enter the anterior horn and connect up with the motor cells there—the tract therefore becomes progressively smaller as it descends.

2. *The direct pyramidal (anterior cerebrospinal or uncrossed motor) tract* is a small tract descending without medullary decussation. At each segment, however, fibres pass from it to the motor cells of the opposite side.

## Ascending tracts (Fig. 230)

1. *The posterior and anterior spino-cerebellar tracts* ascend on the same side of the cord and enter the cerebellum through the inferior and superior cerebellar peduncles respectively.

2. *The lateral* and *anterior spino-thalamic tracts.* Pain and temperature fibres enter the posterior roots, ascend a few segments, relay in the substantia gelatinosa, then cross to the opposite side to ascend in these tracts to the thalamus, where they are relayed to the sensory cortex.

3. *The posterior columns* comprise a medial and lateral tract, termed respectively the *fasciculus gracilis* (of Goll) and *fasciculus cuneatus* (of Burdach). They convey sensory fibres subserving fine touch and proprioception (position sense), mostly uncrossed, to the gracile and cuneate nuclei in the medulla where, after synapse, the fibres decussate, pass to the thalamus and are thence relayed to the sensory cortex. Some fibres pass from the medulla to the cerebellum along the inferior cerebellar peduncle.

### Blood supply

The *anterior* and *posterior spinal arteries* descend in the pia from the intracranial part of the vertebral artery. They are reinforced serially by branches from the ascending cervical, the cervical part of the vertebral, the intercostal and the lumbar arteries.

### CLINICAL FEATURES

1. Complete transection of the cord is followed by total loss of sensation in the regions supplied by the cord segments below the level of injury together with flaccid muscle paralysis. Voluntary sphincter control is lost but reflex emptying of bladder and rectum subsequently return providing that the cord centres situated in the sacral zone of the cord are not destroyed.

2. Destruction of the centre of the cord, as occurs in syringomyelia and in some intramedullary tumours, first involves the decussating

spino-thalamic fibres so that initially there is bilateral loss of pain and temperature sense below the lesion; proprioception and fine touch are preserved till late in the uncrossed posterior columns.

3. Hemisection of the cord is followed by the *Brown-Sequard syndrome*; there is paralysis on the affected side below the lesion (pyramidal tract) and also loss of proprioception and fine discrimination (dorsal columns). Pain and temperature senses are lost on the *opposite* side below the lesion, because the affected spino-thalamic tract carries fibres which have decussated below the level of cord hemisection.

4. *Tabes dorsalis,*which is a syphilitic degenerative lesion of the poster-ior columns and posterior nerve roots, is characterized by loss of pro-prioception; the patient becomes ataxic, particularly if he closes his eyes, because he has lost his position sense for which he can partially compens-ate by visual knowledge of his spatial relationship (Romberg's sign).

5. Intractable pain can be treated in selected cases by cutting the appropriate posterior nerve roots (posterior rhizotomy) or by division of the spinothalamic tract on the side opposite the pain (cordotomy). A knife passed 3 mm into the cord anterior to the denticulate ligament and then swept forward from this point will sever the spinothalamic tract but preserve the pyramidal tract lying immediately behind it.

### The membranes of the cord (Fig. 231)

The spinal cord, like the brain, is closely ensheathed by the *pia mater*. This is thickened on either side between the nerve roots to form the *denticulate ligament*, which passes laterally to adhere to the dura. Infer-iorly the pia continues as the *filum terminale* which pierces the distal extremity of the dural sac and becomes attached to the coccyx.

The *arachnoid mater* lines the *dura mater*, leaving an extensive *sub-arachnoid space*, containing C.S.F., between it and the pia. Both pia and arachnoid are continued along the spinal nerve roots.

The dura itself forms a tough sheath to the cord. It ends distally at the level of the 2nd sacral vertebra. It also continues along each nerve root and blends with the sheaths of the peripheral nerves.

CLINICAL FEATURES

*Lumbar puncture* to withdraw C.S.F. from the spinal subarachnoid space must be performed well clear of the termination of the cord. A line joining the iliac crests passes through the 4th lumbar vertebra and there-fore the intervertebral spaces immediately above or below this landmark

EXTRA DURAL
SUB DURAL
SUB ARACHNOID
CSF. 358

← 3cm of 3 spinal canal.
← DURA
ARACH
← PIA

EPIDURAL →
Spinal Anath. —

## The Central Nervous System

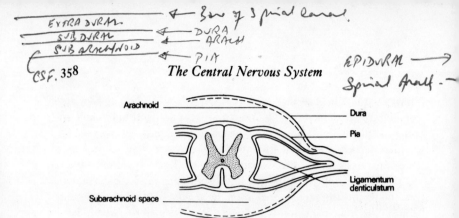

Arachnoid

Dura

Pia

Ligamentum
denticulatum

Subarachnoid space

Fig. 231.   The membranes of the spinal cord.

can be used with safety. The spine must be fully flexed (with the patient either on his side or seated) so that the vertebral interspinous spaces are opened to their maximum extent (Fig. 232). The needle is passed inwards and somewhat cranially exactly in the mid-line and at right angles to the spine; the supraspinous and interspinous ligaments are traversed and then the dura is penetrated, the latter with a distinct 'give'. Occasionally root pain is experienced if a root of the cauda equina is impinged upon, but usually these float clear of the needle.

At spinal puncture C.S.F. can be obtained for examination, antibiotics, radio-opaque dye or anaesthetics may be injected into the subarachnoid space and the C.S.F. pressure estimated. A block in the spinal canal above the point of puncture, produced, for example, by a spinal tumour, can be revealed by *Queckenstedt's test*:

Pressure is applied to the neck in order to compress the jugular veins;

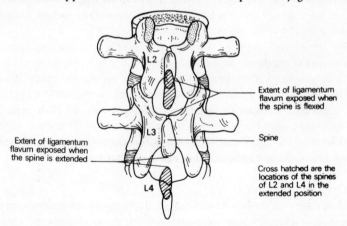

L2

Extent of ligamentum
flavum exposed when
the spine is flexed

L3

Spine

Extent of ligamentum
flavum exposed when
the spine is extended

L4

Cross hatched are the
locations of the spines
of L2 and L4 in the
extended position

Fig. 232.   The lumbar interlaminar gap; this anatomical fact makes lumbar puncture possible.

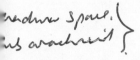 

this reduces venous outflow from the cranium and raises the intracranial pressure. C.S.F. is displaced into the spinal sac and the C.S.F. pressure, as determined by lumbar puncture and manometry, rises briskly. This rise in pressure is not seen if a spinal block is present.

# The Brain

## The medulla

The medulla is 1 in (25 mm) in length and about $\frac{3}{4}$ in (18 mm) in diameter. It is continuous below, through the foramen magnum, with the spinal cord and above with the pons; posteriorly it is connected with the cerebellum by way of the inferior cerebellar peduncle.

### EXTERNAL FEATURES (Fig. 233)

The anterior surface of the medulla is grooved by an anteromedian fissure on either side of which are the swellings due to the pyramidal tracts. These *pyramids*, in turn, are separated from the *olivary eminences* by the antero-lateral sulcus along which the rootlets of the XIIth cranial nerve emerge. Between the olive and the inferior cerebellar peduncle there is yet another groove corresponding to the postero-lateral sulcus of the spinal cord; emerging from this groove are the rootlets of cranial nerves IX, X and XI (*see* Fig. 235). The postero-median fissure of the cord is continued half-way up the medulla, where it widens out to form the posterior part of the IVth ventricle. On either side of the fissure the posterior columns of the spinal cord expand to form two distinct tubercles corresponding to the gracile and cuneate nuclei.

### DEEP STRUCTURE

The deep structure of the medulla is best shown by reference to diagrams representing the cross-sectional appearance of the medulla at the level of the sensory decussation and the lower part of the IVth ventricle (Figs. 234, 235).

The *blood supply* of the medulla is derived from the vertebral artery directly and from its posterior inferior cerebellar branch.

### CLINICAL FEATURES

The medulla contains the respiratory, cardiac and vasomotor centres— the 'vital centres'. The respiratory centre is particularly vulnerable to compression, injury or poliomyelitis with consequent respiratory failure.

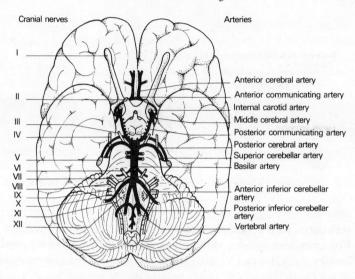

Cranial nerves

Arteries

Anterior cerebral artery
Anterior communicating artery
Internal carotid artery
Middle cerebral artery
Posterior communicating artery
Posterior cerebral artery
Superior cerebellar artery
Basilar artery

Anterior inferior cerebellar artery
Posterior inferior cerebellar artery
Vertebral artery

**Fig. 233.** The base of the brain showing the cranial nerve roots and their relationships to the circle of Willis.

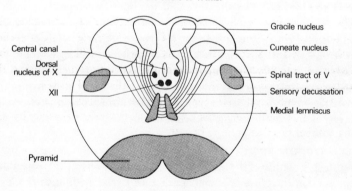

Central canal
Dorsal nucleus of X
XII
Pyramid

Gracile nucleus
Cuneate nucleus
Spinal tract of V
Sensory decussation
Medial lemniscus

**Fig. 234.** The medulla—level of the sensory decusation.

## The pons

EXTERNAL FEATURES (Fig. 233)

The pons lies between the medulla and the mid-brain and is connected to the cerebellum by the middle cerebellar peduncle. It is 1 in (25 mm) in length and $1\frac{1}{2}$ in (38 mm) in width. Its ventral surface presents a shallow median groove and numerous transverse ridges which are continuous laterally with the middle cerebellar peduncle. The dorsal surface of the

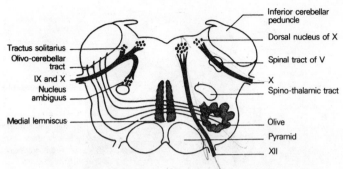

Fig. 235. The medulla—level of the lower part of the IVth ventricle.

pons forms the upper part of the floor of the IVth ventricle. Its junction with the medulla is marked close to the mid-line by the emergence of the VIth cranial nerves and, in the angle between the pons and the cerebellum, by the VIIth and VIIIth nerves. Both the motor and sensory roots of V leave the lateral part of the pons near its upper border.

### INTERNAL STRUCTURE
The pons consists for the most part of a number of cell masses (the *pontine nuclei*), scattered amongst the long ascending and descending pathways and the decussating ponto-cerebellar fibres, the pontine *tegmentum* (the pontine component of the reticular formation) and the central connections of the Vth, VIth and VIIth cranial nerves.

A typical cross-section through the pons is shown in Fig. 236.

The *blood supply* of the pons is derived from the basilar artery by way of a number of small pontine branches.

## The cerebellum

### EXTERNAL FEATURES
The cerebellum is the largest part of the hind-brain and occupies most of the posterior cranial fossa. It is made up of two lateral *cerebellar hemispheres* and a median *vermis*. Inferiorly the vermis is clearly separated from the two hemispheres and lies at the bottom of a deep cleft, the *vallecula*; superiorly it is only marked off from the hemispheres as a low median elevation. A small ventral portion of the hemisphere lying on the middle cerebellar peduncle is almost completely separated from the rest of the cerebellum as the *flocculus*. The surface of the cerebellum is divided into innumerable narrow *folia* and, by a few deep fissures, into a number

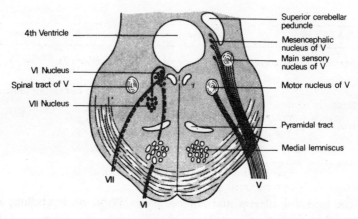

4th Ventricle

VI Nucleus

Spinal tract of V

VII Nucleus

Superior cerebellar peduncle

Mesencephalic nucleus of V

Main sensory nucleus of V

Motor nucleus of V

Pyramidal tract

Medial lemniscus

VII

V

VI

**Fig. 236.** The pons—level of the right VIth nerve nucleus and the intrapontine course of the facial nerve and, on the left, of the nuclei of V.

of lobules. The effect of this fissuring is to give the cerebellum in section the appearance of a many-branched tree (the *arbor vitae*).

### INTERNAL STRUCTURE
The structure of the cerebellum is remarkably uniform. It consists of a *cortex* of grey matter (in which all the afferent fibres terminate) covering a mass of white matter, in which deep nuclei of grey matter are buried. Of these, the *dentate nucleus* is by far the largest and occupies the central area of each hemisphere.

The cerebellum is connected to the brain-stem by way of three pairs of *cerebellar peduncles*. The *inferior peduncles* connect it to the dorso-lateral aspect of the medulla; the *middle cerebellar peduncles* to the dorsum of the pons, and the *superior peduncles* to the back of the mid-brain. Anteriorly, the cerebellum is related to the fourth ventricle and to the medulla and pons, laterally, to the sigmoid sinus and the mastoid antrum and air cells, while above it is separated from the cerebral hemisphere by the *tentorium cerebelli*.

The *blood supply* of the cerebellum is derived from three pairs of arteries; the *posterior inferior cerebellar* branches of the vertebral arteries supply the inferior aspect of the vermis and hemispheres, and the *anterior inferior* and *superior cerebellar branches of the basilar* artery supply the antero lateral part of the under surface and the superior aspect of the cerebellum respectively.

**Table 5**

| Peduncle | Afferent Pathway | Efferent Pathway |
|---|---|---|
| Superior | Anterior spino-cerebellar (uncrossed) | From dentate nucleus to:<br>1. Thalamus<br>2. Cerebral cortex<br>3. Red nucleus (crossed) |
| Middle | Ponto-cerebellar (crossed<br>—relays from cerebral cortex via pontine nuclei | |
| Inferior | 1. Vestibulo-cerebellar (uncrossed)<br>2. Posterior spino cerebellar (uncrossed)<br>3. Olivo-cerebellar (uncrossed)—function unknown | |

CONNECTIONS OF THE CEREBELLUM
The principal afferent and efferent pathways of the cerebellum are set out in Table 5.

CLINICAL FEATURES
1. The cerebellum is principally concerned with the regulation of posture, muscle tone and muscular co-ordination; consequently cerebellar lesions result in some disturbance of one or more of these motor functions in the form of an unsteady gait, hypotonia, tremor, nystagmus and dysarthria. Lesions of the cerebellum give rise to symptoms and signs on the *same side of the body*. Destruction of the dentate nucleus or the superior cerebellar peduncle results in almost as severe a disability as ablation of the entire cerebellar hemisphere.

2. Thrombosis of the posterior inferior cerebellar artery gives rise to a characteristic syndrome marked by ataxia and hypotonia of the homolateral limbs due to involvement of the inferior cerebellar peduncle and cortex, signs of cranial nerve involvement (V to X) and contralateral loss of pain and thermal sensibility (spino-thalamic involvement).

## The mid-brain

The mid-brain is the shortest part of the brain stem (it is just under 1 in (25 cm) long) and connects the pons and cerebellum to the diencephalon. It lies in the gap in the tentorium cerebelli and is largely hidden by the surrounding structures.

EXTERNAL FEATURES (Fig. 233)
The only parts of the mid-brain visible from the ventral aspect of the brain are the two *cerebral peduncles* which emerge from the substance of

the cerebral hemisphere and pass downwards and medially, connecting the internal capsule to the pons. The fibres of the IIIrd nerves emerge between the two cerebral peduncles in the *interpeduncular fossa*. Viewed from the lateral aspect, the mid-brain can be seen to consist of three distinct portions: the basis pedunculi ventrally, the mid-brain *tegmentum* centrally and the *tectum* dorsally. The trochlear nerve, the optic tract and the posterior cerebral artery wind around this aspect of the mid-brain. The dorsal surface of the mid-brain presents the four *corpora quadrigemina* (or colliculi) and the superior *medullary velum* between the two superior cerebellar peduncles. *The pineal gland* rests between the two superior corpora quadrigemina.

INTERNAL STRUCTURE

The internal structure of the mid-brain is again best described by reference to cross-sectional diagrams at representative levels: viz. at the level of the inferior and the superior corpora quadrigemina (Figs. 237 and 238).

Observe that these sections pass through the mid-brain at the level of the decussation of the superior cerebellar peduncle and the nucleus of the IVth nerve, on the one hand, and through the red nucleus and the nucleus of III on the other. The three subdivisions of the mid-brain are also clearly seen in these figures. Above the level of the cerebral aqueduct lies the *tectum* and between the aqueduct and the basis pedunculi is the grey matter of the *tegmentum* separated from basis pedunculi by the deeply pigmented lamina of the *substantia nigra*.

CLINICAL FEATURES

When calcified, the pineal gland is easily identified on skull radiographs.

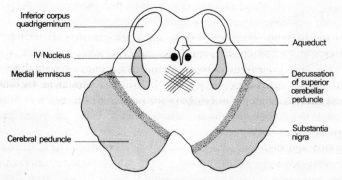

**Fig. 237.** The midbrain—level of the inferior corpus quadrigeminum and decussation of the superior cerebellar peduncle.

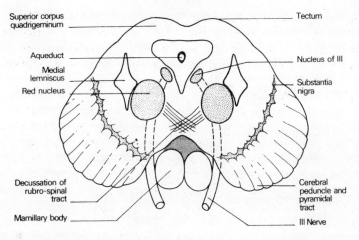

**Fig. 238.** The mid-brain—level of the superior corpus quadrigeminum and the red nucleus.

It may then give the important radiological sign of lateral displacement by a space-occupying lesion of the cerebral hemisphere.

## The diencephalon

The diencephalon comprises the hypothalamus and thalamus. It is that part of the brain surrounding the IIIrd ventricle (Fig. 239).

## The hypothalamus

The hypothalamus forms the floor of the IIIrd ventricle. It includes, from before backwards, the optic chiasma, the tuber cinereum, the infundibular stalk (leading down to the posterior lobe of the pituitary), the mamillary bodies and the posterior perforated substance. In each of these there is a number of cell masses or nuclei surrounded by a fibre pathway—the medial fore-brain bundle—which runs throughout the length of the hypothalamus and serves to link it with the mid-brain posteriorly and the basal fore-brain areas anteriorly. From a functional point of view the hypothalamus is largely concerned with autonomic activity and can be divided into a postero-medial sympathetic area and an antero-lateral area concerned with parasympathetic activity.

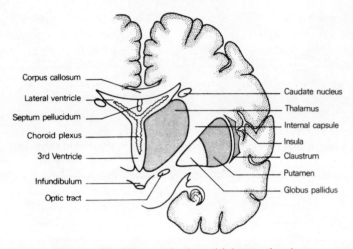

Fig. 239.　The thalamus and 3rd ventricle in coronal section.

CLINICAL FEATURES
1. Lesions of the hypothalamus may result in a variety of autonomic disturbances, e.g. somnolence, disturbances of temperature regulation and obesity.
2. Damage to the supraoptic nuclei or the infundibular stalk leads to diabetes insipidus.

## The pituitary gland (hypophysis cerebri)

This is an example of a 'two in one' organ of which Nature is so keen; compare the two glandular components of the adrenal cortex and medulla, and the exocrine and endocrine parts of the pancreas and testis. The pituitary comprises a larger anterior and smaller posterior lobe, the latter connected by the hollow infundibulum (pituitary stalk) to the tuber cinereum in the floor of the 3rd ventricle. The two lobes are connected by a narrow zone termed the pars intermedia.

The pituitary lies in the cavity of the pituitary fossa covered over by the diaphragma sellae which is a fold of dura mater. This fold has a central aperture through which passes the infundibulum. Below is the body of the sphenoid, laterally lies the cavernous sinus and its contents separated by dura mater (Fig. 210), with inter-cavernous sinuses communicating in front, behind and below. The optic chiasma lies above, immediately in front of the infundibulum.

*Structure*

The anterior lobe is extremely cellular and consists of chromophobe, eosinophilic and basophilic cells. The pars intermedia contains large colloid vesicles reminiscent of the thyroid. The posterior lobe is made up of nerve fibres whose cell stations lie in the hypothalamus.

*Development*

*III ventricle*

The posterior lobe is a cerebral diverticulum. The anterior lobe and the pars intermedia develop from Rathke's pouch in the roof of the primitive buccal cavity. Occasionally a tumour grows from remnants of the epithelium of this pouch (craniopharyngioma). These tumours are often cystic and calcified.

CLINICAL FEATURES

Tumours of the pituitary, as well as forming intracranial space-occupying lesions, may have two special features; their endocrine disturbances and their relationship to the optic chiasma.

*Chromophobe adenoma* is the commonest pituitary tumour. As it enlarges it expands the pituitary fossa (sella turcica) and this may be demonstrated radiologically. Compression of the optic chiasma produces the very typical bitemporal hemianopia (*see* optic nerve, p. 384). The tumour is itself non-secretory and gradually destroys the normally functioning gland. The patient develops hypopituitarism with loss of sex characteristics, hypothyroidism and hypoadrenalism. In childhood there is arrest of growth. As the tumour extends there may be involvement of the hypothalamus with diabetes insipidus and obesity.

*The eosinophil adenoma* secretes the pituitary growth hormones. If it occurs before puberty, which is unusual, it produces gigantism; after puberty it results in acromegaly.

The *basophil adenoma* is small, produces no pressure effects and may be associated with Cushing's syndrome, although this more often results from hyperplasia or tumour of the adrenal cortex.

Pituitary tumours may be approached through a frontal bone flap. Occasionally when access is difficult the surgeon may have to divide the optic nerve and he therefore chooses to attack the tumour on the side with the more serious visual defect.

The close relationship of the pituitary to the sphenoid sinus makes it possible to insert a trocar into the pituitary gland by a trans-nasal, trans-transphenoidal approach. Radioactive material can then be introduced into the pituitary fossa to destroy the gland in the treatment of advanced malignant disease of the breast. At present, most centres perform open transphenoidal hypophysectomy using the operating microscope.

## The thalamus (Figs. 239 and 241)

The thalamus is an oval mass of grey matter which forms the lateral wall of the third ventricle; it extends from the interventricular foramen in front to the mid-brain behind. Laterally it is related to the internal capsule (and through it to the basal ganglia), and above to the body of the lateral ventricle. Medially it is frequently connected with its fellow of the opposite side through the *massa intermedia* (interthalamic connexus). Posteriorly it presents three distinct eminences, the *pulvinar*, and the *medial* and *lateral geniculate bodies*.

The thalamus is largely composed of relay nuclei, which forward impulses either from the main sensory pathways or from supra-segmental levels on to the cerebral cortex. This it does by way of a number of *thalamic radiations* in the internal capsule.

The *blood supply* of the thalamus is derived principally from the posterior cerebral artery through its thalamostriate branches, which pierce the posterior perforated substance to supply also the posterior part of the internal capsule.

## The cerebral hemispheres

The cerebral hemispheres, which in man have developed out of all proportion to the rest of the brain, comprise the cerebral cortex, the basal ganglia, their afferent and efferent connections and the lateral ventricles.

## Cerebral cortex

The cortex of the cerebral hemispheres is divided on topographical and functional grounds into four lobes (Fig. 240).

1. *Frontal lobe*

This includes all the cortical areas in front of the *central sulcus of Rolando*. The important cortical areas of the frontal lobe are:

*a Motor cortex*. The motor area occupies a large part of the pre-central gyrus; it receives afferents from the thalamus and cerebellum and is concerned with voluntary movement.

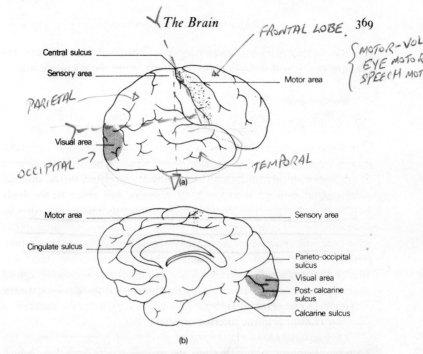

*(handwritten annotations on figure: FRONTAL LOBE; {MOTOR-VOL. / EYE MOTOR / SPEECH MOTOR}; PARIETAL; OCCIPITAL →; TEMPORAL)*

**Fig. 240.** Localization of function in the cerebral cortex. (a) Lateral aspect. (b) Medial aspect.

*b*  *Pre-motor cortex.* This occupies the anterior part of the pre-central gyrus and the adjoining lower part of the frontal gyri. It too is concerned with voluntary movement.

*c*  *Eye motor field.* This lies in the cortex anterior to the premotor area; lesions in this area result in impaired eye movement.

*d*  *Broca's speech area.* Lesions of the area around the posterior part of the inferior frontal gyrus of the dominant hemisphere were shown by Broca to affect the motor element in speech.

*e*  *Frontal association cortex* (clinically called the prefrontal cortex). This comprises a considerable part of the frontal lobe and is one of the remarkable developments of the human brain. Its afferents are derived from the thalamus and also from other cortical areas; it probably sends efferents to the thalamus and hypothalamus. From a functional point of view the lateral aspect of the frontal lobe appears to be related to 'intellectual activity', the medial surface to affective or emotional behaviour and the orbital surface to the control of autonomic activity. In the classical type of prefrontal leucotomy the connections of all these areas with the rest of the brain are severed. Recently more restricted operations, usually confined to the medial or inferior aspect of the frontal lobe, have been performed.

### 2. Parietal lobe

The *parietal lobe* is bounded in front by the central salcus and behind by a line drawn from the parieto-occipital notch to the posterior end of the lateral sulcus. The important cortical areas of the parietal lobe are:

*a  The somatic sensory cortex.* The post-central gyrus receives afferent fibres from the thalamus and is concerned with all forms of somatic sensation.

*b  The parietal association cortex,* comprising the remainder of the parietal lobe, is concerned largely with the recognition of somatic sensory stimuli and their integration with other forms of sensory information. It receives its afferents from the thalamus and, when damaged, gives rise to a peculiar inability to recognize somatic stimuli called astereognosis; put a key or a penny into the patient's hand—he is aware of the object but cannot identify it.

### 3. The temporal lobe

This is arbitrarily separated from the occipital lobe by a line drawn vertically downwards from the upper end of the lateral sulcus.

The important cortical areas of the temporal lobe are:

*a  The auditory cortex.* This lies in the upper part of the temporal lobe. Its afferent fibres are from the medial geniculate body and it is concerned with the reception of auditory stimuli.

*b  The temporal association cortex.* The area surrounding the auditory cortex is responsible for the recognition of auditory stimuli and for their integration with other sensory modalities. Lesions of this area result in auditory agnosia, i.e. the inability to recognize or to understand the significance of meaningful sounds. The cortical region just behind and above this area on the dominant hemisphere is of considerable importance in the sensory aspect of the speech mechanism.

*c  The rest of the temporal cortex* is unique in that it does not appear to receive afferent fibres from the thalamus. Its significance is unknown but there is some evidence that it may be concerned with memory.

*d  The hippocampal gyrus.* The cortex of the most medial part of the under-surface of the temporal lobe is known as the *hippocampal gyrus.* Anteriorly it is related to the olfactory cortex of the uncus but it is not known whether the two are connected. Medially it is in direct continuity with the layer of in-rolled cortex which is the *hippocampus* and which is one of the most important sources of afferents to this obscure cortical structure. The hippocampus occupies the whole length of the floor of the inferior horn of the lateral ventricle and sends its efferents into the overlying layer of white matter known as the *alveus*. The fibres of the alveus collect near the posterior end of the hippocampus to form a compact bundle, the *fimbria*, which, as it arches under the corpus callosum, becomes known as the *fornix*. The fornix passes forwards and then downwards in front of the interventricular foramen and finally backwards into the hypothalamus to terminate in the *mammillary body*. It also gives fibres to the thalamus and the hypothalamus. The function of this complex system is entirely unknown, and although it is often referred to as the '*rhinencephalon*' there is little evidence that it is concerned with olfaction.

### 4. *Occipital lobe*

*The occipital lobe* lies behind the parietal and temporal lobes. On its medial aspect it presents the Y-shaped *calcarine* and *post-calcarine sulci* (Fig. 240). The following cortical areas are noteworthy:

*a* The *visual cortex* surrounds the calcarine and post-calcarine sulci and receives its afferent fibres from the lateral geniculate body of the thalamus of the same side; it is concerned with vision of the opposite half field of sight.

*b* The *occipital association cortex* lies anteriorly to the visual cortex. This area is particularly concerned with the recognition and integration of visual stimuli.

### The insula (Fig. 239)

If the lips of the lateral sulcus are separated it is seen that there is a considerable area of cortex buried in the floor of this sulcus. This area is known as the *insula of Reil*. It is divided into a number of small gyri and is crossed by the middle cerebral artery. Apart from its upper part, which abuts on the sensory cortex and probably represents the taste area of the cerebral cortex, the function of the insula is unknown.

### The connections of the cerebral cortex

As has been indicated, most areas of the cerebral cortex receive their main afferent input from the thalamus, but, in addition to this, there are well-established *commissural connections* with the corresponding area of the opposite hemisphere by way of the *corpus callosum. Associational connections* also link neighbouring cortical areas on the same side and, in some cases, connect distant cortical areas, thus the frontal, occipital and temporal lobes are directly connected by long association pathways.

CLINICAL FEATURES

It is convenient to summarize here the clinical effects of lesions affecting the principal cortical areas.

1.   *Frontal cortex*—impairment of higher mental functions and emotions.

2.   *Precental (motor) cortex*—weakness of the opposite side of the body; lesions low down the cortex affecting the face and arm, high lesions affecting the leg. Mid-line lesions (meningioma, sagittal sinus thrombosis or a gun-shot wound) may produce paraplegia by involving both leg areas.

3.   *Sensory cortex*—contralateral hemianaesthesia (distributed in the same pattern as the motor cortex) affecting especially the higher sensory modalities such as stereognosis and two-point position sense.

4.   *Occipital cortex*—homonymous hemianopia.

5.   Lesions adjacent to the lateral sulcus in either the frontal, parietal or temporal lobes of the dominant hemisphere result in dysphasia.

## The basal ganglia (Figs. 239 and 241)

These compact masses of grey matter are situated deep in the substance of the cerebral hemisphere and comprise: the *corpus striatum* (composed of the *caudate nucleus*, the *putamen* and the *globus pallidus*), the *claustrum* and the *amygdala*.

The *corpus striatum*. The *caudate nucleus* is a large homogeneous mass of grey matter consisting of a *head*, anterior to the interventricular foramen and forming the lateral wall of the anterior horn of the lateral ventricle, a *body*, forming the lateral wall of the body of the ventricle, and an elongated *tail*, which forms the roof of the inferior (temporal) horn of the ventricle. It is largely separated from the putamen by the *internal capsule*, but the two structures are connected anteriorly. The *putamen* is a roughly ovoid mass closely applied to the lateral aspect of the *globus pallidus* with which it is sometimes described as the *lentiform nucleus*. The basal ganglia receive their afferent connections from the cerebral cortex and the thalamus and send their efferents to the globus pallidus. Efferent fibres pass to the hypothalamus, red nucleus, substantia nigra and olivary nucleus.

homonymous.
bitemporal

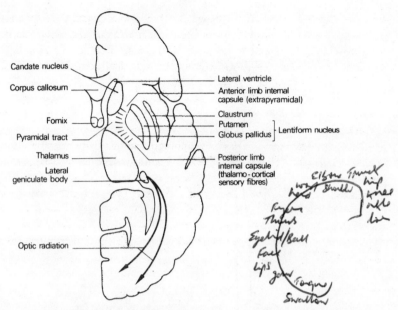

Fig. 241. The basal ganglia and internal capsule shown in horizontal section through the cerebrum.

## The long ascending and descending pathways

A. THE SOMATIC AFFERENT PATHWAYS (Fig. 242)

1. Proprioceptive and tactile impulses pass uninterruptedly through the posterior root ganglia, through the ipsilateral *posterior columns* of the spinal cord to the *gracile and cuneate nuclei* in the lower part of the medulla. In the posterior columns there is a fairly precise organization of the afferent fibres; those from sacral and lumbar segments are situated medially in the tracts while fibres from thoracic and cervical levels are successively added to their lateral aspect. This arrangement according to body segments is maintained in the gracile and cuneate nuclei and in the efferents from these nuclei to the contralateral thalamus. The fibres from the dorsal column nuclei immediately cross over to the opposite side in the *sensory decussation* of the medulla (Fig. 234) and continue up to the thalamus as a compact bundle—the *medial lemniscus*.

2. Dorsal root fibres subserving *pain and temperature*, together with some *tactile afferents*, end in the *substantia gelatinosa* of the posterior horn. From the posterior horn afferent fibres are relayed up to the contralateral

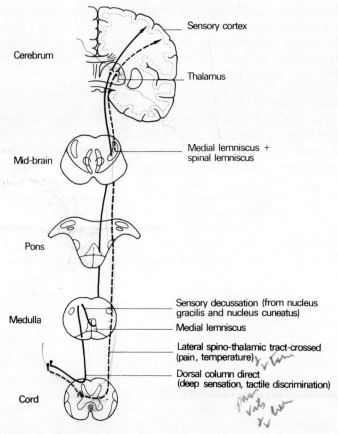

Sensory cortex

Cerebrum

Thalamus

Mid-brain

Medial lemniscus +
spinal lemniscus

Pons

Sensory decussation (from nucleus
gracilis and nucleus cuneatus)

Medulla

Medial lemniscus

Lateral spino-thalamic tract-crossed
(pain, temperature)

Dorsal column direct
(deep sensation, tactile discrimination)

Cord

**Fig. 242.** The long ascending pathways of the dorsal columns and spino-thalamic tracts.

thalamus in the *anterior and lateral spino-thalamic tracts* of the opposite side, the fibres crossing in the anterior white commissure of the spinal cord. In the brain stem these fibres come to lie immediately lateral to the medial lemniscus and are sometimes known as the *spinal lemniscus*. They terminate in the thalamus.

These somatic afferents are relayed from the thalamus, through the posterior limb of the internal capsule (Fig. 241) to the somatic sensory cortex of the *post-central gyrus*. In the internal capsule the fibres are arranged in the sequence 'face, arm, trunk and leg' from before backwards, and this segregation persists in the sensory cortex, where the leg is represented on the upper part of the cortex, the trunk and arm in its middle portion and the face most inferiorly. Since the cortical representation reflects the density of the peripheral innervation rather than the size of the receptive field, there is a good deal of distortion of the body image in the cortex, the cortical representation of the face and hand being much greater than that of the limbs and trunk.

CLINICAL FEATURES

1. Lesions of the sensory pathway most commonly occur in the internal capsule following some form of cerebro-vascular accident. If complete, these result in a total hemianaesthesia of the opposite side of the body. In partial lesions the area of sensory loss will be determined by the site of the injury in the internal capsule and, from a knowledge of the sensory (and motor) loss, it is usually possible to determine with some degree of accuracy the site of a lesion in the capsule.

2. Since there is some degree of modality segregation below the level of the thalamus, lesions of the sensory pathways at these levels may result in, say, an area of analgesia without much impairment of tactile sensibility.

The auditory, visual and olfactory pathways are dealt with under the appropriate cranial nerves.

B. THE MOTOR PATHWAYS (Fig. 243)

It is customary to divide the motor pathways of the brain and spinal cord into pyramidal and extra-pyramidal systems, although the latter is anatomically rather ill-defined.

1. *The pyramidal tract*

The pyramidal system, by definition, is the motor pathway which forms the medullary pyramids. It receives fibres from a rather wide area of cortex both in front of and behind the central sulcus, but the great majority of its fibres are derived from the pyramidal cells of the '*motor cortex*' of the *precentral gyrus* and the *premotor area* immediately in front of it. In both these areas there is an organization comparable to that seen in the sensory area. Again, the body is inverted so that the 'leg area' is situated in the upper part of the precentral gyrus encroaching on the medial surface of the hemisphere, the 'face area' is near the foot of the gyrus while the 'arm area' occupies a central position. Again, the body-form is greatly distorted so that the area representing the hand, lips, eyes and foot are exaggerated out of all proportion to the rest of the body.

From the cortex, the motor fibres pass through the genu and posterior

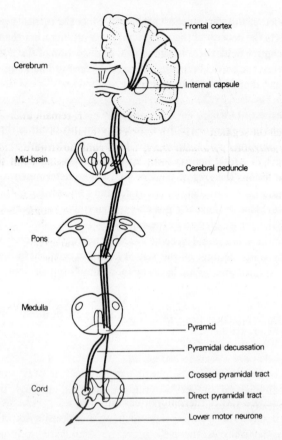

**Fig. 243.** The long descending pathway of the pyramidal tract.

limb of the internal capsule (Fig. 241) where they are again organized in the sequence of 'face, arm, leg', from before backwards. From the internal capsule the fibres form a compact bundle which occupies the central third of the cerebral peduncle. Hence they pass through the pons, where they are broken up into a number of small bundles between the cells of the pontine nuclei and the transversely disposed ponto-cerebellar fibres. Near the lower end of the pons they again collect to form a single bundle which comes to lie on the ventral surface of the medulla and forms the elevation known as the 'pyramid'. As it passes through the brain stem the pyramidal system gives off, at regular intervals, contributions to the somatic efferent nuclei of the cranial nerves. Most of these *cortico-bulbar fibres* cross over in the brain stem, but many of the cranial nerve nuclei are bilaterally innervated.

Near the lower end of the medulla the great majority of the pyramidal tract fibres cross over to the opposite side and come to occupy a central position in the lateral white column of the spinal cord. This is the so-called *crossed*, or *indirect*, *pyramidal tract* shown in Fig. 230. A small proportion of the fibres of the medullary pyramid, however, remain uncrossed until they reach the segmental level at which they finally terminate. This is the *direct* or *uncrossed pyramidal tract*, which runs downwards close to the anteromedian fissure of the cord with fibres passing from it at each segment to the opposite side.

In view of the frequent involvement of the pyramidal tract in cerebrovascular accidents, its blood supply is listed here in some detail.

*Motor cortex:* leg area: anterior cerebral artery;
                face and arm areas: middle cerebral artery;
*Internal capsule:* branches of the middle cerebral artery;
*Cerebral peduncle:* posterior cerebral artery;
*Pons:* pontine branches of basilar artery;
*Medulla:* medullary branches of vertebral artery and its posterior inferior cerebellar branch;
*Spinal cord:* segmental branches of anterior and posterior spinal arteries.

CLINICAL FEATURES

1.   It is important to remember that, in the motor cortex, movements are represented rather than individual muscles; lesions of this pathway result in paralysis of voluntary movement on the opposite side of the body although the muscles themselves are not paralysed and may cause involuntary movements. This is the essential difference between an 'upper motor neurone' lesion (i.e. a lesion of the central motor pathway) and a 'lower motor neurone' lesion (i.e. a lesion affecting the cranial nerve nuclei, or the anterior horn cells or their processes). In the latter type of lesion muscular paralysis results and, since reflex activity is abolished, flaccidity and muscular atrophy follow, whereas, in pyramidal lesions, there is spasticity, increased tendon reflexes and an extensor plantar response.

2.   Experimental lesions strictly confined to the pyramidal tract are not followed by increased muscular tone in the affected part (spasticity), but clinically this is a feature of upper motor neurone lesions; it is attributable to concomitant involvement of the extra-pyramidal system.

3.   The pyramidal tract is most frequently involved in cerebrovascular

accidents where it passes through the internal capsule. Indeed, the
artery supplying this area—the largest of the perforating branches of the
middle cerebral artery—has been termed the '*artery of cerebral haemorrhage*'.

4.   A list of the more important related signs is here given for involvement of the pyramidal tract at each level.

•Cortex: isolated lesions may occur here, resulting in loss of voluntary
movement in, say, only one limb, but often the sensory cortex is also
involved. Dysphasia in dominant hemisphere lesions is common.

•Internal capsule: usually all parts of the tract are involved, giving a
complete hemiplegia with associated sensory loss. The lesion may extend
back to involve the visual radiation, giving a homonymous field defect
(hemianopia).

•Cerebral peduncle and mid-brain: the fibres from the IIIrd nerve are
often concomitantly involved so that there are the associated signs of a
IIIrd nerve palsy.

•Pons: here the VIth nerve is often involved, alone or together with VII.
There may then be a hemiplegia affecting the arm and leg of the opposite
side and an abducens and a facial palsy of the lower motor neurone type
on the same side as the lesion.

•Medulla: because of the proximity of the pyramids to one another,
medullary lesions often affect both sides of the body. Paralysis of the
tongue on the side of the lesion is due to involvement of the XIIth nerve
or its nucleus. The respiratory, vasomotor and swallowing centres may
also be affected.

•Spinal cord: the paralysis following lesions of the spinal cord accurately
depends on the level at which the pyramidal tract is involved. The proximity of the pyramidal tracts to the ascending sensory pathways accounts
for the concomitant sensory changes usually found.


*The extra-pyramidal system*

This, by definition, should include all motor mechanisms whose fibres
do not pass through the medullary pyramids. However, it is customary
to restrict the term to non-pyramidal motor areas of the cerebral cortex,
the basal ganglia and certain brain stem motor mechanisms. The cortical
areas concerned include the premotor area and possibly some of the
sensory areas which send efferents to the corpus striatum. From the
corpus striatum efferent fibres pass to a number of brain stem structures,
including the substantia nigra, red nucleus, the thalamus and subthalamic

nucleus. These nuclei connect with the nuclei of the brain stem reticular formation from which derive the reticulo-spinal tracts.

Animal experiments have shown that this complex extra-pyramidal system receives afferent fibres from nearly all the sensory centres (by way of collateral fibres), from the corpus striatum, the cerebellum and possibly directly from the cerebral cortex. It can exert a remarkably widespread effect on the activity of the cortex on the one hand and on segmental motor activity on the other. Thus it is probable that the brain stem reticular formation serves as a sort of final common path for all the non-pyramidal motor mechanisms of the brain. Efferent pathways from this system ultimately reach segmental levels by way of fibres in the ventral white column, which may be called the *reticulo-spinal pathways*. Their effect on the anterior horn cells is not such as to produce actual movements but rather to 'set the background' for subsequent pyramidal activity. In keeping with this, it is well known that lesions affecting the extra-pyramidal pathways result in disturbance (but not loss) of voluntary movement, certain involuntary movements and disturbance in muscular tone. For a detailed description of these disturbances in different clinical conditions reference should be made to a textbook of clinical neurology; here it will suffice to list the site of the lesion in some of the more common conditions involving this system.

•Parkinsonism—globus pallidus.
•Wilson's Disease—especially putamen; caudate nucleus to lesser extent.
•Sydenham's chorea ⎫ basal ganglia.
•Huntington's chorea ⎭
•Athetosis—anterior segment of putamen.

It should be noted that precise correlation of these clinical syndromes with anatomical sites is difficult since many lesions are diffuse.

The localization of function in the basal ganglia has now become of intense practical importance since, with the advent of stereotactic coagulation, it has been found that the symptoms of Parkinson's disease can frequently be completely relieved; the rigidity by a lesion in the globus pallidus and the tremor by one in the ventrilateral nucleus of the thalamus.

## The membranes of the brain

The three membranes surrounding the spinal cord, the *dura mater*, *arachnoid mater* and *pia mater*, are continued upwards as coverings to the brain.

*The dura* is a dense membrane which, within the cranium, is made up of two layers. The outer layer is intimately adherent to the skull; the inner layer is united to the outer layer except where separated by the great dural venous sinuses and where it projects to form four sheets (Fig. 209):

•the falx cerebri;
•the falx cerebelli;
•the tentorium cerebelli and
•the diaphragma sellae.

*The arachnoid* is a delicate membrane separated from the dura by the capillary *subdural space*. It projects only into the longitudinal fissure and the stem of the lateral fissure.

*The pia* is closely moulded to the outline of the brain; it dips down into the cerebral sulci leaving the *subarachnoid space* between it and the arachnoid. This space is broken up by trabeculae of fine fibrous strands and contains the cerebro-spinal fluid.

## The ventricular system and the cerebro-spinal fluid circulation

The cerebro-spinal fluid (c.s.f.) is formed by the secretory and filtering activity of the ependymal lining of the choroid plexuses in the lateral, third and fourth ventricles; it circulates through the ventricular system of the brain and the subarachnoid space before being re-absorbed into the dural venous system.

The general appearance of the ventricular system is indicated in Fig. 244. The two *lateral ventricles*, which are by far the largest components of the system, occupy a considerable part of the cerebral hemispheres. Each has an *anterior horn* (in front of the interventricular

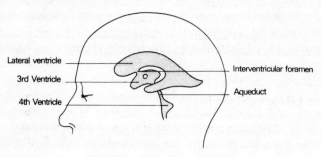

Fig. 244. The ventricular system.

foramen), a *body*, above and medial to the body of the caudate nucleus, a *posterior horn* in the occipital lobe and an *inferior horn* reaching down into the temporal lobe. The choroid plexuses of the lateral ventricles, which are responsible for the production of most of the c.s.f., extend from the inferior horn, through the body, to the interventricular foramen where they become continuous with the plexus of the IIIrd ventricle (Fig. 239).

The *third ventricle* is a narrow slit-like cavity between the two thalami. Its floor is formed by the hypothalamus. From the IIIrd ventricle the c.s.f. passes through the narrow *cerebral aqueduct* (of Sylvius) in the mid-brain to reach the IVth ventricle.

The *fourth ventricle* is diamond-shaped when viewed from above and tent-shaped as seen from the side. Its floor is formed below by the medulla and above by the pons. Its roof is formed by the cerebellum and the superior and inferior medullary vela. The c.s.f. escapes from the IVth ventricle into the subarachnoid space by way of the *median* and *lateral apertures* (of *Magendie* and *Luschka* respectively) and then flows over the surface of the brain and spinal cord.

In certain areas the subarachnoid space is considerably enlarged to form distinct cisterns. The most important of these are: the *cisterna magna* between the cerebellum and the back of the medulla; the *cisterna pontis* over the front of the pons; the *interpeduncular cistern* between the two cerebral peduncles and the *chiasmatic cistern* around the optic chiasma. Re-absorption of c.s.f. is principally by way of the superior longitudinal and the other dural sinuses, the modified arachnoid of the arachnoid granulations piercing the dura and bringing the c.s.f. into direct contact with the sinus endothelium. Along the superior sagittal sinus these granulations (or arachnoid villi) clump together to form the Pacchionian bodies which produce the pitted erosions readily seen along the median line of the inner aspect of the skull cap.

About one-fifth of the c.s.f. is absorbed along similar spinal villi or escapes along the nerve sheaths into the lymphatics. This absorption of c.s.f. is passive, depending on its higher hydrostatic pressure than that of the venous blood.

CLINICAL FEATURES

1. The general form of the ventricular system and its distortion by space-occupying lesions can be visualized radiographically by injecting a small quantity of air into the posterior horn of a lateral ventricle through a brain needle introduced through a burr hole. The head is then posi-

tioned so that air occupies, in turn, the different parts of the system (*ventriculography*).

Nowadays, air encephalography is commoner by the lumbar route; air readily enters the ventricles when introduced by lumbar puncture with the patient in the sitting position.

2.   The c.s.f. probably serves two purposes; a protective water-jacket and a regulating mechanism of intracranial pressure with changing cerebral blood flow.

3.   The total capacity of the c.s.f. in the adult is about 150 ml, of which some 25 ml is contained within the spinal theca; it is normally under a pressure of about 100 mm of water in the lateral horizontal position. The dural theca acts as a simple hydrostatic system, so that when the patient sits up, the c.s.f. pressure in the lumbar theca rises to between 350 and 550 mm, whereas the ventricular fluid pressure falls to below atmospheric.

4.   Certain parts of the c.s.f. pathway are narrow and easily obstructed. These sites are the interventricular foramina, the IIIrd ventricle, the aqueduct, the exit foramina of the IVth ventricle and the subarachnoid space around the mid-brain in the tentorial notch. Obstruction to the system causes increased intracranial pressure and ventricular dilatation (hydrocephalus).

5.   The subarachnoid space is prolonged around the optic nerve; raised c.s.f. pressure can therefore interfere with the venous drainage of the orbit and produce *papilloedema*.

6.   *Lumbar puncture*—see p. 357.

# The cranial nerves

## 1.   The olfactory nerve

The fibres of the olfactory nerve, unlike other afferent fibres, are unique in being the central processes of the olfactory cells and not the peripheral processes of a central group of ganglion cells.

The central processes of the olfactory receptors pass upwards from the olfactory mucosa in the upper part of the superior nasal concha and septum, through the cribriform plate of the ethmoid bone to end by synapsing with the dendrites of mitral cells in the olfactory bulb. The mitral cells in turn send their axons back in the olfactory tract to terminate in the cortex of the uncus and the region of the anterior perforated space. The further course of the olfactory pathway is unknown, but it is now clear that the hippocampus-fornix system is not directly concerned with olfaction.

CLINICAL FEATURES

1. The sense of smell is not highly developed in man and is easily disturbed by conditions affecting the nasal mucosa generally (e.g. the common cold). However, unilateral anosmia may be an important sign in the diagnosis of frontal lobe tumours. Tumours in the region of the uncus may give rise to the so-called uncinate type of fit, characterized by olfactory hallucinations associated with impairment of consciousness and involuntary chewing movements.

2. Bilateral anosmia due to interruption of the 1st nerve is common after head injuries, particularly in association with anterior cranial fossa fractures.


2. The optic nerve and the visual pathway

(For a description of the eye itself see the section on special senses.)

The optic nerve is the nerve of vision. It is not a true cranial nerve but should be thought of as a brain tract which has become drawn out from the cerebrum. Embryologically it is developed, together with the retina, as a lateral diverticulum of the fore-brain. Devoid of neurilemmal sheaths, its fibres, like other brain tissues, are incapable of regeneration after division.

From a functional point of view the retina can be regarded as consisting of three cellular layers; a layer of receptor cells—the *rods* and *cones*— an intermediate layer of *bipolar cells*, and a layer of *ganglion cells*, whose axons form the optic nerve (Fig. 245). From all parts of the retina these axons converge on the *optic disc* whence they pierce the sclera to form the optic nerve.

The optic nerve passes backwards and medially to the optic foramen through which it reaches the optic groove on the dorsum of the body of the sphenoid. Here, all the fibres from the medial half of the retina (i.e. those concerned with the temporal visual field) cross over in the *optic chiasma* to the optic tract of the opposite side, while the fibres from the lateral half of the retina (nasal visual field) pass back in the optic tract of the same side. The great majority of the fibres in the optic tract end in the six-layered *lateral geniculate body* of the thalamus, but a small proportion, subserving pupillary and ocular reflexes, by-pass the geniculate body to reach the superior corpus quadrigeminum. From the lateral geniculate

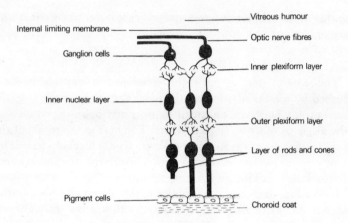

**Fig. 245.** The layers of the retina.

body the fibres of the optic radiation sweep laterally and backwards to the occipital visual cortex (the striate area surrounding the calcarine fissure) where they terminate in such a way that the upper and lower halves of the retina are represented on the upper and lower lips of the fissure respectively (Figs. 240 and 246).

CLINICAL FEATURES

1.   Lesions of the retina or optic nerve result in *unilateral blindness* in the affected segment, but lesions of the optic tract and central parts of visual pathway result in *homonymous defects*. Similarly lesions of the optic chiasma (e.g. from an expanding pituitary tumour) will give rise to a *bi-temporal hemianopia*, i.e. there will be a loss of vision in both temporal eye-fields.

2.   The lesion responsible for the *Argyll Robertson pupil* is thought to be in the vicinity of the pretectal area, but there is no satisfactory explanation why the pupillary reaction to light should be abolished while the convergence-accommodation reflex is preserved.

3.   **The oculomotor nerve**    $LR_6 (SO_4)_3.$

In addition to supplying most of the extrinsic eye muscles, the oculomotor nerve conveys the preganglionic parasympathetic fibre for the sphincter of the pupil via the ciliary ganglion. Its nucleus of origin lies in the floor of the cerebral aqueduct at the level of the superior corpus quadrigeminum and consists essentially of two components: the *somatic efferent nucleus*, which supplies the ocular muscles, and the *Edinger-Westphal nucleus* from which the parasympathetic fibres are derived.

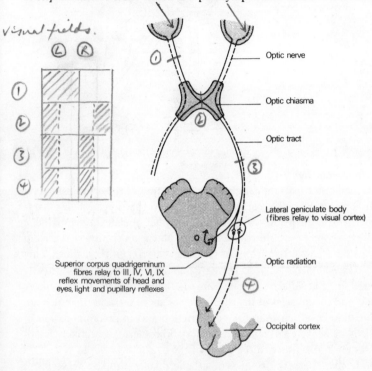

Fig. 246.   Schematic drawing of the optic pathway.

From these nuclei, fibres pass vertically through the mid-brain tegmentum to emerge just medial to the cerebral peduncle. Passing forwards between the superior cerebellar and posterior cerebral arteries, the nerve pierces the dura mater to run in the lateral wall of the cavernous sinus (Fig. 247) as far as the superior orbital fissure. Before entering the fissure it divides into a superior and inferior branch; both branches enter the orbit through the tendinous ring from which the recti arise (Fig. 251). The superior branch passes lateral to the optic nerve to supply the superior rectus muscle and levator palpebrae superioris; the inferior branch supplies three muscles, the medial rectus, the inferior rectus and the inferior oblique, the nerve to the last conveying the parasympathetic fibres to the ciliary ganglion.

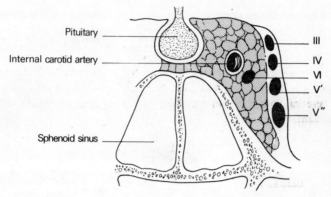

**Fig. 247.** The cavernous sinus—showing the relations of the IIIrd, IVth, Vth and VIth cranial nerves.

## The ciliary ganglion

This small but important ganglion lies in the apex of the orbit just lateral to the optic nerve. It receives, in addition to the preganglionic para-sympathetic fibres from the Edinger-Westphal nucleus, a sympathetic (post-ganglionic) root ultimately from the plexus on the internal carotid artery, and a *sensory root* from the nasociliary nerve. Of these fibres, only the *parasympathetic* synapse in the ganglion, the others pass directly through it. The efferent fibres from the ganglion pass to the ciliary muscle and the muscles of the iris by way of about 10 *short ciliary nerves*. Stimulation results in pupillary constriction and in accommodation of the lens. The sympathetic and sensory fibres are respectively vaso-constrictor and sensory to the globe of the eye.

(Note that the majority of sympathetic dilator pupillae nerve fibres are transmitted to the eye in the long ciliary branches of the nasociliary nerve.)

CLINICAL FEATURES

1.  Complete division of the IIIrd nerve results in a characteristic group of signs:
- *ptosis*—due to paralysis of the levator palpebrae superioris;
- *a divergent squint*—due to the unopposed action of the superior oblique and lateral rectus muscles, rotating the eyeball laterally;
- *dilatation of the pupil*—the dilator action of the sympathetic fibres being unopposed;
- *loss of the accommodation-convergence* and *light reflexes*—due to con-strictor pupillae paralysis;
- *double vision.*

## 4. The trochlear nerve

The trochlear nerve is the most slender of the cranial nerves and supplies only one eye muscle, the *superior oblique*. Its nucleus of origin lies in a similar position to that of the IIIrd nerve at the level of the inferior corpus quadrigeminum, but from here its fibres pass dorsally around the cerebral aqueduct and decussate in the superior medullary vellum (Fig. 248).

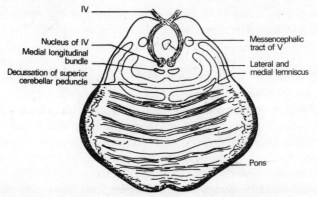

IV

Nucleus of IV
Medial longitudinal bundle
Decussation of superior cerebellar peduncle

Messencephalic tract of V

Lateral and medial lemniscus

Pons

**Fig. 248.** Section through the upper pons to show the nucleus of the IV nerve.

Emerging on the dorsum of the brain, the nerve winds round the cerebral peduncle and then passes forwards between the superior cerebellar and posterior cerebral arteries to pierce the dura. It then runs forwards in the lateral wall of the cavernous sinus (Fig. 247) between the oculomotor and ophthalmic nerves to enter the orbit through the superior orbital fissure, lateral to the tendinous ring from which the recti take origin. It then passes medially over the optic nerve to enter the superior oblique muscle.

#### CLINICAL FEATURES
A lesion of the trochlear nerve results in paralysis of the superior oblique muscle with the result that diplopia occurs when the patient attempts to turn the eye downwards and laterally.

## 5. The trigeminal nerve

As the name suggests, this nerve consists of three divisions. Together they supply sensory fibres to the greater part of the skin of the head and face, the mucous membranes of the mouth, nose and paranasal air sinuses and, by way of a small motor root, the muscles of mastication. In addition it is associated with four autonomic ganglia, the ciliary, sphenopalatine, otic and submandibular.

### The trigeminal ganglion

This ganglion, which is also termed the semilunar or Gasserian ganglion, is equivalent to the dorsal sensory ganglion of a spinal nerve. It is crescent shaped and is situated within an invaginated pocket of dura immediately inferior to the anterior attachment of the tentorium cerebelli. The ganglion lies near the apex of the petrous temporal bone, which is somewhat hollowed for it. The motor root of the trigeminal nerve and the greater superficial petrosal nerve both pass deep to the ganglion. Above lies the hippocampal gyrus of the temporal lobe of the cerebrum; medially lies the internal carotid artery and the posterior part of the cavernous sinus. The trigeminal ganglion represents the first cell station for the sensory fibres of the trigeminal nerve.

### $V_I$: *The ophthalmic division*

This is the smallest division of the trigeminal nerve; it is wholly sensory and is responsible for the innervation of the skin of the forehead, the upper eyelid and most of the nose. Passing forwards from the trigeminal ganglion, it immediately enters the lateral wall of the cavernous sinus where it lies beneath the trochlear nerve (Fig. 247). Just before entering the orbit it divides into three branches, *frontal*, *lacrimal* and *nasociliary*. The *frontal nerve* runs forward just beneath the roof of the orbit for a short distance before dividing into its two terminal branches, the *supratrochlear* and *supra-orbital* nerves, which supply the upper eyelid and the scalp as far back as the lambdoid suture.

The *lacrimal nerve* supplies the lacrimal gland (with post-ganglionic parasympathetic fibres from the sphenopalatine ganglion which reach it by way of the maxillary nerve) and the lateral part of the conjunctiva and upper lid.

The *nasociliary nerve* gives branches to the ciliary ganglion, the eyeball, the medial half of the lower eyelid, the dura of the anterior cranial fossa, and to the mucosa and skin of the nose.

### $V_2$: *The maxillary nerve*

The maxillary nerve is again purely sensory. Passing forwards from the central part of the trigeminal ganglion, close to the cavernous sinus, it leaves the skull by way of the *foramen rotundum* and emerges into the upper part of the pterygo-palatine fossa. Here it gives off a number of branches before continuing through the inferior orbital fissure and the infra-orbital canal as the *infra-orbital nerve* which supplies the skin of the cheek.

The maxillary nerve hás the following named branches:

1. the *zygomatic nerve*, whose zygomatico-temporal and zygomatico-facial branches supply the skin of the temple and cheek respectively;
2. *superior dental branches* to the teeth of the upper jaw; and
3. the *branches from the sphenopalatine ganglion*, which run a descending course and are distributed as follows: *the greater and lesser palatine nerves*, which pass through the corresponding palatine foramina to supply the mucous membrane of the hard and soft palates, the uvula and the tonsils, *nasal branches* to the mucous membrane of the nose and a *pharyngeal branch* supplying the mucosa of the nasopharynx. The *long and short sphenopalatine nerves* supply the nasal septum. The former also supplies the incisive gum of the hard palate.

### The sphenopalatine ganglion

Associated with the maxillary division of V as it lies in the pterygopalatine fossa is the relatively large sphenopalatine ganglion. This receives its *parasympathetic* or secreto-motor root from the greater superficial petrosal branch of VII, its sensory component from two sphenopalatine branches of the maxillary nerve and its sympathetic root from the internal carotid plexus. Its parasympathetic efferents pass to the lacrimal gland through a communicating branch to the lacrimal nerve. Sensory and and sympathetic (vaso-constrictor) fibres are distributed to nose, naso-pharynx, palate and orbit.

### $V_3$: *the mandibular nerve*

This is the largest of the three divisions of the trigeminal nerve. In addition to supplying the skin of the temporal region, part of the auricle and the lower face, the mucous membrane of the anterior two-thirds of the tongue and the floor of the mouth, it also conveys the motor root to the muscles of mastication and secreto-motor fibres to the parotid gland.

Passing forwards from the trigeminal ganglion, it almost immediately enters the foramen ovale through which it reaches the infratemporal fossa. Here it divides into a small anterior and a larger posterior trunk, but before doing so it gives off the *nervus spinosus* to supply the dura mater and the *nerve to the medial pterygoid muscle* from which the *otic ganglion* is suspended and through which motor fibres are transmitted to tensor palati and tensor tympani.

The anterior trunk gives off:

1. A sensory branch, the *buccal nerve*, which supplies part of the skin of the cheek and the mucous membrane on its inner aspect and

2. Motor branches to the masseter, temporalis and lateral pterygoid muscles.

The posterior trunk, which is principally sensory, divides into three branches:

1. The *auriculo-temporal nerve*, which conveys sensory fibres to the skin of the temple and auricle and secreto-motor fibres from the otic ganglion to the parotid gland.

2. The *lingual nerve*, which passes downwards under cover of the ramus of the mandible to the side of the tongue (Fig. 192), where it supplies the mucous membrane of the floor of the mouth, the anterior two-thirds of the tongue (including the taste buds by way of fibres which join it from the chorda tympani) and the sublingual and submandibular salivary glands.

3. The *inferior dental nerve*, which passes down into the mandibular canal and supplies branches to the teeth of the lower jaw. It then emerges from the mental foramen to supply the skin of the chin and lower lip. This branch also conveys the only motor component of the posterior trunk; the *nerve to the mylohyoid*, supplying the muscle of that name and the anterior belly of the digastric.

*The otic ganglion*

The otic ganglion is unique among the four ganglia associated with the trigeminal nerve in having a motor as well as parasympathetic, sympathetic and sensory components. It lies immediately below the foramen ovale as a close medial relationship to the mandibular nerve.

Its *parasympathetic* fibres reach the ganglion by the lesser superficial petrosal branch of the glossopharyngeal nerve; these relay in the ganglion and pass via the auriculo-temporal nerve to the parotid gland, and are its secretomotor supply. The *sympathetic* fibres are derived from the superior cervical ganglion along the plexus which surrounds the middle meningeal artery and the *sensory* fibres arrive from the auriculo-temporal nerve; they are respectively vasoconstrictor and sensory to the parotid gland.

*Motor* fibres pass through the ganglion from the nerve to the medial pterygoid (a branch of the mandibular nerve) and supply the tensor tympani and tensor palati muscles.

### The submandibular ganglion

This is suspended from the lower aspect of the lingual nerve. Its *para-sympathetic* supply is derived from the chorda tympani branch of the facial nerve (see Fig. 252) by which it is conveyed to the lingual nerve; it carries the secreto-motor supply to the submandibular and sublingual salivary glands.

*Sympathetic* fibres are transmitted from the superior cervical ganglion via the plexus on the facial artery and supply vasoconstrictor fibres to these same two salivary glands. The *sensory* component is contributed by the lingual nerve itself, which provides sensory fibres to these salivary glands and also to the mucous membrane of the floor of the mouth.

### The central connections of the trigeminal nerve

The central processes of the trigeminal ganglion cells enter the lateral aspect of the pons and divide into ascending and descending branches which terminate in one or other component of the sensory nucleus of V (Figs. 236 and 249). This nucleus consists of three parts each of which appears to subserve different sensory modalities; a chief sensory nucleus in the pontine tegmentum concerned with touch, a descending, or spinal, nucleus subserving pain and temperature and a mesencephalic nucleus receiving proprioceptive afferents. *The motor root of the trigeminal nerve* lies just medial to the sensory nucleus in the upper part of the pons; its efferents pass out with the sensory fibres and are distributed by way of the mandibular division of the nerve.

CLINICAL FEATURES

1.   Section of the whole trigeminal nerve results in unilateral anaesthesia of the face and anterior part of the scalp, the auricle and the mucous membranes of the nose, mouth and anterior two-thirds of the tongue, together with paralysis and wasting of the muscles of mastication on the affected side. Lesions of separate divisions give rise to corresponding sensory and motor deficits in the area of distribution of the affected nerve.

2.   Trigeminal neuralgia may affect any one or more of the three divisions, giving rise to the characteristic pain over the appropriate area (Fig. 250). Surgical treatment for this condition may involve:

•alcohol injection of the trigeminal ganglion;

•section of the central root of the nerve or of the appropriate division or

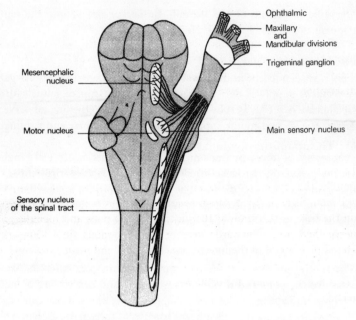

Ophthalmic

Maxillary
and
Mandibular divisions

Trigeminal ganglion

Mesencephalic
nucleus

Motor nucleus

Main sensory nucleus

Sensory nucleus
of the spinal tract

**Fig. 249.** Plan of the trigeminal nerve and its nuclei in dorsal view.

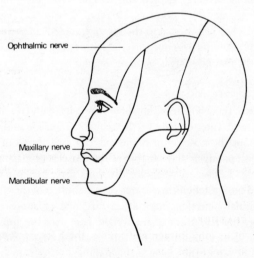

Ophthalmic nerve

Maxillary nerve

Mandibular nerve

**Fig. 250** Areas of the face supplied by the three divisions of the trigeminal nerve.

•section of the descending tract of V as this lies superficially in the medulla—a procedure known as medullary tractotomy.

3. Pain is frequently referred from one segment to another. Thus a patient with a carcinoma of the tongue (lingual nerve) frequently complains bitterly of earache (auriculo-temporal nerve). The classical description of such a case is an old gentleman sitting in out-patients spitting blood and with a piece of cotton wool in his ear.

## 6. The abducent nerve

Like the trochlear nerve, the abducent nerve supplies only one eye muscle, the *lateral rectus*. Its nucleus lies in the upper part of the pons and from there its fibres pass through the pontine tegmentum to emerge on the base of the brain at the junction of the pons and medulla. The nerve then passes forwards to enter the cavernous sinus (Fig. 247). Here it lies lateral to the internal carotid artery and medial to the IIIrd, IVth and Vth nerves. Passing through the tendinous ring just below the IIIrd nerve, it enters the orbit to pierce the deep surface of the lateral rectus (Fig. 251).

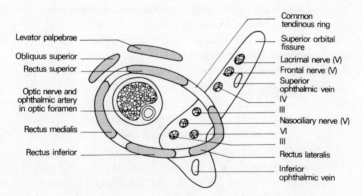

**Fig. 251.** The superior orbital fissure and tendinous ring of origin of the extrinsic orbital muscles, showing the relations of the cranial nerves as they enter the orbit.

CLINICAL FEATURES

On account of its long intracranial course, the VIth nerve is frequently involved in injuries to the base of the skull. When damaged, it gives rise to diplopia and a convergent squint.

## 7. The facial nerve

In addition to supplying the muscles of facial expression, the facial nerve conveys secreto-motor fibres to the sublingual and submandibular salivary glands and the lacrimal gland; it also carries taste fibres from the anterior two-thirds of the tongue.

The fibres innervating the facial muscles have their nucleus of origin in the ventral part of the lower pons; the secreto-motor fibres for the salivary glands are derived from the superior salivary nucleus. The sensory fibres associated with the nerve have their cells of origin in the facial ganglion.

From the motor nucleus, fibres of the facial nerve run a devious course over the nucleus of the abducent nerve (Fig. 236), where they form an elevation on the floor of the fourth ventricle known as the *facial colliculus*, then downwards and forwards to emerge from the lateral aspect of the pons between the olive and the inferior cerebellar peduncle.

The sensory and motor fibres pass together into the internal auditory meatus at the bottom of which they leave the VIIIth nerve and enter the facial canal. Here they run laterally over the vestibule before bending sharply backwards over the promontory of the middle ear. This bend, or genu of the facial nerve, as it is called, marks the site of the facial ganglion and the point at which the secreto-motor fibres for the lacrimal gland leave to form the greater superficial petrosal nerve. The facial nerve then passes downwards, medial to the middle ear, to reach the stylomastoid foramen (Fig. 252).

Just before entering this foramen it gives off the branch, known as the *chorda tympani*, which runs back through the middle ear to the infra-temporal fossa where it joins the lingual nerve. Hence its taste fibres reach the anterior two-thirds of the tongue and its secreto-motor fibres are conveyed to the submandibular ganglion.

On emerging from the stylomastoid foramen the nerve supplies the stylohyoid and the posterior belly of digastric muscle. It then enters the parotid gland where it divides into five divisions for the supply of the facial muscles; the temporal, zygomatic, buccal, mandibular, and cervical branches (see Figs. 203, 204 and 253).

### CLINICAL FEATURES

1. It is important to distinguish between 'nuclear' and 'infranuclear' facial palsies on the one hand and 'supranuclear' palsies on the other. Both nuclear and infranuclear palsies result in a facial paralysis which is

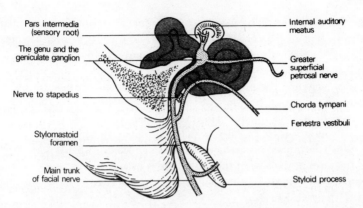

Pars intermedia
(sensory root)

The genu and the
geniculate ganglion

Nerve to stapedius

Stylomastoid
foramen

Main trunk
of facial nerve

Internal auditory
meatus

Greater
superficial
petrosal nerve

Chorda tympani

Fenestra vestibuli

Styloid process

Fig. 252. Distribution of the facial nerve within the temporal bone.

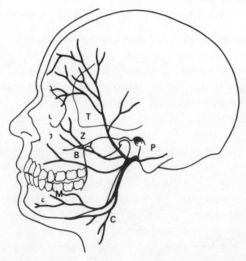

Fig. 253. Distribution of the facial nerve: T=temporal, Z=zygomatic, B=buccal, M=mandibular, C=cervical and P=posterior auricular branch.

complete and which affects *all* the muscles on one side of the face. In supranuclear palsies there is no involvement of the muscles above the palpebral fissure since the portion of the facial nucleus supplying these muscles receives fibres from both cerebral hemispheres. Furthermore, in such cases the patient may involuntarily use the facial muscles but will be unable to do so on request.

2. Supranuclear facial palsies most frequently result from vascular involvement of the cortico-bulbar pathways, e.g. in cerebral haemorrhage. Nuclear palsies may occur in poliomyelitis or other forms of bulbar paralysis, while infranuclear palsies may result from a variety of causes including compression in the cerebello-pontine angle (as by an acoustic neuroma), fractures of the temporal bone and invasion by a malignant parotid tumour. However, by far the commonest cause of facial paralysis is Bell's Palsy, which is of unknown aetiology.

When the intracranial part of the nerve is affected or when it is involved in fractures of the base of the skull there is usually loss of taste over the anterior two-thirds of the tongue and an associated loss of hearing (VIIIth nerve damage).

## 8. The auditory nerve (Fig. 254)

The VIIIth nerve consists of two sets of fibres: *cochlear* and *vestibular*. The *cochlear fibres* (concerned with hearing) represent the central processes of the bipolar spiral ganglion cells of the cochlea which traverse the internal auditory meatus to reach the lateral aspect of the medulla, where they terminate in the *dorsal* and *ventral cochlear nuclei*. The majority of the efferent fibres from these nuclei cross to the opposite side, those from the dorsal nucleus forming the *auditory striae* in the floor of the fourth ventricle, those from the ventral nucleus forming the *trapezoid body* in the ventral part of the pons. Most of these efferent fibres terminate in nuclei associated with the trapezoid body, either on the same or the opposite side, and then ascend in the lateral lemniscus to the *inferior corpus quadrigeminum* and the *medial geniculate body*; from the former, fibres reach the motor nuclei of the cranial nerves and form the pathway of auditory reflexes; from the latter, fibres sweep laterally in the *auditory radiation* to the *auditory cortex* on the superior temporal gyrus.

The *vestibular fibres* (concerned with equilibrium) enter the medulla just medial to the cochlear division and terminate in the *vestibular nuclei*. Many of the efferent fibres from these nuclei pass to the cerebellum in the inferior cerebellar peduncle together with fibres by-passing the vestibular nuclei and passing directly to the cerebellum.

Other vestibular connections are to the nuclei of III, IV and VI, the nucleus of XI and the upper cervical cord (via the vestibulo-spinal tract). These connections bring the eye and neck muscles under reflex vestibular control.

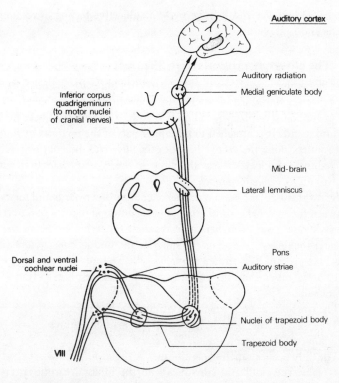

Auditory cortex

Auditory radiation

Medial geniculate body

Inferior corpus quadrigeminum (to motor nuclei of cranial nerves)

Mid-brain

Lateral lemniscus

Pons

Dorsal and ventral cochlear nuclei

Auditory striae

Nuclei of trapezoid body

Trapezoid body

VIII

**Fig. 254.** The central connections of the auditory pathway of VIII.

### CLINICAL FEATURES

**1.** Lesions of the cochlear division result in deafness which may, or may not, be accompanied by tinnitus.

The differential diagnosis between middle ear deafness and cochlear (inner ear) or auditory nerve lesions can be made clinically by the use of a tuning fork. Air conduction (the fork being held beside the ear) is normally louder than bone conduction (the fork being held against the mastoid process). If the middle ear is damaged, the reverse will hold true.

**2.** Apart from injury to the cochlear nerve itself, unilateral lesions of the auditory pathway do not greatly affect auditory acuity because of the bilaterality of the auditory projections.

**3.** Temporal lobe tumours may give rise to auditory hallucinations if they encroach upon the auditory radiation or superior temporal gyrus.

**4.** Lesions of the vestibular division of the labyrinth or of the vestibulo-cerebellar pathway result in vertigo—a subjective feeling of rotation—ataxia and nystagmus.

### 9. The glossopharyngeal nerve (Fig. 205)

The glossopharyngeal nerve contains sensory fibres for the pharynx and the posterior one-third of the tongue (including the taste buds), motor fibres for the stylopharyngeus muscle and secreto-motor fibres for the parotid gland. It is attached to the upper part of the medulla by four or five rootlets along the groove between the olive and the inferior cerebellar peduncle and leaves the skull by way of the jugular foramen in which it gives off its *tympanic branch*.

Below the jugular foramen the nerve courses downwards and forwards between the internal carotid artery and the internal jugular vein to reach the styloid process. From here it passes along the stylopharyngeus muscle to enter the pharynx between the superior and middle constrictors. Here it breaks up into its terminal branches which supply the posterior one-third of the tongue and the mucous membrane of the pharynx (including the tonsil).

The *tympanic branch*, which is continued as the *lesser superficial petrosal nerve*, conveys the preganglionic parasympathetic fibres to the otic ganglion (parotid secreto-motor fibres).

The only other branch of significance is the *carotid nerve* which arises just below the skull and runs down on the internal carotid artery to supply the carotid body and sinus. This small twig serves as the afferent limb of the presso-receptor and chemo-receptor reflexes from the carotid sinus and body respectively.

CLINICAL FEATURES

Complete section of the glossopharyngeal nerve results in:

Sensory loss in the pharynx, loss of taste and common sensation over the posterior one-third of the tongue, some pharyngeal weakness and loss of salivation from the parotid gland. However, such lesions are frequently difficult to detect and rarely occur as isolated phenomena since there is so often associated involvement of the vagus or its nuclei.

### 10. The vagus nerve

The vagus has the most extensive distribution of all the cranial nerves, innervating the heart and the major part of the respiratory and alimentary tracts.

*Central connections*

The *dorsal nucleus* of the vagus (Fig. 235) is a mixed visceral afferent and efferent nucleus. It receives sensory fibres from the heart, the lower respiratory tract and the alimentary tract down to the transverse colon; in addition it gives rise to motor fibres to the heart and the smooth muscles of the bronchi and gut.

From the *nucleus ambiguus* efferent fibres pass to the striped muscles of the pharynx.

*Distribution*

The nerve is connected to the side of the medulla by about 10 filaments lying in series with the glossopharyngeal nerve along the groove between the olive and the inferior cerebellar peduncle. These filaments unite to form a single bundle which passes beneath the cerebellum to the jugular foramen and is ensheathed in the same dural covering as the accessory nerve. Two sensory ganglia are associated with this part of the nerve; a superior within the jugular foramen and an inferior immediately beneath the skull.

The vagus then passes vertically downwards to the root of the neck, lying in the posterior part of the carotid sheath between the internal jugular vein and the internal and then common carotid arteries (Fig. 208). There are a number of important branches in the neck: *pharyngeal* to the pharyngeal and palatal musculature by way of the pharyngeal plexus; *superior laryngeal,* supplying the interior of the larynx above the vocal folds and the cricothyroid and inferior constrictor muscles, and the superior and inferior *cardiac branches* which are inhibitory to the heart.

Below the level of the subclavian arteries the course and relations of the nerve on the two sides differ:

On the *right side* the *recurrent laryngeal* branch is given off as it crosses the subclavian artery; beyond this the nerve descends through the superior mediastinum in close association with the great veins. Behind the root of the lung it takes part in the formation of the *pulmonary plexus* and then passes on to the oesophagus to form, with its fellow, the *oeso-phageal plexus.*

The *left vagus* enters the thorax in close association with the great arteries, lying at first lateral to the common carotid and then crossing the arch of the aorta (*see* Fig. 38). The *left recurrent laryngeal* branch, which is given off as the vagus crosses the aortic arch, passes below the ligamentum arteriosum, behind the arch and then ascends in the groove between the trachea and oesophagus (Fig. 34a). The vagus then passes behind the root of the lung, enters into the formation of the pulmonary plexus and passes onto the oesophagus.

The two vagi then enter the abdomen through the oesophageal opening in the diaphragm, the left vagus passing onto the anterior surface and the right passing to the posterior aspect of the stomach (Fig. 53). Beyond this it is difficult to trace the course of the nerves, but branches are given to the *coeliac*, *hepatic* and *renal plexuses* and, by way of these plexuses, are distributed to the fore and mid-gut and to the kidneys.

CLINICAL FEATURES
1. Isolated lesions of the vagus nerve are uncommon but it may be involved in injuries or disease of related structures.
2. Vagotomy—see p. 77.
3. Injuries to the recurrent laryngeal nerve—see larynx, p. 306.

11. **The accessory nerve** (Fig. 205)

The accessory nerve comprises a small cranial root, which is distributed by way of the vagus to the musculature of the palate, pharynx and larynx, and a larger spinal root which supplies the sterno-mastoid and trapezius muscles.

The fibres of the *cranial root* are derived from the lower part of the nucleus ambiguus and leave the medulla, below the vagus, in 4 or 5 rootlets. Just before leaving the skull through the jugular foramen it is joined by the *spinal root*; this is formed by the union of fibres from an elongated nucleus in the anterior horn of the upper 5 cervical segments which leave the cord midway between the anterior and posterior roots and pass upwards through the foramen magnum. The two roots are united for only a short distance, the cranial root joining the inferior ganglion of the vagus immediately below the skull. The fibres of this root may be regarded as a caudal component of the vagus and are probably the source of the motor fibres to the palatal, pharyngeal and laryngeal musculature which leave the main trunk of the vagus in its pharyngeal and recurrent laryngeal branches. The spinal root passes backwards over the internal jugular vein to the sterno-mastoid muscle which it pierces (and supplies) and then crosses the posterior triangle of the neck to enter and supply the deep surface of the trapezius.

CLINICAL FEATURES
1. Isolated lesions of the cranial root of the accessory nerve are in-

frequent; more commonly it is involved concomitantly with the vagus when it gives rise to paresis of the laryngeal and pharyngeal muscles, resulting in dysphonia and dysphagia.

2.   Division of the fibres of the spinal root (or lesions affecting their cells of origin) results in paresis of the sterno-mastoid and trapezius muscles. This follows, for example, most block dissections of the lymph nodes of the neck, the nerve being sacrificed in clearing the posterior triangle.

### 12.   The hypoglossal nerve

The hypoglossal nerve supplies all the intrinsic and extrinsic muscles of the tongue (with the exception of the palatoglossus). From its nucleus, which lies in the floor of the 4th ventricle (Fig. 235), a series of about a dozen rootlets leave the side of the medulla in the groove between the pyramid and the olive. These rootlets unite to leave the skull by way of the anterior condylar, or hypoglossal, canal. Lying at first deep to the internal carotid artery and the jugular vein, the nerve passes downwards between these two vessels to just above the level of the angle of the mandible. Here it passes forwards over the internal and external carotid

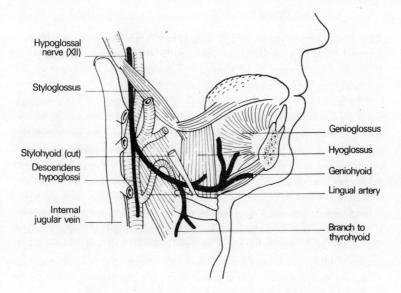

**Fig. 255.**   The distal course of the hypoglossal nerve.

arteries, and gives off its descending and thyrohyoid branches. It then crosses the hyglossus and genioglossus muscles to enter the tongue (Fig. 255). Its *descending branch* passes more or less vertically downwards upon the internal carotid artery to join the descending cervical nerve (C2 and 3) to form a loop known as *ansa hypoglossi* just above the omohyoid muscle. From this loop branches are given to three infrahyoid muscles—sterno-thyoid, sternohyoid and omohyoid.

CLINICAL FEATURES

1. Division of the hypoglossal nerve, or lesions involving its nucleus, result in an ipsilateral paralysis and wasting of the muscles of the tongue. This is detected clinically by deviation of the protruded tongue to the *affected side*.
2. Supranuclear paralysis (due to involvement of the cortico-bulbar pathways) leads to paresis but not atrophy of the muscles of the contra-lateral side.

# The special senses

## THE NOSE (see also accessory nasal sinuses)

The *external nose* consists of a bony and cartilaginous framework closely overlaid by skin and fibro-fatty tissues. The bones are the two nasal bones and the frontal processes of the maxilla. The *ala* is composed solely of fatty tissue at its lower free edge.

The *nasal cavity* is divided into right and left halves by a median *nasal septum* formed by the *perpendicular plate* of the *ethmoid bone*, the *septal cartilage* and the *vomer* (Fig. 256). Each cavity extends from the nostril (or anterior nares) in front to the posterior nasal aperture behind. communicating through the latter with the nasopharynx. The *lateral wall* is very irregular, due to the projection of the three *conchae* (superior, middle and inferior) and the underlying meatuses (see Figs. 218 and 219).

The superior meatus receives the opening of the posterior ethmoidal air cells. Opening into the middle meatus are (from before backwards) the frontal and maxillary sinuses and the anterior and middle ethmoidal air cells. Only the naso-lacrimal duct opens into the inferior meatus. The roof of the cavity is horizontal in its central portion, where it is formed by the cribriform plate of the ethmoid, but slopes downwards both anteriorly (the frontal and nasal bones) and posteriorly (the sphen-

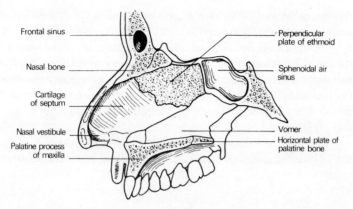

Frontal sinus

Nasal bone

Cartilage
of septum

Nasal vestibule

Palatine process
of maxilla

Perpendicular
plate of ethmoid

Sphenoidal air
sinus

Vomer

Horizontal plate of
palatine bone

Fig. 256.   The septum of the nose.

oid). The *floor* corresponds to the roof of the mouth; it comprises the palatine process of the maxilla, the horizontal process of the palatine bone and the soft palate.

*Mucous membrane*
The olfactory portion, which is confined to the superior concha and the adjacent upper part of the septum, is thin and dull yellow in colour; it contains the olfactory receptors and supporting cells. The remaining *respiratory portion* is thick, vascular and moist with secretions of mucous glands; its epithelium is ciliated.

The upper part of the nasal cavity receives its arterial supply from the ethmoidal branches of the ophthalmic artery, a terminal of the internal carotid. The sphenopalatine branch of the maxillary artery, a terminal of the external carotid, supplies the lower part of the cavity. Just within the vestibule of the nose, on the antero-inferior part of the septum, it links with a septal branch of the facial artery and it is from this zone, Little's area, that 90 per cent of nose-bleeds occur. The veins drain downwards into the facial vein and upwards to the ethmoidal tributaries of the ophthalmic veins.

CLINICAL FEATURES
1.   The skin of the external nose and its surrounds contains many sebaceous glands and hair follicles which may become blocked and infected. The significance of this fact is that the facial veins, which may become infected, communicate directly with the ophthalmic veins and hence with

the cavernous sinus. For this reason, this zone is often known as the 'danger area of the face'.

2. The extensive relations of the nasal cavity are important in the spread of infection. Observe that it is in direct continuity with (i) the anterior cranial fossa (via the cribriform plate of the ethmoid bone); (ii) the nasopharynx and, through the pharyngo-tympanic tube, the middle ear; (iii) the paranasal air sinuses; (iv) the lacrimal apparatus and conjunctiva.

3. The septum is frequently deviated to one or other side, interfering both with inspiration and with drainage of the nose and accessory sinuses.

## THE EAR

### 1. The external ear (Fig. 257)

This comprises the auricle and the external auditory meatus. The *auricle*, for the most part, consists of a cartilaginous framework to which the skin is closely applied. The intrinsic and extrinsic muscles described for the ear are of no significance in man.

The *external auditory meatus* extends inwards to the tympanic membrane. It is about $1\frac{1}{2}$ in (37 cm) long, and has a peculiar S-shaped course, being directed first medially upwards and forwards, then medially and backwards and, finally, medially forwards and downwards. The outer

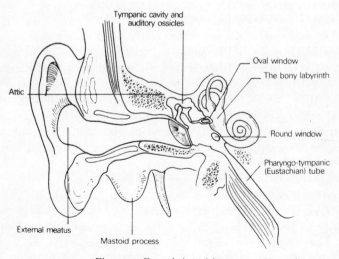

**Fig. 257.** General view of the ear.

third of the canal is cartilaginous and somewhat wider than the medial osseous portion. The whole canal is lined by skin, which is closely adherent to the osseous portion but is separated from the cartilaginous part by the ceruminous glands in the subcutaneous tissue.

The *tympanic membrane*, or *ear drum* (Fig. 259), separates the middle ear from the external auditory meatus. It is made up of an outer cutaneous layer, continuous with the skin of the external auditory meatus, a middle fibrous layer and an inner mucous layer continuous with the muco-periosteum of the rest of the tympanic cavity. It is oval in outline, a little less than $\frac{1}{2}$ in (12 mm) in its greatest (vertical) diameter, and faces laterally, downwards and forwards; it is slightly concave outwards. Since it is translucent (except at its margin where it is attached to the medial aspect of the external auditory meatus), it is possible on examination to see the underlying malleus and part of the incus. The greater part of the membrane is taut and is known as the *pars tensa*, but above the lateral process of the malleus there is a small triangular area where the membrane is thin and lax—the *pars flaccida*. This area is bounded by two distinct malleolar folds which reach down to the lateral process of the malleus. The point of greatest concavity of the membrane is known as the *umbo*; this marks the attachment of the handle of the malleus to the membrane.

## 2. The middle ear

The *middle ear*, or *tympanic cavity*, is the narrow slit-like cavity in the petrous part of the temporal bone containing the three auditory ossicles (Fig. 257).

The walls of the cavity and its important relations are as follows:

The *lateral wall* is formed mainly by the tympanic membrane, which divides it from the external auditory meatus, and above this by the squamous part of the temporal bone; the part of the cavity above the tympanic membrane is known as the *epi-tympanic recess* or *attic*; this part of the cavity contains the incus and the head of the malleus.

The *medial wall*, which separates the cavity from the internal ear, presents the *fenestra cochleae* (round window), closed by the secondary tympanic membrane; the *fenestra vestibuli* (oval window), occupied by the base of the stapes; the *promontory*, formed by the first turn of the cochlea; and the *prominence* caused by the underlying canal for the facial nerve (see Fig. 252).

The *floor* is a thin plate of bone separating the cavity from the bulb of the jugular vein.

The *roof* is formed by the thin sheet of bone known as the *tegmen tympani*, which separates it from the middle cranial fossa and the temporal lobe of the brain.

*Anteriorly*, the cavity communicates with the pharynx by way of the *pharyngo-tympanic* or *Eustachian tube*.

*Posteriorly* it communicates with the tympanic antrum and the mastoid air cells.

The *mastoid antrum* is a small cavity in the posterior part of the petrous temporal bone connected to the epi-tympanic recess of the middle ear by way of the narrow *aditus*. Its importance is two-fold; it is in communication with the mastoid air cells (hence the portal through which infection may spread to these spaces from the middle ear) and it is intimately related posteriorly to the sigmoid sinus and the cerebellum, both of which may be involved from a middle ear infection.

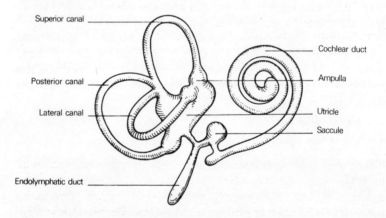

Superior canal

Posterior canal

Lateral canal

Endolymphatic duct

Cochlear duct

Ampulla

Utricle

Saccule

**Fig. 258.** Detail of the membranous labyrinth.

The *mastoid air cells* arise post-natally as diverticula from the tympanic antrum. They may invade not only the mastoid process but also the squamous part of the temporal bone. They are lined by a mucoperiosteum continuous anteriorly with that of the tympanic cavity.

The *pharyngo-tympanic* tube reaches downwards, forwards and medially from the anterior part of the tympanic cavity to the lateral walls of the nasopharynx. In all it is about $1\frac{1}{2}$ in (37 mm) long, the first $\frac{1}{2}$ in (12 mm) being bony while the rest is cartilaginous. It is lined by a ciliated columnar epithelium. The mucous membrane is thin in its bony part but the cartilaginous segment contains numerous mucous glands and, near its pharyngeal orifice, a considerable collection of lymphoid tissue termed the *tubal tonsil*. This may become swollen in infection, producing blockage of the tube. The tube is widest at its pharyngeal end and narrowest at the junction of the bony and cartilaginous portions.

Conduction of sound through the middle ear is by way of the malleus, incus and stapes. The *malleus* is the largest of the three and is described as having a handle, attached to the tympanic membrane, a rounded head, which articulates with the incus, and a lateral process, which can be seen through the tympanic membrane and from which the malleolar folds radiate. The *incus* comprises a body, which articulates with the malleus, and two processes, a short process attached to the posterior wall of the middle ear and a long process for articulation with the stapes. The shadow of the long process can often be seen through an auroscope running downwards behind the handle of the malleus. The *stapes* has a head for articulation with the incus, a neck and a base, which is firmly fixed in the fenestra vestibuli (the oval window).

Two small muscles are associated with these ossicles; the *stapedius*, which is attached to the neck of the stapes and is supplied by the facial nerve, and the tensor tympani, which is inserted into the handle of the malleus and is supplied by the mandibular division of V. Both serve to damp high-frequency vibrations.

### 3. The internal ear (Fig. 258)

The *internal ear* consists essentially of a complicated *bony labyrinth* made up of a central *vestibule*, which communicates posteriorly with three semicircular ducts and anteriorly with the spiral *cochlea*. This cavity contains a fluid known as perilymph and encloses the *membranous laby-*

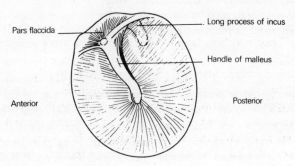

**Fig. 259.** The tympanic membrane as seen through an auroscope.

*rinth*, comprising the *utricle* and *saccule*, which communicate respectively with the *semicircular canals* and the *cochlear canal*. The duct system is filled with endolymph.

In each component of the membranous labyrinth there are specialized sensory receptor areas known as the *maculae* of the utricle and saccule, the *ampullary crests* of the semicircular canals and the *spiral organ of Corti* in the cochlea.

The disposition of the semicircular canals in three planes at right-angles to each other renders this part of the labyrinth particularly well suited to signal changes in position of the head. The organ of Corti is adapted to record the sound vibrations transmitted by the stapes at the round window.

# The eye and associated structures

(For optic nerve and visual pathway see pages 383–5).

### The eyeball (Fig. 260)

The eyeball, which is just under 25 mm in all diameters, is formed by segments of two spheres of different size; a prominent anterior segment, which is transparent and forms about one-sixth of the eyeball, and a larger posterior segment, which is opaque and comprises five-sixths of a sphere. A line joining the points of greatest curvature of these two segments (the anterior and posterior poles of the eye) is known as the *optical axis* of the eye. The optic nerve enters the eye about $\frac{1}{8}$ in (3 mm) to the nasal (medial) side of the posterior pole.

The eyeball is formed by three coats: a fibrous outer coat, a vascular middle coat and an inner neural coat—the retina.

### The fibrous coat
The *fibrous coat* comprises a transparent anterior part, *the cornea*, and an opaque posterior portion, *the sclera*. Peripherally the cornea is continuous with the sclera at the *sclerocorneal junction*. *The sclera* is a tough, fibrous membrane which is responsible for the maintenance of the shape of the eyeball and which receives the insertion of the extraocular muscles. Posteriorly it is pierced by the optic nerve, with whose dural sheath it is continuous.

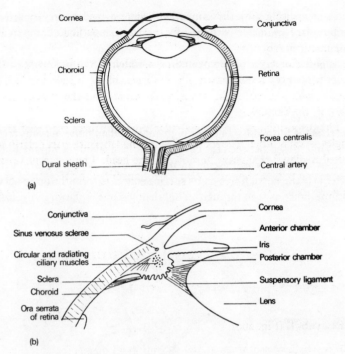

Fig. 260. (a) The eyeball in section. (b) Detail of the ciliary region.

## The vascular coat

This is made up of the choroid, the ciliary body and the iris.

*The choroid* is a thin but highly vascular membrane lining the inner surface of the sclera. Posteriorly it is pierced by the optic nerve and anteriorly it is connected to the iris by the ciliary body.

*The ciliary body* includes the *ciliary ring*, a fibrous ring continuous with the choroid, the *ciliary processes*, a group of sixty to eighty folds arranged radially between the ciliary ring and the iris and connected posteriorly to the suspensory ligament of the lens, and the *ciliary muscles*, an outer radial and inner circular layer of smooth muscle responsible for the changes in convexity of the lens in accommodation.

*The iris* is the contractile disc surrounding the pupil. It consists of four layers:

1.   an anterior mesothelial lining;
2.   a connective tissue stroma containing pigment cells;
3.   a group of radially arranged smooth muscle fibres—the dilator of

the pupil (supplied by the sympathetic system) and a circular group, the pupillary sphincter (supplied by the parasympathetic fibres in the oculomotor nerve).

4. a posterior layer of pigmented cells which is continuous with the ciliary part of the retina.

### The neural coat

*The retina* is formed by an outer pigmented and an inner nervous layer, and is interposed between the choroid and the hyaloid membrane of the vitreous. Anteriorly it presents an irregular edge, *the ora serrata*, while posteriorly the nerve fibres on its surface collect to form the optic nerve. Its appearance as seen through an ophthalmoscope is shown in Fig. 261.

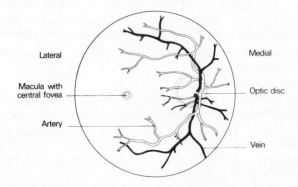

**Fig. 261.**   The right fundus oculi as seen through an ophthalmoscope.

Near its posterior pole there is a pale yellowish area, the *macula lutea*, the site of central vision, and just medial to this is the pale *optic disc* formed by the passage of nerve fibres through the retina, corresponding to the '*blind-spot*'. The *central artery of the retina* emerges from the disc and then divides into upper and lower branches; each of these in turn divides into a nasal and temporal branch. Histologically the retina consists of a number of layers but from a functional point of view only three need be considered: an inner receptor cell layer—the layer of rods and cones—an intermediate layer of bipolar neurones and the layer of ganglion cells, whose axons form the superficial layer of optic nerve fibres (Fig. 245).

## Contents of the eyeball

Within the eyeball are found: the lens, the aqueous humour and the vitreous body. *The lens* is biconvex and is placed between the vitreous and the aqueous humour, just behind the iris. *The aqueous humour* is a filtrate of plasma secreted by the vessels of the iris and ciliary body into the *posterior chamber* of the eye (i.e. the space between the lens and the iris). From here it passes through the pupillary aperture into the *anterior chamber* (between the cornea and the iris) and is re-absorbed into the ciliary veins by way of the *sinus venosus sclerae* (or canal of Schlemm). The *vitreous body*, which occupies the posterior four-fifths of the eyeball, is a thin transparent gel contained within a delicate membrane—*the hyaloid membrane*—and pierced by the lymph-filled *hyaloid canal*. The anterior part of the hyaloid membrane is thickened, receives attachments from the ciliary processes and gives rise to the *suspensory ligament of the lens*. This ligament is attached to the capsule of the lens in front of its equator and serves to retain it in position. It is relaxed by contraction of the radial fibres of the ciliary muscle and so allows the lens to assume a more convex form.

*The orbital muscles* (Fig. 251)

These are the levator palpebrae superioris and the extra-ocular muscles; the medial, lateral, superior and inferior recti and the superior and inferior obliques. The *four recti* arise from a tendinous ring around the optic foramen and the medial part of the superior orbital fissure and are inserted into the sclera anterior to the equator of the eyeball. The lateral

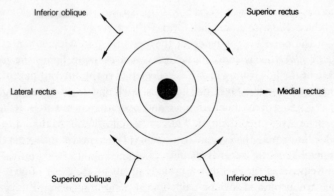

**Fig. 262.** The muscles acting on the eyeball from the primary position (i.e. looking directly forwards).

rectus is supplied by the VIth nerve, the others by III. The *superior oblique* arises just above the tendinous ring and is inserted by means of a long tendon which loops around a fibrous pulley on the medial part of the roof of the orbit into the sclera just lateral to the insertion of the superior rectus. It is supplied by the IVth nerve. The *inferior oblique* passes like a sling from its origin on the medial side of the orbit around the under-surface of the eye to insert into the sclera between the superior and lateral recti; it is supplied by III.

The actions of these muscles are shown diagrammatically in Fig. 262.

*The fascial sheath of the eye* (Tenon's capsule) is the membrane enclosing the eyeball from the optic nerve behind to the sclero-corneal junction in front. It is pierced by the vessels and nerves of the eye and by the tendons of the extra-ocular muscles. It is thickened inferiorly, where it forms the suspensory ligament.


### The eyelids and conjunctiva

Of the two eyelids, the upper is the larger and more mobile, but apart from the presence of the levator palpebrae superioris in this lid, the structure of the two eyelids is essentially the same. Each consists of the following layers from without inwards; skin, loose connective tissue, fibres of the orbicularis oculi muscle, the tarsal plates, of very dense fibrous tissue, tarsal glands and conjunctiva. The eyelashes arise along the muco-cutaneous junction and immediately behind the lashes there are the openings of the *tarsal* (*Meibomian*) *glands*. These are large sebaceous glands whose secretion helps to seal the palpebral fissure when the eyelids are closed and forms a thin layer over the exposed surface of the open eye; if blocked, they distend into Meibomian cysts.

*The conjunctiva* is the delicate mucous membrane lining the inner surface of the lids from which it is reflected over the anterior part of the sclera to the cornea. Over the lids it is thick and highly vascular, but over the sclera it is much thinner and over the cornea it is reduced to a single layer of epithelium. The line of reflection from the lid to the sclera is known as the conjunctival fornix; the superior fornix receives the openings of the lacrimal glands.

Movements of the eyelids are brought about by the contraction of the orbicularis oculi and levator palpebrae superioris muscles. The width of the palpebral fissure at any one time depends on the tone of these muscles and the degree of protrusion of the eyeball.

## The lacrimal apparatus

The *lacrimal gland* is situated in the upper, lateral part of the orbit in what is known as the lacrimal fossa. The main part of the gland is about the size and shape of an almond, but it is connected to a small terminal process which extends into the posterior part of the upper lid. The gland is drained by a series of 8–12 small ducts which open into the lateral part of the superior conjunctival fornix whence its secretion is spread over the surface of the eye by the action of the lids.

The tears are drained by way of the *lacrimal canaliculi* whose openings, the *lacrimal puncta*, can be seen on the small elevation near the medial margin of each eyelid known as the *lacrimal papilla*. The two canaliculi, superior and inferior, open into the *lacrimal sac*, which is situated in a small depression on the medial surface of the orbit. This in turn drains through the *naso-lacrimal duct* into the anterior part of the inferior meatus of the nose. The naso-lacrimal duct, which not uncommonly becomes obstructed, is about $\frac{1}{2}$ in (12 mm) in length and lies in its own bony canal in the medial wall of the orbit.

## THE AUTONOMIC NERVOUS SYSTEM

The nervous system is divided into two great sub-groups; the cerebro-spinal system, made up of the brain, spinal cord and the peripheral cranial and spinal nerves, and the autonomic system (also termed the vegetative, visceral or involuntary system), comprising the autonomic ganglia and nerves. Broadly speaking, the cerebrospinal system is concerned with the responses of the body to the external environment. In contrast, the autonomic system is concerned with the control of the internal environment, exercised through the inervation of the non-skeletal muscle of the heart, blood vessels, bronchial tree, gut and the pupils and the secretormotor supply of many glands, including those of the alimentary tract and its outgrowths, the sweat glands, and, as a rather special example, the adrenal medulla.

The two systems should not be regarded as being independent of each other for they are linked anatomically and functionally. Anatomically, autonomic nerve fibres are transmitted in all of the peripheral and some of the cranial nerves; moreover, the higher connections of the autonomic system are situated within the spinal cord and brain. Functionally the two systems are closely linked within the brain and cord.

The characteristic feature of the autonomic system is that its efferent nerves emerge as medullated fibres from the brain and spinal cord, are interrupted in their course by a synapse in a peripheral ganglion and are then relayed for distribution as fine non-medullated fibres. In this respect they differ from the cerebro-spinal efferent nerves, which pass without interruption to their terminations (Fig. 263).

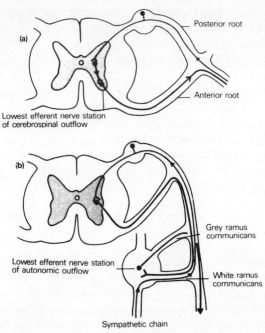

**Fig. 263.** The essential difference between the cerebrospinal and autonomic outflows: (a) the cerebrospinal system has its lowest efferent nerve cell stations within the C.N.S.; (b) the autonomic system has its lowest efferent cell stations in a peripheral ganglion (here illustrated by a typical sympathetic nerve ganglion).

The autonomic system is sub-divided into the *sympathetic* and *para-sympathetic* systems on anatomical, functional, and to a considerable extent, pharmacological grounds.

*Anatomically* thé sympathetic nervous system has its motor cell station in the lateral grey column of the thoracic and upper two lumbar segments of the spinal cord. The parasympathetic system is less neatly defined anatomically since it is divided into a cranial outflow, which passes along the cranial nerves III, VII, IX and X, and a sacral outflow, with cell stations in the 2nd, 3rd and sometimes 4th sacral segments of the cord.

*Functionally* the sympathetic system is concerned principally with stress reactions of the body. When stimulated the puṭ is dilate, peripheral blood vessels constrict, the force, rate and oxygen consumption of the heart increase, the bronchial tree dilates, visceral activity is diminished by inhibition by peristalsis and increase of sphincter tone, glycogenolysis takes place in the liver, the adrenal medulla is stimulated to secrete, and there is cutaneous sweating and pilo-erection. The sympathetic pelvic nerves inhibits bladder contraction, are motor to the internal vesical sphincter and innervate the uterine musculature.

Although the coronary blood flow is increased it is doubtful whether this is a direct sympathetic effect on the coronary arteries, but is more likely to be due to indirect factors including more vigorous cardiac contraction, reduced systole, relatively increased diastole and an increased concentration of vasodilator metabolites.

The parasympathetic system tends to be antagonistic to the sympathetic system. Its stimulation results in constriction of the pupils, diminution in the rate, conduction and excitability of the heart, an increase in gut peristalsis with sphincter relaxation and enhanced alimentary glandular secretion. In addition the pelvic parasympathetic nerves inhibit the vesical internal sphincter and are motor to the detrusor muscle of the bladder

The sympathetic system tends to have a "mass action" effect; stimulation of any part of it results in a widespread response. In contrast, parasympathetic activity is usually discrete and localized. This difference can be explained, at least in part, by differences in anatomical peripheral connections of the two systems, as will be shown below.

It is perhaps useful to think of the two systems as acting synergistically. For example, reflex slowing of the heart is effected partly from increased vagal and partly from decreased sympathetic stimulation. In addition, some organs receive their autonomic innervation from one system only; for example, the adrenal medulla and the cutaneous arterioles receive only sympathetic fibres, whereas neurogenic gastric secretion is entirely under parasympathetic control via the vagus nerve.

*Pharmacologically* the sympathetic post ganglionic terminals release adrenaline and nor-adrenaline; this with a single exception of the sweat glands, which, in common with all the parasympathetic postganglionic terminations, releases acetylcholine.

## Autonomic afferents

As well as the efferent system, there are afferent autonomic fibres which are concerned with the afferent arc of autonomic reflexes and with the conduction of visceral pain stimuli. These nerves have their cell stations in the dorsal root ganglia of the spinal nerves or of the ganglia of the cranial nerves concerned with the autonomic system. The fibres from the viscera ascend in the autonomic plexuses; those from the body wall are conveyed in the peripheral spinal nerves. The afferent course from any structure is therefore along the same pathway as the efferent autonomic fibres which supply the part.

The afferent fibres ascend centrally to the hypothalamus and thence to the orbital and frontal gyri of the cerebral cortex along as yet indeterminate pathways. Normally we are unaware of the afferent impulses from the viscera unless they become sufficiently great to exceed the pain threshold when they are perceived as visceral pain, e.g. the pain of coronary ischaemia or intestinal colic.

### The sympathetic system

The efferent fibres of the sympathetic system arise in the lateral grey column of the spinal cord (see Fig. 263) from segments T.1 to L2. From each of these segments small medullated axons emerge into the corresponding anterior primary ramus and pass via a white ramus cummunicans into the sympathetic trunk.

The spinal segments responsible for the sympathetic innervation of the various parts of the body are approximately as follows:
•head and neck T.1.–2.
•upper limb T2–7
•thoracic viscera T1–4
•abdominal viscera T4–L2
•lower limb T11–L2

### The sympathetic trunk

The sympathetic trunk on each side is a ganglionated nerve chain which extends from the base of the skull to the coccyx in close relationship to the vertebral column and maintaining a distance of about one inch from the mid-line throughout its course. Commencing in the superior cervical ganglion beneath the skull base, the chain descends closely behind the posterior wall of the carotid sheath, enters the thorax anterior to the neck of the first rib, descends over the heads of the upper ribs and then on the sides of the bodies of the last three or four thoracic vertebrae, passes into the abdomen behind the medial arcuate ligament of the diaphragm and descends in a groove between psoas major and the sides of the lumbar vertebral bodies, overlapped by the abdominal aorta on the left and the inferior vena cava on the right. The chain then passes behind the common iliac vessels to enter the pelvis anterior to the ala of the sacrum and then descends medial to the anterior sacral foramena. The sympathetic trunks end below by meeting each other at the *ganglion impar* on the anterior face of the coccyx.

The details of the cervical, thoracic and lumbar portions of the trunk are given on pages 327, 49 and 164 respectively.

The sympathetic trunk bears a series of ganglia along its course which contain motor cells with which pre-ganglionic medullated fibres enter into synapse and from which non-medullated post-ganglionic axons originate. Developmentally there was originally one ganglion for each peripheral nerve, but by a process of fusion these have been reduced in man to three cervical, twelve or less thoracic, two to four lumbar and four sacral ganglia. Only the ganglia of $T_1$ to $L_2$ receive white rami directly; the higher and lower ganglia must receive their pre-ganglionic supply from medullated nerves which travel through their corresponding ganglia without relay and which then ascend or descend in the sympathetic chain. Still other pre-ganglionic fibres pass intact through the ganglia to peripheral visceral ganglia for relay.

There are thus three fates which may befall white rami (Fig. 264):

1.  to enter into synapse from the corresponding sympathetic ganglion (this applies only to the $T_1$ to $L.2$ segments);

2.  to ascend or descend in the sympathetic chain with relay in higher or lower ganglia;

3.  to traverse the ganglia intact and relay in peripheral ganglia.

Stimulation of a single white ramus communicans would thus obviously have widespread effects—the anatomical basis of the "mass action" response of sympathetic stimulation.

The branches of the sympathatic ganglionic chain have somatic and visceral distribution:

SOMATIC
Each spinal nerve receives one or more grey rami from a sympathetic ganglion which distributes postganglionic non-medullated sympathetic fibres to the segmental skin area supplied by the spinal nerve. These fibres are vasoconstrictor to the skin arterioles, sudo-motor to sweat glands and pilo-motor to the cutaneous hairs.

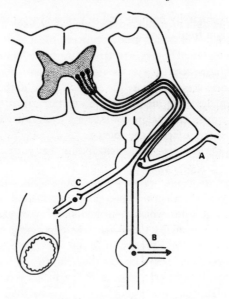

**Fig. 264.** The three fates of sympathetic white rami. These may (a) relay in their corresponding ganglion and pass to their corresponding spinal nerve for distribution (b) ascend or descend in the sympathetic chain and relay in higher or lower ganglia or (c) pass without synapse to a peripheral ganglion for relay.

**VISCERAL**

Postganglionic fibres to the head and neck and to the thoracic viscera arise from the ganglion cells of the sympathetic chain. Those to the head ascend along the internal carotid and vertebral arteries, whereas those to the thoracic organs are distributed by the cardiac, pulmonary and oesophageal plexuses.

The abdominal and pelvic viscera, however, are supplied by postganglionic fibres which have their cell stations in peripheral ganglia—the coelic, hypogastric and pelvic plexuses—which receive their preganglionic fibres from the splanchnic nerves.

The adrenal medulla has a unique nerve supply comprising a rich plexus of preganglionic fibres which pass without relay from the coeliac ganglion to the gland. These fibres end in direct contact with the chromaffin medullary cells, and liberate acetylcholine (as in all autonomic ganglia) which stimulates the secretion of adrenaline and noradrenaline by the adrenal medulla.

The chromaffin cells of the adrenal medulla may thus be regarded as sympathetic cells which have not developed postganglionic fibres;

indeed, embryologically both the medulla and sympathetic nerves have a common origin from the neural crest.

## The parasympathetic system

As already stated this system has a cranial and a sacral component. Its medullated preganglionic fibres synapse with ganglion cells which lie close to, or actually in the walls of, the viscera supplied. Postganglionic fibres therefore have only a short and direct course to their effector cells and there is thus the anatomical pathway of a local discrete response to parasympathetic stimulation (Fig. 265).

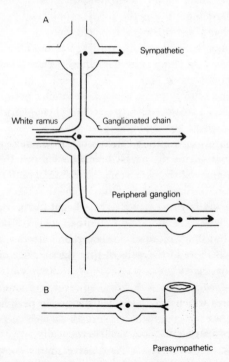

**Fig. 265.** The anatomical basis of widespread sympathetic and local parasympathetic response. (a) the widespread distribution of postganglionic fibres from a single sympathetic white ramus, (b) the localized distribution of postganglionic parasympathetic fibres.

### Cranial outflow

The cranial component of the parasympathetic system is conveyed in cranial nerves III, VII, IX and X, of which X (the vagus) is the most important and the most widely distributed. The functions of this group of nerves can be summarised as follows:

1.  pupils—constrictor to pupil, motor to ciliary muscle (accommodation);
2.  salivary glands—secreto-motor;
3.  lacrimal glands—secreto-motor;
4.  heart—inhibitor of cardiac conduction, contraction, excitability and impulse formation (with consequent slowing of the heart and diminution of its contraction force).
5.  lungs—broncho-constrictor, secreto-motor to mucous glands;
6.  alimentary canal—motor to gut muscles as far as the region of the ascending colon; inhibitor to the pyloric sphincter. Secreto-motor to the glands and adnexae of the stomach and intestine.

The parasympathetic distribution of III, VII and IX is carried out via four ganglia from which postganglionic fibres relay. These ganglia also transmit (without synapse and therefore without functional connection) sympathetic and sensory fibres which have similar peripheral distribution. These ganglia are the ciliary (page 386) sphenophalatine (page 389), submandibular (page 391) and otic (page 390).

The Xth (vagal) distribution conveys by far the most important and largest contributions of the parasympathetic system. It is responsible for all the functions of the parasympathetic cranial outflow enumerated above apart from the innervation of the eye and the secreto-motor supply to the salivary and lacrimal glands. The efferent fibres are derived from the dorsal nucleus of X and are distributed widely in the cardiac, pulmonary and alimentary plexuses. Postganglionic fibres are relayed from tiny ganglia which lie in the walls of the viscera concerned; in the gut these constitute the submucosal *plexus of Meissner* and the myenteric *plexus of Auerbach.*

### The sacral outflow

The anterior primary rami of S2, 3 and occasionally 4 give off nerve fibres termed the *pelvic splanchnic nerves* or *nervi erigentes*, which join the sympathetic pelvic plexuses for distribution to the pelvic organs. Tiny ganglia in the walls of the viscera then relay postganglionic fibres.

The sacral parasympathetic system has been termed by Cannon "the mechanism for emptying". It supplies visceromotor fibres to the muscles of the rectum and inhibitor fibres to the internal anal sphincter, motor fibres to the bladder wall and inhibitor fibres to the internal vesical sphincter. In addition, vasodilator fibres supply the erectile cavernous sinuses of the penis and the clitoris.

### Afferent parasympathetic fibres

Visceral afferent fibres from the heart, lung and the alimentary tract are conveyed in the vagus nerve. Sacral afferents are conveyed in the pelvic splanchnic nerves and are responsible for visceral pain experienced in the bladder, prostate, rectum and uterus. The reference of pain from these structures to the sacral area, buttocks and posterior aspect of the thighs is explained by the similar segmental supply of the sacral dermatomes.

Note that although afferent fibres are conveyed in both sympathetic and parasympathetic nerves they are completely independent of the autonomic system. They do not relay in the autonomic ganglia and terminate, just like somatic sensory fibres, in the dorsal ganglia of the spinal and cranial nerves. They simply use the autonomic nerves as a convenient anatomical conveyor system from the periphery to the brain.

# Index

# Index